Chemical and Biosensors Based on Metal-Organic Frames (MOFs)

Chemical and Biosensors Based on Metal-Organic Frames (MOFs)

Guest Editors

Chunsheng Wu
Liping Du
Wei Chen

Basel • Beijing • Wuhan • Barcelona • Belgrade • Novi Sad • Cluj • Manchester

Guest Editors

Chunsheng Wu
Institute of Medical Engineering
School of Basic Medical Sciences
Xi'an Jiaotong University
Xi'an
China

Liping Du
Institute of Medical Engineering
School of Basic Medical Sciences
Xi'an Jiaotong University
Xi'an
China

Wei Chen
Institute of Medical Engineering
School of Basic Medical Sciences
Xi'an Jiaotong University
Xi'an
China

Editorial Office
MDPI AG
Grosspeteranlage 5
4052 Basel, Switzerland

This is a reprint of the Special Issue, published open access by the journal *Chemosensors* (ISSN 2227-9040), freely accessible at: https://www.mdpi.com/journal/chemosensors/special_issues/R4TBO8TO49.

For citation purposes, cite each article independently as indicated on the article page online and as indicated below:

Lastname, A.A.; Lastname, B.B. Article Title. *Journal Name* **Year**, *Volume Number*, Page Range.

ISBN 978-3-7258-3539-3 (Hbk)
ISBN 978-3-7258-3540-9 (PDF)
https://doi.org/10.3390/books978-3-7258-3540-9

© 2025 by the authors. Articles in this book are Open Access and distributed under the Creative Commons Attribution (CC BY) license. The book as a whole is distributed by MDPI under the terms and conditions of the Creative Commons Attribution-NonCommercial-NoDerivs (CC BY-NC-ND) license (https://creativecommons.org/licenses/by-nc-nd/4.0/).

Contents

About the Editors vii

Chunsheng Wu, Liping Du and Wei Chen
Chemical Sensors and Biosensors Based on Metal–Organic Frameworks (MOFs)
Reprinted from: *Chemosensors* **2025**, *13*, 72, https://doi.org/10.3390/chemosensors13020072 . . . 1

Sara Maira Mohd Hizam and Mohamed Shuaib Mohamed Saheed
Facile Electrochemical Approach Based on Hydrogen-Bonded MOFs-Derived Tungsten Ethoxide/Polypyrrole-Reduced GO Nanocrystal for ppb Level Ammonium Ions Detection
Reprinted from: *Chemosensors* **2023**, *11*, 201, https://doi.org/10.3390/chemosensors11030201 . . 7

Wenyue Ma, Zijian Gu, Guocui Pan, Chunjuan Li, Yu Zhu, Zhaoyang Liu, et al.
Dual-Response Photofunctional Covalent Organic Framework for Acid Detection in Various Solutions
Reprinted from: *Chemosensors* **2023**, *11*, 214, https://doi.org/10.3390/chemosensors11040214 . . 25

Xiangxiang Fan, Susu Yang, Chun Huang, Yujie Lu and Pan Dai
Preparation and Enhanced Acetone-Sensing Properties of ZIF-8-Derived Co_3O_4@ZnO Microspheres
Reprinted from: *Chemosensors* **2023**, *11*, 376, https://doi.org/10.3390/chemosensors11070376 . . 36

Shaikha S. AlNeyadi, Mohammed T. Alhassani, Ali S. Aleissaee and Ibrahim AlMujaini
Pyrene-Derived Covalent Organic Framework Films: Advancements in Acid Vapor Detection
Reprinted from: *Chemosensors* **2024**, *12*, 37, https://doi.org/10.3390/chemosensors12030037 . . . 48

Yu Zhang, Chang Liu, Rongqiu Yan and Chenghong Lei
Bipyridyl Ruthenium-Decorated Ni-MOFs on Carbon Nanotubes for Electrocatalytic Oxidation and Sensing of Glucose
Reprinted from: *Chemosensors* **2024**, *12*, 39, https://doi.org/10.3390/chemosensors12030039 . . . 59

Wen Si, Yang Jiao, Xianchao Jia, Meng Gao, Yihao Zhang, Ye Gao, et al.
A Host–Guest Platform for Highly Efficient, Quantitative, and Rapid Detection of Nitroreductase
Reprinted from: *Chemosensors* **2024**, *12*, 145, https://doi.org/10.3390/chemosensors12080145 . . 74

Yuyu Peng, Chunyan Wang, Gen Li, Jianguo Cui, Yina Jiang, Xiwang Li, et al.
The Efficient and Sensitive Detection of Serum Dopamine Based on a MOF-199/Ag@Au Composite SERS Sensing Structure
Reprinted from: *Chemosensors* **2024**, *12*, 187, https://doi.org/10.3390/chemosensors12090187 . . 91

Junrong Li, Yihao Ding, Yuxuan Shi, Zhiying Liu, Jun Lin, Rui Cao, et al.
A Zinc Oxide Nanorod-Based Electrochemical Aptasensor for the Detection of Tumor Markers in Saliva
Reprinted from: *Chemosensors* **2024**, *12*, 203, https://doi.org/10.3390/chemosensors12100203 . . 105

Sofía V. Piguillem, Germán E. Gomez, Gonzalo R. Tortella, Amedea B. Seabra, Matías D. Regiart, Germán A. Messina and Martín A. Fernández-Baldo
Paper-Based Analytical Devices Based on Amino-MOFs (MIL-125, UiO-66, and MIL-101) as Platforms towards Fluorescence Biodetection Applications
Reprinted from: *Chemosensors* **2024**, *12*, 208, https://doi.org/10.3390/chemosensors12100208 . . 117

Miloš Ognjanović, Milena Marković, Vladimír Girman, Vladimir Nikolić, Sanja Vranješ-Đurić, Dalibor M. Stanković and Branka B. Petković
Metal–Organic Framework-Derived CeO$_2$/Gold Nanospheres in a Highly Sensitive Electrochemical Sensor for Uric Acid Quantification in Milk
Reprinted from: *Chemosensors* **2024**, *12*, 231, https://doi.org/10.3390/chemosensors12110231 . . **132**

About the Editors

Chunsheng Wu

Chunsheng Wu received his Ph.D. in Biomedical Engineering from Zhejiang University, Hangzhou, China, in 2009. From 2009 to 2009, he was a joint Ph.D. student of Micro Systems Laboratories at the University of California, Los Angeles. From 2010 to 2012, he worked as a research associate at the Biosensor National Special Laboratory of Zhejiang University. From 2012 to 2015, he worked as a postdoctoral fellow at the Institute of Nano- and Biotechnologies (INB) at Aachen University of Applied Sciences, Germany. At present, he works as a professor at Xi'an Jiaotong University, China. His research primarily focuses on the development and application of biomimetic organoid-, cell- and molecule-based biosensors, instruments, and microsystems.

Liping Du

Liping Du received her Ph.D. in Biomedical Engineering from Zhejiang University, Hangzhou, China, in 2013. From 2013 to 2015, she worked as a postdoctoral researcher at Research Center Jülich, Germany. At present, she works as an associate professor at Xi'an Jiaotong University, China. Her main research interests focus on the development and application of biomimetic olfactory and taste biosensors.

Wei Chen

Wei Chen obtained his B.Sc. and master's degrees in Applied Chemistry from Xi'an Jiaotong University in 2005 and 2008, respectively. He obtained his Ph.D. in Materials Science from City University of Hong Kong, Hong Kong SAR, in 2016. In 2017, he joined the Institute of Medical Engineering at Xi'an Jiaotong University as a lecturer. At present, he works as an associate professor at Xi'an Jiaotong University, China. His research focuses on nanomaterial-based biosensors, drug delivery, and composites.

Editorial

Chemical Sensors and Biosensors Based on Metal–Organic Frameworks (MOFs)

Chunsheng Wu *, Liping Du and Wei Chen

Institute of Medical Engineering, Department of Biophysics, School of Basic Medical Sciences, Health Science Center, Xi'an Jiaotong University, Xi'an 710061, China; duliping@xjtu.edu.cn (L.D.); weiwcchen@xjtu.edu.cn (W.C.)
* Correspondence: wuchunsheng@xjtu.edu.cn

1. Introduction

Metal–organic frameworks (MOFs), also referred to as porous coordination polymers [1–8], have experienced significant advancements in recent decades [9–17]. These structures, which are composed of organic linkers and metal ions or clusters, present unique opportunities for the innovation of novel biosensors due to their notable characteristics, including a substantial surface area, high porosity, customizable structures, and adaptability in terms of tailored properties [18–26]. Sensitive elements can be integrated into MOFs in situ by incorporating bioactive molecules during the synthesis phase [27–36]. Furthermore, the adjustable dimensions and extensive surface area, along with channels of varying sizes, render MOFs ideal candidates for the development of hybrid composite materials that can serve as sensitive components in chemical sensing and biosensing applications [37–46]. The growing interest in the utilization of MOFs in chemical sensing and biosensing technologies has spurred the creation of innovative structures and functionalities in MOF-based sensors [47–52], characterized by enhanced stability, sensitivity, flexibility, and specificity [53–59]. Consequently, we are pleased to present this Special Issue, which focuses on the latest advancements and applications of MOF-based chemical sensors and biosensors, with particular attention to the synthesis and modification of MOFs and their subsequent applications.

2. The Special Issue

This Special Issue comprises a compilation of eight high-quality original research articles and two communications, all focused on the advancement and application of metal–organic framework (MOF)-based chemical sensors and biosensors.

In their study, Ognjanović et al. employed spherical gold nanoparticles (AuNPs) to functionalize MOF-derived nanoceria (MOFdNC), which was subsequently utilized to modify a carbon paste electrode [60]. This modification enhanced the performance of the non-enzymatic electrode material for the highly sensitive detection of the significant biomolecule uric acid (UA) [60]. The results of the measurements demonstrated linear operating ranges of 0.05 to 1 µM and 1 to 50 µM, with a detection limit of 0.011 µM. Furthermore, this method has been successfully applied for the quantification of UA in milk.

In another study, Piguillem et al. developed three innovative platforms utilizing metal–organic frameworks (MOFs) to create paper-based analytical devices (PADs) that enable sensitive and portable monitoring of alkaline phosphatase (ALP) enzyme activity through laser-induced fluorescence (LIF) [61]. The findings suggest that amino-derived

Received: 7 February 2025
Accepted: 13 February 2025
Published: 17 February 2025

Citation: Wu, C.; Du, L.; Chen, W. Chemical Sensors and Biosensors Based on Metal–Organic Frameworks (MOFs). *Chemosensors* 2025, 13, 72. https://doi.org/10.3390/chemosensors13020072

Copyright: © 2025 by the authors. Licensee MDPI, Basel, Switzerland. This article is an open access article distributed under the terms and conditions of the Creative Commons Attribution (CC BY) license (https://creativecommons.org/licenses/by/4.0/).

MOF platforms possess considerable potential for integration into biosensor PADs, offering essential characteristics that enhance their efficacy and applicability in the realms of analytical chemistry and diagnostics.

Li et al. created an aptamer-based biosensor for the label-free detection of carcinoembryonic antigen (CEA) in saliva, which holds significant promise for the diagnosis and early screening of oral squamous cell carcinoma (OSCC) [62]. The measurement outcomes demonstrated that this biosensor exhibits high performance in detecting CEA, characterized by a broad linear response range and a low detection limit. This aptamer-based biosensor represents a novel approach for the label-free detection of CEA in saliva, with potential applications in clinical diagnostics and early screening for OSCC.

Peng et al. successfully synthesized a MOF-199/Ag@Au surface-enhanced Raman scattering (SERS) sensing structure, which integrates MOFs with SERS technology for the detection of dopamine (DA) in serum [63]. The results reveal a strong linear correlation between the SERS signal at 1140 cm^{-1} and DA concentration (ranging from 0.001 M to 1 pM). This work establishes a solid technical foundation for the application of SERS technology in the screening of clinical neurological diseases.

Si et al. developed a host–guest complex designed for the detection and imaging of nitroreductase (NTR), wherein a fluorescent substrate, GP-NTR, was encapsulated within a metal–organic capsule, Zn-MPB [64]. The findings indicate a linear correlation between NTR concentration and a low detection limit of 6.4 ng/mL. This complex has also been utilized for the rapid imaging of NTR in tumor cells that overexpress NTR, as well as in tumor-bearing animal models. It facilitates not only swift and quantitative detection of NTR activity but also serves as an imaging tool for the early diagnosis of hypoxia-related tumors.

Zhang et al. developed a glucose sensor based on Bipyridyl Ruthenium-decorated nickel metal–organic frameworks (MWCNT-RuBpy@Ni-MOF), which can also be employed for electrocatalytic oxidation [65]. The results obtained demonstrate the high performance of this glucose sensor, characterized by a broad range of linear responses, as well as high sensitivity and selectivity. This sensor is applicable for enzymeless glucose detection and can also be utilized for electrocatalytic oxidation.

AlNeyadi et al. synthesized three distinct types of pyrene-based covalent organic frameworks (COFs) for the detection of volatile acid vapors [66]. Notably, one specific type of COF exhibited exceptional sensitivity and rapid response as a fluorescent chemosensor for the detection of hydrochloric acid (HCl) in solution. This COF shows significant potential for use as a reliable and reusable sensor for the detection of harmful acid vapors, thereby addressing environmental concerns associated with industrial activities.

Fan et al. synthesized Co_3O_4@ZnO microspheres derived from ZIF-8 and utilized them for the enhanced detection of acetone [67]. Their findings revealed that the response of Co_3O_4@ZnO microspheres to 50 ppm acetone was 4.5 times greater than that of pure Co_3O_4, with a detection limit as low as 0.5 ppm. The mechanisms underlying the enhanced sensing capabilities for acetone were also examined.

Ma et al. introduced a novel class of covalent organic frameworks (COFs) designed for acid detection in various solutions [68]. This framework is synthesized through the Schiff-base condensation of N, N, N′, N′-tetrakis(4-aminophenyl)-1,4-phenylenediamine (TPBD) and terephthalaldehyde (TA). The COFs exhibit a dual response to acid, characterized by a color transition from red to dark red and fluorescence quenching as the acid concentration increases.

Mohd Hizam et al. developed an electrochemical sensor for the detection of ammonium ions, utilizing a metal–organic framework-derived tungsten ethoxide/polypyrrole-reduced graphene oxide (MOFs-W(OCH$_2$CH$_3$)$_6$/Ppy-rGO) [69]. The results indicate that this sensor demonstrates high performance in detecting NH^{4+} ions, suggesting its significant potential as a platform for applications in the agricultural sector.

In summary, the original research articles and communications featured in this Special Issue contribute valuable insights into the development and application of MOF-based chemical sensors and biosensors. The presented examples and advancements highlight the potential for integrating various types of transducers with functional MOFs for both biosensing and chemical sensing applications. This integration is likely to stimulate increased interest in the field, fostering new research initiatives within the scientific community.

Acknowledgments: We extend our gratitude to all authors who submitted their outstanding papers for consideration in this Special Issue. We also appreciate the reviewers for their diligent efforts in evaluating and enhancing the manuscripts throughout the publication process. Furthermore, we express our sincere thanks to the members of the *Chemosensors* Editorial Office for facilitating this opportunity and for their ongoing support in the management and organization of this Special Issue. We also thank the funding support from National Natural Science Foundation of China (Grant Nos. 32271427, 32471433, and 32071370).

Conflicts of Interest: The authors declare no conflicts of interest.

References

1. Zhang, J.; Wang, Z.; Suo, J.; Tuo, C.; Chen, F.; Chang, J.; Zheng, H.; Li, H.; Zhang, D.; Fang, Q.; et al. Morphological Tuning of Covalent Organic Framework Single Crystals. *J. Am. Chem. Soc.* **2024**, *146*, 35090–35097. [CrossRef]
2. Yin, L.; Huang, J.-B.; Yue, T.-C.; Wang, L.-L.; Wang, D.-Z. Synthesis, magnetic and dye adsorption properties of three metal-organic frameworks based on purine carboxylic acid. *J. Mol. Struct.* **2025**, *1319*, 139598. [CrossRef]
3. Patra, K.; Pal, H. Lanthanide-based metal–organic frameworks (Ln-MOFs): Synthesis, properties and applications. *RSC Sustain.* **2025**, *402*, 629–660. [CrossRef]
4. Tian, S.-Y.; Ding, Z.; Yao, M.-L.; Liu, S.-M.; Hou, X.-Y.; Tang, L.; Yue, E.-L.; Wang, X.; Wang, J.-J. A novel highly selective Fe@UiO-67-BDA/GCE sensor for efficient detecting Hg^{2+}. *Microchem. J.* **2024**, *206*, 111670. [CrossRef]
5. Chen, W.; Wu, C. Synthesis, Functionalization, and Applications of Metal-Organic Frameworks in Biomedicine. *Dalton Trans.* **2018**, *47*, 2114–2133. [CrossRef]
6. Zhong, L.; Qian, J.; Wang, N.; Komarneni, S.; Hu, W. Metal–organic frameworks on versatile substrates. *J. Mater. Chem. A* **2023**, *11*, 20423–20458. [CrossRef]
7. Zou, W.; Zhang, L.; Lu, J.; Sun, D. Recent development of metal–organic frameworks in wound healing: Current status and applications. *Chem. Eng. J.* **2024**, *480*, 148220. [CrossRef]
8. Cai, W.; Wang, J.; Chu, C.; Chen, W.; Wu, C.; Liu, G. Metal-organic framework-based stimuli-responsive systems for drug delivery. *Adv. Sci.* **2019**, *6*, 1801526. [CrossRef]
9. Li, H.; Chang, X.; Wang, J.; Zhang, X.; Zhao, W.; Meng, F. Preparation and ferroelectric properties of metal-organic frame lithium ion liquid crystal composites. *Liq. Cryst.* **2024**, *5*, 1–11. [CrossRef]
10. Wu, Z.; Ye, Y.; Guo, Z.; Wu, X.; Zhang, L.; Huang, Z.; Chen, F. Stereoselective reduction of diarylmethanones via a ketoreductase@metal–organic framework. *Org. Biomol. Chem.* **2024**, *22*, 5198–5204. [CrossRef] [PubMed]
11. Molavi, H. Cerium-based metal-organic frameworks: Synthesis, properties, and applications. *Coord. Chem. Rev.* **2025**, *527*, 216405. [CrossRef]
12. Mendes, R.F.; Figueira, F.; Leite, J.P.; Gales, L.; Almeida Paz, F.A. Metal–organic frameworks: A future toolbox for biomedicine. *Chem. Soc. Rev.* **2020**, *49*, 9121–9153. [CrossRef] [PubMed]
13. Chen, W.; Kong, S.; Lu, M.; Chen, F.; Cai, W.; Du, L.; Wang, J.; Wu, C. Comparison of different zinc precursors for the construction of zeolitic imidazolate framework-8 artificial shells on living cells. *Soft Matter* **2020**, *16*, 270–275. [CrossRef] [PubMed]
14. Du, L.; Zhang, T.; Li, P.; Chen, W.; Wu, C. Zeolitic imidazolate framework-8/Bacterial Cellulose Composite for Iodine Loading and Its Antibacterial Performance. *Dalton Trans.* **2022**, *51*, 14317–14322. [CrossRef] [PubMed]
15. Bigham, A.; Islami, N.; Khosravi, A.; Zarepour, A.; Iravani, S.; Zarrabi, A. MOFs and MOF-Based Composites as Next-Generation Materials for Wound Healing and Dressings. *Small* **2024**, *20*, 2133903. [CrossRef]
16. Chen, W.; Chen, F.; Zhang, G.; Kong, S.; Cai, W.; Wang, J.; Du, L.; Wu, C. Fast decomposition of hydrogen peroxide by Zeolitic Imidazolate Framework-67 crystals. *Mater. Lett.* **2019**, *239*, 94–97. [CrossRef]
17. Natarajan, S.; Manna, K. Bifunctional MOFs in Heterogeneous Catalysis. *ACS Org. Inorg. Au* **2024**, *4*, 59–90. [CrossRef] [PubMed]

18. Vizuet, J.P.; Mortensen, M.L.; Lewis, A.L.; Wunch, M.A.; Firouzi, H.R.; McCandless, G.T.; Balkus, K.J. Fluoro-Bridged Clusters in Rare-Earth Metal–Organic Frameworks. *J. Am. Chem. Soc.* **2021**, *143*, 17995–18000. [CrossRef] [PubMed]
19. Chen, W.; Ca, W.; Liu, H.; Fu, L.; Lu, W.; Zhang, C.; ADale, M.; Da, C.; Pan, H.; Kong, S.; et al. Facially-controllable synthesis of zeolitic imidezolate framework-8 nanocrystal and its colloidal stability in phosphate buffered saline. *Mater. Chem. Phys.* **2020**, *245*, 122576. [CrossRef]
20. Chen, W.; Zhu, P.; Chen, Y.; Liu, Y.; Du, L.; Wu, C. UiO-66 Loaded Iodine/Poly(ε-caprolactone) with Low Iodine Content as An Effective Antibacterial Material. *Polymers* **2022**, *14*, 283. [CrossRef]
21. Le, V.N.; Kim, D.; Kim, J.; Othman, M.S. Freeze Granulation of Nanoporous UiO-66 Nanoparticles for Capture of Volatile Organic Compounds. *ACS Appl. Nano Mater.* **2021**, *4*, 8863–8871. [CrossRef]
22. Seal, N.; Palakkal, A.S.; Singh, M.; Goswami, R.; Pillai, R.S.; Neogi, S. Chemically Robust and Bifunctional Co(II)-Framework for Trace Detection of Assorted Organo-toxins and Highly Cooperative Deacetalization–Knoevenagel Condensation with Pore-Fitting-Induced Size-Selectivity. *ACS Appl. Mater. Interfaces* **2021**, *13*, 28378–28389. [CrossRef] [PubMed]
23. Xia, T.; Wan, Y.; Li, Y.; Zhang, J. Highly Stable Lanthanide Metal–Organic Framework as an Internal Calibrated Luminescent Sensor for Glutamic Acid, a Neuropathy Biomarker. *Inorg. Chem.* **2020**, *59*, 8809–8817. [CrossRef] [PubMed]
24. Wang, Q.; Astruc, D. State of the Art and Prospects in Metal–Organic Framework (MOF)-Based and MOF-Derived Nanocatalysis. *Chem. Rev.* **2020**, *120*, 1438–1511. [CrossRef] [PubMed]
25. Yin, Q.; Li, Y.-L.; Li, L.; Lü, J.; Liu, T.-F.; Cao, R. Novel Hierarchical Meso-Microporous Hydrogen-Bonded Organic Framework for Selective Separation of Acetylene and Ethylene versus Methane. *ACS Appl. Mater. Interfaces* **2019**, *11*, 17823–17827. [CrossRef] [PubMed]
26. Wu, C.; Geng, P.; Zhang, G.; Li, X.; Pang, H. Synthesis of Conductive MOFs and Their Electrochemical Application. *Small* **2024**, *20*, 2308264. [CrossRef]
27. Shubhangi; Nandi, I.; Rai, S.K.; Chandra, P. MOF-based nanocomposites as transduction matrices for optical and electrochemical sensing. *Talanta* **2024**, *266*, 125124. [CrossRef] [PubMed]
28. Gao, F.; Zhao, Y.; Dai, X.; Xu, W.; Zhan, F.; Liu, Y.; Wang, Q. Aptamer tuned nanozyme activity of nickel-metal–organic framework for sensitive electrochemical aptasensing of tetracycline residue. *Food Chem.* **2024**, *430*, 137041. [CrossRef] [PubMed]
29. Wei, W.; Zhou, S.; Ma, D.-D.; Li, Q.; Ran, M.; Li, X.; Wu, X.-T.; Zhu, Q.L. Ultrathin Conductive Bithiazole-Based Covalent Organic Framework Nanosheets for Highly Efficient Electrochemical Biosensing. *Adv. Funct. Mater.* **2023**, *33*, 2302917. [CrossRef]
30. Liu, J.; Li, B.; Lu, G.; Wang, G.; Zheng, J.; Huang, L.; Feng, Y.; Xu, S.; Jiang, Y.; Liu, N. Toward Selective Transport of Monovalent Metal Ions with High Permeability Based on Crown Ether-Encapsulated Metal–Organic Framework Sub-Nanochannels. *ACS Appl. Mater. Interfaces* **2024**, *16*, 26634–26642. [CrossRef] [PubMed]
31. Wang, J.; Wang, C.; Liu, S.; Zhang, Y.; Zhang, J.; Dang, W.; Wu, W.; Wang, J. Laminar Composite Electrolytes with Nanoporous Sulfonated Covalent Organic Framework-Confined Crown Ether for Solid-State Lithium–Sulfur Batteries. *ACS Appl. Nano Mater.* **2024**, *7*, 3774–3781. [CrossRef]
32. Wu, Q.; Liang, J.; Wang, D.; Wang, R.; Janiak, C. Host molecules inside metal–organic frameworks: Host@MOF and guest@host@MOF (Matrjoschka) materials. *Chem. Soc. Rev.* **2025**, *123*, 601–622. [CrossRef] [PubMed]
33. Peng, X.; Wu, X.; Zhang, M.; Yuan, H. Metal–Organic Framework Coated Devices for Gas Sensing. *ACS Sens.* **2023**, *8*, 2471–2492. [CrossRef] [PubMed]
34. Lu, X.; Jayakumar, K.; Wen, Y.; Hojjati-Najafabadi, A.; Duan, X.; Xu, J. Recent advances in metal-organic framework (MOF)-based agricultural sensors for metal ions: A review. *Microchim. Acta* **2024**, *191*, 58. [CrossRef] [PubMed]
35. Zhan, F.; Zhao, Y.; Dai, X.; Zeng, J.; Wang, Q. Electrochemically synthesized polyanine@Cu-BTC MOF as a bifunctional matrix for aptasensing of tetracycline in aquatic products. *Microchem. J.* **2024**, *196*, 109512. [CrossRef]
36. Yang, Z.-W.; Li, J.-J.; Wang, Y.-H.; Gao, F.-H.; Su, J.-L.; Liu, Y.; Wang, H.-S.; Ding, Y. Metal/covalent-organic framework-based biosensors for nucleic acid detection. *Coord. Chem. Rev.* **2023**, *491*, 215249. [CrossRef]
37. Mohan, B.; Priyanka; Singh, G.; Chauhan, A.; Pombeiro, A.J.L.; Ren, P. Metal-organic frameworks (MOFs) based luminescent and electrochemical sensors for food contaminant detection. *J. Hazard. Mater.* **2023**, *453*, 131324. [CrossRef] [PubMed]
38. Ameen, S.M.; Bedair, A.; Hamed, M.; Mansour, F.R.; Omer, K.M. Recent Advances in Metal–Organic Frameworks as Oxidase Mimics: A Comprehensive Review on Rational Design and Modification for Enhanced Sensing Applications. *ACS Appl. Mater. Interfaces* **2025**, *17*, 110–129. [CrossRef] [PubMed]
39. Cui, L.; Li, H.; Shi, W.; Jing, Y.; Sun, S.; Ai, S.; Guo, Z. The coordination effect of organic ligands in Ce-MOF brings about atomically dispersed Fe in CeO2 for TAC detection in commercial samples. *Talanta* **2025**, *285*, 127405. [CrossRef] [PubMed]
40. Karrat, A.; Benssbihe, J.; Ameen, S.M.; Omer, K.M.; Amine, A. Development of a Silver-Based MOF Oxidase-Like nanozyme modified with molecularly imprinted polymer for sensitive and selective colorimetric detection of quercetin. *Spectrochim. Acta Part A Mol. Biomol. Spectrosc.* **2025**, *330*, 125735. [CrossRef]
41. Chen, W.; Kong, S.; Wang, J.; Du, L.; Cai, W.; Wu, C. Enhanced fluorescent effect of graphitic C3N4@ZIF-8 nanocomposite contribute to its improved sensing capabilities. *RSC Adv.* **2019**, *9*, 3734–3739. [CrossRef] [PubMed]

42. Ameen, S.M.; Omer, K.M. Multifunctional MOF: Cold/hot adapted sustainable oxidase-like MOF nanozyme with ratiometric and color tonality for nitrite ions detection. *Food Chem.* **2025**, *462*, 141027. [CrossRef] [PubMed]
43. Bedair, A.; Hamed, M.; Ameen, S.M.; Omer, K.M.; Mansour, F.R. Polyphenolic antioxidant analysis using metal organic Frameworks: Theoretical foundations and practical applications. *Microchem. J.* **2024**, *205*, 111183. [CrossRef]
44. Du, L.; Chen, W.; Tian, Y.; Zhu, P.; Chen, Y.; Wu, C. Applications of functional metal-organic frameworks in biosensors. *Biotechnol. J.* **2020**, *16*, 1900424. [CrossRef]
45. Hua, Y.; Kukkar, D.; Brown, R.J.; Kim, K.H. Recent advances in the synthesis of and sensing applications for metal-organic framework-molecularly imprinted polymer (MOF-MIP) composites. *Crit. Rev. Environ. Sci. Technol.* **2022**, *53*, 258–289. [CrossRef]
46. Anik, U.; Timur, S.; Dursun, Z. Metal organic frameworks in electrochemical and optical sensing platforms: A review. *Microchim. Acta* **2019**, *186*, 196. [CrossRef] [PubMed]
47. Du, L.; Chen, W.; Wang, J.; Cai, W.; Kong, S.; Wu, C. Folic acid-functionalized zirconium metal-organic frameworks based electrochemical impedance biosensor for the cancer cell detection. *Sens. Actuators B Chem.* **2019**, *301*, 127073. [CrossRef]
48. Wang, J.; Yang, Z.; Chen, W.; Du, L.; Jiao, B.; Krause, S.; Wang, P.; Wei, Q.; Zhang, D.-W.; Wu, C. Modulated light-activated electrochemistry at silicon functionalized with metal-organic frameworks towards addressable DNA chips. *Biosens. Bioelectron.* **2019**, *146*, 111750. [CrossRef] [PubMed]
49. Hang, T.; Zhang, C.; Pei, F.; Yang, M.; Wang, F.; Xia, M.; Hao, O.; Lei, W. Magnetism-Functionalized Lanthanide MOF-on-MOF with Plasmonic Differential Signal Amplification for Ultrasensitive Fluorescence Immunoassays. *ACS Sens.* **2024**, *9*, 6779–6788. [CrossRef]
50. Chen, Y.; Tian, Y.; Zhu, P.; Du, L.; Chen, W.; Wu, C. Electrochemical activated conductive Ni-based MOFs for non-enzymatic sensors towards long-term glucose monitoring. *Front. Chem.* **2020**, *8*, 602752. [CrossRef] [PubMed]
51. Ghahari, A.; Raissi, H. Enhanced Antibiotic Pollutant Capture: Coupling Carbon Nanotubes with Covalent Organic Frameworks. *J. Phys. Chem. C* **2024**, *128*, 17141–17152. [CrossRef]
52. Zhang, W.; Sun, Q.; Zhu, Y.; Sun, J.; Wu, Z.; Tian, N. High-Performance Trimethylamine Sensor Based on an Imine Covalent Organic Framework. *ACS Sens.* **2024**, *9*, 3262–3271. [CrossRef] [PubMed]
53. Liu, Y.; Wang, M.; Hui, Y.; Tian, J.; Xu, J.; Lu, Z.; Yang, Z.; Guo, H.; Yang, W. Eu3+-Modified Covalent Organic Frameworks for the Detection of a Vinyl Chloride Monomer Exposure Biomarker. *ACS Sens.* **2024**, *9*, 315–324. [CrossRef]
54. Zhou, L.; Yang, R.; Li, X.; Dong, N.; Zhu, B.; Wang, J.; Lin, X.; Su, B. COF-Coated Microelectrode for Space-Confined Electrochemical Sensing of Dopamine in Parkinson's Disease Model Mouse Brain. *J. Am. Chem. Soc.* **2023**, *145*, 23727–23738. [CrossRef] [PubMed]
55. Zhou, K.; Jia, Z.; Zhou, Y.; Ding, G.; Ma, X.-Q.; Niu, W.; Han, S.-T.; Zhao, J.; Zhou, Y. Covalent Organic Frameworks for Neuromorphic Devices. *J. Phys. Chem. Lett.* **2023**, *14*, 7173–7192. [CrossRef] [PubMed]
56. Lv, Y.; Zhou, Y.; Li, C.; Lv, C.; Dong, H.; Xu, M.; Zhang, J.; Yan, M. In situ formation of boronic acid-based covalent organic frameworks for specific and ultra-sensitive electrochemical assay of glycosylated amyloid-beta proteins. *Talanta* **2025**, *285*, 127435. [CrossRef] [PubMed]
57. Li, Y.; Zhao, L.; Bai, Y.; Feng, F. Applications of covalent organic frameworks (COFs)-based sensors for food safety: Synthetic strategies, characteristics and current state-of-art. *Food Chem.* **2025**, *469*, 142495. [CrossRef] [PubMed]
58. Chen, Y.; Chen, W.; Tian, Y.; Zhu, P.; Liu, S.; Du, L.; Wu, C. DNA and RhoB-functionalized metal-organic framework for the sensitive fluorescent detection of liquid alcohol. *Microchem. J.* **2021**, *170*, 106688. [CrossRef]
59. Chen, W.; Tan, Y.; Zheng, H.; Wang, Z.; Qu, Z.; Wu, C. Advance in metal–organic frameworks hybrids-based biosensors. *Microchem. J.* **2024**, *206*, 111441. [CrossRef]
60. Ognjanović, M.; Marković, M.; Girman, V.; Nikolić, V.; Vranješ-Đurić, S.; Stanković, D.M.; Petković, B.B. Metal–Organic Framework-Derived CeO$_2$/Gold Nanospheres in a Highly Sensitive Electrochemical Sensor for Uric Acid Quantification in Milk. *Chemosensors* **2024**, *12*, 231. [CrossRef]
61. Piguillem, S.V.; Gomez, G.E.; Tortella, G.R.; Seabra, A.B.; Regiart, M.D.; Messina, G.A.; Fernández-Baldo, M.A. Paper-Based Analytical Devices Based on Amino-MOFs (MIL-125, UiO-66, and MIL-101) as Platforms towards Fluorescence Biodetection Applications. *Chemosensors* **2024**, *12*, 208. [CrossRef]
62. Li, J.; Ding, Y.; Shi, Y.; Liu, Z.; Lin, J.; Cao, R.; Wang, M.; Tan, Y.; Zong, X.; Qu, Z.; et al. A Zinc Oxide Nanorod-Based Electrochemical Aptasensor for the Detection of Tumor Markers in Saliva. *Chemosensors* **2024**, *12*, 203. [CrossRef]
63. Peng, Y.; Wang, C.; Li, G.; Cui, J.; Jiang, Y.; Li, X.; Wang, Z.; Zhou, X. The Efficient and Sensitive Detection of Serum Dopamine Based on a MOF-199/Ag@Au Composite SERS Sensing Structure. *Chemosensors* **2024**, *12*, 187. [CrossRef]
64. Si, W.; Jiao, Y.; Jia, X.; Gao, M.; Zhang, Y.; Gao, Y.; Zhang, L.; Duan, C. A Host–Guest Platform for Highly Efficient, Quantitative, and Rapid Detection of Nitroreductase. *Chemosensors* **2024**, *12*, 145. [CrossRef]
65. Zhang, Y.; Liu, C.; Yan, R.; Lei, C. Bipyridyl Ruthenium-Decorated Ni-MOFs on Carbon Nanotubes for Electrocatalytic Oxidation and Sensing of Glucose. *Chemosensors* **2024**, *12*, 39. [CrossRef]

66. AlNeyadi, S.S.; Alhassani, M.T.; Aleissaee, A.S.; AlMujaini, I. Pyrene-Derived Covalent Organic Framework Films: Advancements in Acid Vapor Detection. *Chemosensors* **2024**, *12*, 37. [CrossRef]
67. Fan, X.; Yang, S.; Huang, C.; Lu, Y.; Dai, P. Preparation and Enhanced Acetone-Sensing Properties of ZIF-8-Derived Co_3O_4@ZnO Microspheres. *Chemosensors* **2023**, *11*, 376. [CrossRef]
68. Ma, W.; Gu, Z.; Pan, G.; Li, C.; Zhu, Y.; Liu, Z.; Liu, L.; Guo, Y.; Xu, B.; Tian, W. Dual-Response Photofunctional Covalent Organic Framework for Acid Detection in Various Solutions. *Chemosensors* **2023**, *11*, 214. [CrossRef]
69. Mohd Hizam, S.M.; Mohamed Saheed, M.S. Facile Electrochemical Approach Based on Hydrogen-Bonded MOFs-Derived Tungsten Ethoxide/Polypyrrole-Reduced GO Nanocrystal for ppb Level Ammonium Ions Detection. *Chemosensors* **2023**, *11*, 201. [CrossRef]

Disclaimer/Publisher's Note: The statements, opinions and data contained in all publications are solely those of the individual author(s) and contributor(s) and not of MDPI and/or the editor(s). MDPI and/or the editor(s) disclaim responsibility for any injury to people or property resulting from any ideas, methods, instructions or products referred to in the content.

Article

Facile Electrochemical Approach Based on Hydrogen-Bonded MOFs-Derived Tungsten Ethoxide/Polypyrrole-Reduced GO Nanocrystal for ppb Level Ammonium Ions Detection

Sara Maira Mohd Hizam [1,2] and Mohamed Shuaib Mohamed Saheed [1,3,*]

[1] Centre of Innovative Nanostructures and Nanodevices (COINN), Universiti Teknologi PETRONAS, Seri Iskandar 32610, Perak, Malaysia
[2] Department of Fundamental and Applied Sciences, Universiti Teknologi PETRONAS, Seri Iskandar 32610, Perak, Malaysia
[3] Department of Mechanical Engineering, Universiti Teknologi PETRONAS, Seri Iskandar 32610, Perak, Malaysia
* Correspondence: shuaib.saheed@utp.edu.my

Abstract: Ammonium (NH_4^+) ions are a primary contaminant in the river and along the waterside near an agricultural area, therefore, necessitating sensitive detection of pollutants before irreversibly damaging environment. Herein, a new approach of metal-organic framework-derived tungsten ethoxide/polypyrrole-reduced graphene oxide (MOFs-W(OCH_2CH_3)$_6$/Ppy-rGO) electrochemical sensors are introduced. Through a simple hydrothermal process, Ppy-rGO is linked to tungsten ethoxide as an organic linker. This creates the MOFs-W(OCH_2CH_3)$_6$/Ppy-rGO nanocrystal through hydrogen bonding. The synergistic combination of tungsten ethoxide and Ppy-rGO provides three-fold advantages: stabilization of Ppy-rGO for extended usage, enabling detection of analytes at ambient temperature, and availability of multiple pathways for effective detection of analytes. This is demonstrated through excellent detection of NH_4^+ ions over a dynamic concentration range of 0.85 to 3.35 µM with a ppb level detection limit of 0.278 µM (9.74 ppb) and a quantitation limit of 0.843 µM (29.54 ppb). The increment in the concentration of NH_4^+ ions contributes to the increment in proton (H^+) concentration. The increment in proton concentration in the solution will increase the bonding activity and thus increase the conductivity. The cyclic voltammetry curves of all concentrations of NH_4^+ analytes at the operating potential window between −1.5 and 1.5 V exhibit a quasi-rectangular shape, indicating consistent electronic and ionic transport. The distinctive resistance changes of the MOFs-W(OCH_2CH_3)$_6$/Ppy-rGO to various NH_4^+ ion concentrations and ultrasensitive detection provide an extraordinary platform for its application in the agriculture industry.

Keywords: graphene; metal-organic framework; nanostructured materials; sensor design; water contamination

Citation: Mohd Hizam, S.M.; Mohamed Saheed, M.S. Facile Electrochemical Approach Based on Hydrogen-Bonded MOFs-Derived Tungsten Ethoxide/Polypyrrole-Reduced GO Nanocrystal for ppb Level Ammonium Ions Detection. *Chemosensors* **2023**, *11*, 201. https://doi.org/10.3390/chemosensors11030201

Academic Editors: Chunsheng Wu, Liping Du and Wei Chen

Received: 8 February 2023
Revised: 9 March 2023
Accepted: 15 March 2023
Published: 21 March 2023

Copyright: © 2023 by the authors. Licensee MDPI, Basel, Switzerland. This article is an open access article distributed under the terms and conditions of the Creative Commons Attribution (CC BY) license (https://creativecommons.org/licenses/by/4.0/).

1. Introduction

Ammonia or ammonium ions (NH_4^+) can be found largely in agricultural vegetation sites. Additionally, the major usage of ammonia in industry, especially fertilizer production for plantations, is a leading source of NH_4^+ ion discharge in running water sources such as rivers. NH_4^+ ions are the main part of the nitrogen cycle and the metabolic product of many substances [1]. At a low concentration of approximately <3 mM, NH_4^+ ions are an important nitrogen source for plants. However, it will be detrimental to plants and affect the balance of the ecosystem when it is above the threshold. NH_4^+ ions introduced to terrestrial and aquatic systems are a source of acidity in the water through the biological uptake of NH_4^+ ions contributed by pollution and the nitrification process, which produces hydrogen ions [2,3]. The abundance of NH_4^+ ions in ecosystems can be determined by many factors, including the chemical nature of the soil, the accumulation of organic compounds, temperature, pH, light, CO_2, and oxygenation [4]. Soils with low pH and anoxic conditions indicate higher ammonification than nitrification rates, which are rich in NH_4^+ ions. Therefore, comprehensive monitoring of NH_4^+ ion content in the environment especially close to industrial areas is crucial for early screening and subsequent preventive measures [5].

Numerous methods have been introduced for monitoring NH_4^+ ions in water surroundings. Ions are naturally unstable, so immediate quantification is crucial after collection. In the earliest methods of analysis, the most common classical methods used for NH_4^+ ion detection are spectrophotometric, such as ultraviolet-visible (UV-Vis), infrared (IR), and fluorimetric, and electroanalytical analyses, including potentiometric based on the use of ion-selective electrodes, amperometric, coulometric, and voltamperometric [6]. The method for the determination of NH_4^+ ions or NH_3 was proposed by Nessler in 1856. In this method, Nessler reagent or an alkaline solution of potassium tetraiodomercurate (II) (K_2HgI_4) would react with the ions to give a color complex [7,8]. Unfortunately, this method suffers from poor selectivity, and the elimination of other ions' interference is difficult. NH_4^+ ions were often determined by a colorimetric indophenol titration method and a potentiometric method with an ion-selective electrode [9]. Meanwhile, the ion chromatography method offered simultaneous determination of a few nitrogen-contained ions in small-volume samples and the possibility of using different detectors such as UV-vis and mass spectrometry [10]. However, this method has low selectivity and sensitivity with poor repeatability in the analyte detections.

Although the aforementioned techniques have been in practice for over a decade, they are complicated and require a long time for data processing. Amperometric and potentiometric methods are simpler methods that require selective electrodes for NH_4^+ ion detection due to ions' interference, particularly Fe^{2+}, Ni^{2+}, P^+, K^+, and Na^+ [11]. Though the sensor signal is not predicted to be highly affected by the presence of certain ions, the selectivity test is essential to conduct for the possibility of ion interference. This is to ensure the optimum sensing performance of the sensor for real time monitoring as these ions present in the natural water resources. Meanwhile, other techniques such as spectrophotometric, colorimetric, and fluorometric methods are straightforward and cost-effective. However, it requires elaborate sample pre-treatment to eliminate background interferences and the use of several reagents for color formation. Among the approaches to the detection of ammonium ions quantitatively, electrochemical detection gives the possibility of real-time monitoring with the advantages of high sensitivity and selectivity, low cost, and independence of the sample color and turbidity. Various types of ammonium ion sensors have been developed, with the main obstacle primarily focused on choosing suitable ion-selective components and enzymes.

The high potential of conducting polymers, 2-dimensional layered materials, and metal-organic frameworks (MOFs) has been widely recognized for sensing applications. MOFs consist of inorganic metal centers and organic linkers with a rich hydrogen bonding network that exhibits various unique features such as high porosity, tunable pore size, and large surface area. MOFs are an attractive proposition for sensing applications due to their ability to be tuned to detect specific analytes, thus offering good adsorption or active sites and thereby promoting lower detection limits. The choice of ligand or organic linker is highly crucial when designing MOFs with excellent electronic conductivity. MOFs with a large pore size are not enough for the efficient detection of toxic liquids. The active functionalizations on the pore surface can contribute to unique and specific molecular recognition between host porous frameworks and guest analytes, which then enhances the sensing performance of MOFs. Common ligands used as organic linkers are 1,3,5-benzenetricarboxylate (BTC) [12], tetrakis-4-carboxyphenyl ethylene (TCPE) [13], 2,5-dihydrobenzene-1,4-dicarboxylate (H_2HBDC) [13], hydroquinone [14], 5-(dihydroxyphosphoryl)isophthalic acid (DPPA) [15], and 2,5-dihydroxy terephthalic acid or benzene dicarboxylate (BDH-$(OH)_2$) [16]. The porosity, surface area, conductivity, and functionalization of the complex material can be tuned by the type of ligand.

There are two strategies to design and prepare good, functionalized MOFs with excellent proton-conductive properties. Firstly, by choosing and introducing acidic organic linkers, such as phosphate groups, sulfonic groups, carboxyl groups, and other groups as well, into the MOFs. Secondly, by adding additional protonic molecules or carriers into the pores of the MOFs. In this study, rGO is used as the organic linker, while the additional protonic molecules would be from the polypyrrole chain attached to the graphene oxide. Concerning ammonium ions, the ideal sorbent might have a high density of reactive or adsorption sites for ammonia and thus be prone to having an affinity for NH_4^+ ions. The incorporation of metal oxide and structural defects (rGO) can impart a high density, and thus metal-ammonia bonding can occur.

Graphene is a sp^2-hybridized carbon atom arranged in a two-dimensional hexagonal conjugated structure. The extraordinary physical properties, such as high carrier mobility with remarkable electrical and antimicrobial properties, are undeniably incredible. Graphene is also known for its great refractive index and optical transparency of up to 97.7% for single-layer graphene [17], its high mechanical strength with Young's modulus of 1 TPa, and its fracture strength of 130 GPa [18]. It has a remarkable thermal conductivity of approximately ~50,000 W/mK [19], a large surface area of 2630 m^2/g [20], and is thermally stable. Further, reduced graphene oxide (rGO) has spectacular flexibility that sustains sensing performance stability in devices even under extreme bending stress [21,22]. The GO is well-known for its strong oxidizing properties and can act as an oxidizing agent for the oxidative polymerization of pyrrole during the reduction process. This characteristic of graphene makes it a perfect candidate to be an organic ligand.

Likewise, tungsten oxide is a very promising material because it has a high theoretical capacity, excellent chemical stability, and good conductivity. It is an n-type semiconductor that works best as a negative electrode as it has many oxidation states and is known for its high energy and packing density with massive pseudocapacitance [23]. It has been widely used in many applications, such as secondary batteries [24], photocatalysis [25], electrochemical [26], dye-solar cells [27], sensing devices [28], and flexible and portable supercapacitors [29]. It also shows an exceptionally high surface area, which is good for sensing performance. Thus, the combination between tungsten ethoxide and pyrrole-reduced graphene oxide could enhance the surface area, mobility, and active sites for ion detection, therefore developing a MOF.

In this work, the synergistic combination of tungsten ethoxide and reduced graphene oxide would be beneficial for room temperature sensor operation as compared to the high-temperature operation of tungsten-based sensors as reported in various literature [30,31]. Metal-organic frameworks-derived tungsten ethoxide/polypyrrole-reduced graphene oxide (MOFs-W(OCH$_2$CH$_3$)$_6$/Ppy-rGO) nanocrystals as the sensing elements were developed, and their performance in detecting NH_4^+ ions at different concentrations (0, 0.84, 1.26, 1.68, 2.52, and 3.35 µM) were thoroughly studied. The as-synthesized MOFs-W(OCH$_2$CH$_3$)$_6$/Ppy-rGO nanocrystal is in black powder form, and it was mixed with Nafion and dimethylformamide (DMF) to become a viscous liquid. The MOFs-W(OCH$_2$CH$_3$)$_6$/Ppy-rGO solution is then dropped onto the working electrode of the screen-printed electrode (SPE). The Ppy-rGO is astonishingly well-known for its high mobility, which is due to its excellent electroactive surface sites. The addition of metal oxide, which is tungsten ethoxide, can enhance the performance of graphene. It is expected to be extremely sensitive to the presence of NH_4^+ ions. Consequently, the fabricated electrochemical sensor demonstrates the extremely sensitive determination of NH_4^+ ions over a concentration range of 0.84 µM to 3.35 µM with a limit of detection (LOD) of 0.385 µM, limit of quantification (LOQ) of 1.166 µM. The developed electrochemical sensor demonstrates good reproducibility and stability for the detection of NH_4^+ ions in water.

2. Materials and Methods

2.1. Materials

Freeze-dried powder of GO was received from Universiti Putra Malaysia (Serdang, Selangor, Malaysia) [32]. Pyrrole (C$_4$H$_5$N, 97%), ammonia solution (NH$_3$.H$_2$O), and N,N-dimethylformamide (C$_3$H$_7$NO) were obtained from Merck (Shah Alam, Malaysia). Ultrapure deionized water was obtained from THE Direct-Q 8 UV system (Darmstadt,

German). Tungsten(VI) ethoxide (W(OCH$_2$CH$_3$)$_6$, 99.8%) was received from Alfa Aesar (Ward Hill, MA, United States). Ethanol (CH$_3$CH$_2$OH) was obtained from VWR BDH Chemicals (Rosny-sous-Bois cedex, France). Phosphate buffer solution (PBS) of ultra-pure grade is obtained from 1st Base (Selangor, Malaysia) (containing 137 mM NaCl, 2.7 mM KCl, 10 mM Na$_2$HPO$_4$ and 1.8 mM KH$_2$PO$_4$; pH 7.4 at 25 °C; 0.1 M). Distilled water was obtained from Apex WS 4L (Kuala Lumpur, Malaysia). The screen-printed carbon electrode (SPCE) transducer, equipped with a working electrode with a diameter of 4 mm, a counter electrode made of carbon, a reference electrode or electric contacts made of Ag/AgCl, and a carbon ring, was purchased from Metrohm Sdn. Bhd. (Selangor, Malaysia).

2.2. Preparation of Polypyrrole-Reduced Graphene Oxide

The as-received GO powder (0.1 g) was mixed with pyrrole (0.1 mL) at a 1:1 mass-to-volume ratio and then subsequently subjected to a vaporizing process at 50 °C for a duration of 18 h. This process will reduce the GO, which can be clearly observed by the reduction of the carboxyl group and the hydroxyl group in Fourier transform infrared (FTIR) and X-ray photoelectron spectroscopy (XPS). This was followed by annealing at 120 °C for 5 h in a flow of argon gas at atmospheric pressure.

2.3. Preparation of Metal-Organic Frameworks Derived Tungsten Ethoxide/Polypyrrole-Reduced Graphene Oxide

Ppy-rGO is a polymer chain attached to a graphene structure with an oxygenated-functional group that would act as an organic ligand. Tungsten ethoxide provides metal ions with possible attachment or bonding to oxygenated-functional groups in Ppy-rGO. The well-prepared Ppy-rGO of 30 wt% was added to well-dissolved tungsten ethoxide in N,N-dimethylformamide (DMF). The 30 wt% of Ppy-rGO is chosen after optimization of weight percentage usage explained in the supplementary information referring to Figure S3B and Table S1. The mixture was then sonicated for 5 min before being stirred for 30 min and later poured into a 100 mL Teflon liner. The liner was then placed in a Parr acid digestion vessel and continuously heated to 150 °C for 4 days. The vessel was progressively cooled to room temperature for 24 h. The powder formed after 10 h of drying at 80 °C in the oven, as illustrated in Figure S1. The synthesized product was kept in the desiccator to avoid oxidation and contamination. The synthesis was similar to the previous work [33].

2.4. Characterizations

Fourier transform infrared spectroscopy (FTIR) spectra were obtained using Perkin Elmer (Waltham, MA, United States) with the sample in a KBr pellet. A linear baseline was removed from all spectra. X-ray photoelectron spectroscopy (XPS) spectra were received and analyzed using Thermo Scientific (NE Dawson Creek Drive Hillsboro, OR, United States) (Al Kα radiation, hv = 1486.7 eV). Raman spectra were acquired from Horiba Jobin Yvon Scientific (Kyoto, Japan) (HR800) with wavelengths ranging from 200–4000 nm and powers $\leq$ 300 mW. Brunauer, Emmett, and Teller (BET) data were received by N$_2$ adsorption at 77 K using a surface area and porosimetry analyzer (SAP) from Micromeretics (Tristar 3020) (Norcross, GA, United States) with a degassing temperature of 150 °C for a 240 min period. The BET methods were used to measure the pore size distribution and calculate the surface area of the nanocomposites. The total pore volume was measured at a relative pressure (P/P$_o$) of 0.99. X-ray Diffraction (XRD) spectra were obtained from Xpert3 powder PANalytical with step scan mode between 2θ = 10–90° and a scan rate of 0.01°/min using Cu Kα of 1.5406 Å. Elemental composition, which consists of carbon, hydrogen, nitrogen, sulfur, and oxygen, was analyzed using the CHNS/O elemental analyzer (Thermo Scientific, UKM, Malaysia). The surface morphology, phase structure, and elemental composition of the composites were studied using a field emission scanning electron microscope (FESEM) with energy dispersive X-ray spectroscopy (EDX) (Zeiss, Oberkochen, Germany, Supra 55VP) at 5 keV. The EDX spot pattern and mapping scanning analysis were performed by the FESEM attachment. High-resolution transmission electron microscopic (HRTEM) images were performed using a Hitachi (Tokyo, Japan) (HT7830) at magnifications of 10 to 300,000 times with a virtual depth of field.

2.5. Ammonium Ion Sensing Measurement

The screen-printed electrode (SPE), purchased from DropSens (DRP-110), Metrohm Malaysia Sdn Bhd, was composed of a carbon counter electrode, an Ag pseudo-reference electrode, and a printed graphene working electrode (diameter, Ø = 4 mm). The device was cleaned with 70% ethanol and dried at room temperature. 0.08 g of MOFs-W(OCH$_2$CH$_3$)$_6$/Ppy-rGO and 0.2 mL (2 × 10^{-3} wt%) Nafion were mixed with 1 mL DMF under ultrasonication for 30 min to obtain a black suspension. 1.2 µL of the suspension is dropped onto SPE and allowed to dry at room temperature. Metrohm Autolab/PGSTAT302N Potentiostat/Galvanostat NOVA 2.1 was used to measure the sensitivity using EIS and CV. For stability analysis, both EIS and CV were performed three times at room temperature for four weeks. EIS assays were conducted using five ammonia solutions with different concentrations spiked in distilled water (0.84 µM, 1.26 µM, 1.68 µM, 2.52 µM and 3.35 µM), at a standard potential of +0.10 V, using the sinusoidal amplitude of 10 mV and a frequency of 50 Hz, logarithmically distributed over a frequency range of 10–100 kHz. Impedance data were fitted to a [R (RQ)] circuit to determine the simplified Randles circuit using NOVA 2.1 software. This circuit included solution resistance (R_s), constant phase element (CPE), and charge transfer resistance (R_{ct}) components, where CPE and R_{ct} are in parallel. The 0 µM ammonia concentration means the W-MOF/Ppy-rGO/SPE device was analyzed by EIS and CV analyses in PBS. The solution was prepared by mixing PBS with ultrapure water in a 1:9 ratio. Meanwhile, CV measurements were conducted in the same ammonia solution using a potential scan from −1.0 to +1.0 V at 100 mV/s.

3. Results and Discussion

Thermal and chemical reduction routes are the most commonly utilized techniques to prepare rGO from GO. In this study, chemical reduction is used to synthesize Ppy-rGO nanocomposites with pyrrole as the oxidizing agent as it is a more feasible method that produces a higher percentage yield and gives a better response to the targeted analytes [34]. In addition, hydrothermal is the most preferable method to obtain the MOFs-W(OCH$_2$CH$_3$)$_6$/Ppy-rGO due to the excellent sensing performance reported [35]. GO was synthesized using a simplified version of Hummer's method [32]. When pyrrole was added, the GO sheets became crumpled and wrinkled (Figure 1e) due to the presence of oxygen functional groups. The folded sheets also confirmed the growth of polypyrrole along the GO sheets. The tungsten nanoparticles were observed to be dispersed evenly on the surface of Ppy-rGO wrinkled sheets, which can be seen in Figure 1f. The degradation FESEM images of Ppy-rGO and MOFs-W(OCH$_2$CH$_3$)$_6$/Ppy-rGO are less comparable as shown in Figure S7.

As for the BET linear isotherm, it is observed to be a Type II isotherm based on the linear plot. At lower relative pressure, there is positive isothermal curvature, giving the first isothermal knee. This is due to the fact that the initial absorption occurred on an energetically more favorable surface area before being located on a less favorable one at low pressure. This indicates that most small pores will be filled up with a monolayer in the initial phase. Consequently, the monolayer absorption volume can be calculated, followed by specific surface areas that are to be 11.678, 19.263, and 6.312 m^2/g for GO, Ppy-rGO, and MOFs-W(OCH$_2$CH$_3$)$_6$/Ppy-rGO, respectively. The second region of absorption occurs at higher relative pressure which took place as saturation was approached. Essentially, the increment of absorption volume was asymptotic to the volume axis. At higher relative pressure, the profile showed an increase in adsorption volume. There were many possibilities for this occurrence, such as either multilayer adsorbate on adsorbate absorption of a non-porous sample or incomplete filling of a porous sample for nitrogen absorption at 77 K. This could also occur in a sample with a macroporous structure that possesses large pores to be filled by nitrogen molecules as saturation is approached. However, for this case, the pore sizes of GO, Ppy-rGO, and MOFs-W(OCH$_2$CH$_3$)$_6$/Ppy-rGO are 10.607, 11.152, and 18.886 nm, respectively. Thus, the porosity size range of 2 nm to 50 nm has been defined as mesoporous composites. There was a high probability of inconsistency in porosity size, which caused some pores to remain incompletely filled. This was supported by the appearance of a hysteresis loop at high relative pressure in the gas desorption isotherm linear plot as depicted in Figure 1g–i. Hysteresis was observed when there was more than one pore size. Moreover, hysteresis occurred due to the incomplete evaporation of nitrogen molecules from the poor neck or body of pores. Consequently, this will give less information on pore sizes from the desorption isotherm.

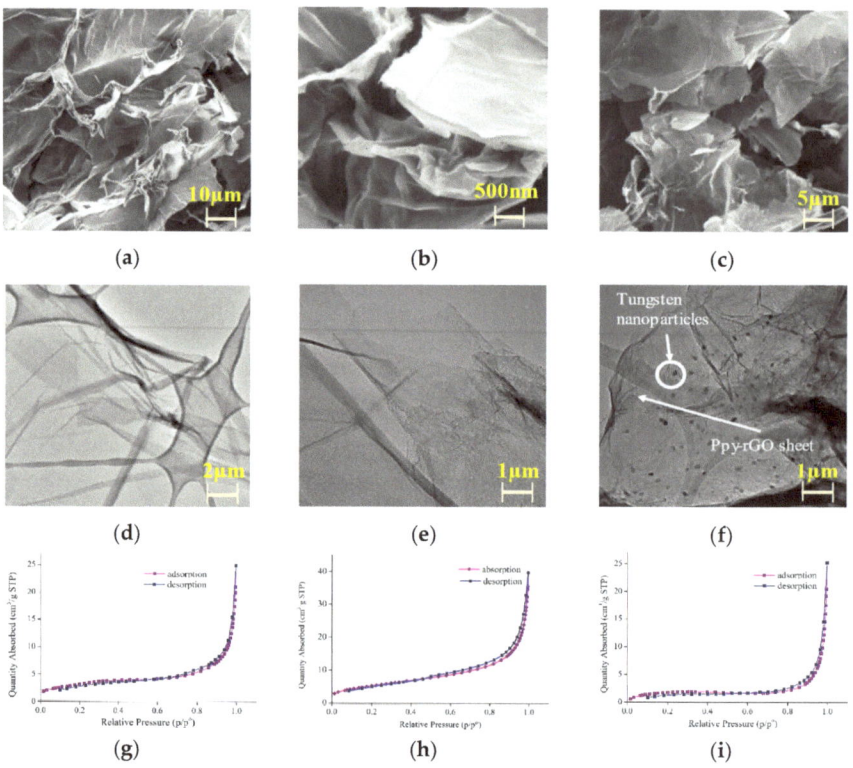

Figure 1. FESEM images of (**a**) GO; (**b**) Ppy-rGO; (**c**) MOFs-W(OCH$_2$CH$_3$)$_6$/Ppy-rGO; HRTEM images of (**d**) GO, (**e**) Ppy-rGO; (**f**) MOFs-W(OCH$_2$CH$_3$)$_6$/Ppy-rGO; and an isotherm linear plot for (**g**) GO and (**h**) Ppy-rGO; (**i**) MOFs-W(OCH$_2$CH$_3$)$_6$/Ppy-rGO.

Based on the calculated percentage of elements obtained from the carbon, hydrogen, nitrogen, sulfur, and oxygen databases (Table S2), it can be clearly seen that the reduction of GO occurs as the calculated number of oxygen atoms is reduced from 41 to 27. Meanwhile, nitrogen atoms were observed after reduction, attributed to the presence of a reducing agent. This result was supported by the hyperchromic shift observed in the Raman spectrum of Ppy-rGO as compared to GO (Figure S3A). Raman spectroscopy is used to distinguish ordered and disordered compound structures of carbon and solid-state materials. The upright shifting was caused by the addition of pyrrole as a reducing agent. Referring to the data in Table 1, the intensity increment in the D peak over the G peak is attributed to the increment in defect concentration present in Ppy-rGO relative to that of GO. The defect could be at the edges, a dislocation, cracks, or vacancies. Furthermore, the intensity ratios (I_D/I_G) of the GO and Ppy-rGO are 0.949 and 1.146, respectively. The I_D/I_G is used to determine the degree of disorder within the graphitic lattice. The increment in I_D/I_G of Ppy-rGO as compared to that of GO is attributed to the formation of the sp^2 domain, or unsaturated carbon [36]. The D and G peaks are left-shifted after the reduction of GO nanocomposites, as shown in Figure 2. The left-shifting occurred due to mechanical strain, polarity change, or charge transfer [37,38].

Table 1. Raman shifts and intensity ratios of GO, Ppy-rGO, and MOFs-W(OCH$_2$CH$_3$)$_6$/Ppy-rGO.

Nanocomposites	(cm^{-1})		I_D/I_G Ratio
	D Band	G Band	
GO	1352.7	1603.0	0.949
Ppy-rGO	1340.5	1572.8	1.146
MOFs-W(OCH$_2$CH$_3$)$_6$/Ppy-rGO	1357.7	1587.9	0.887

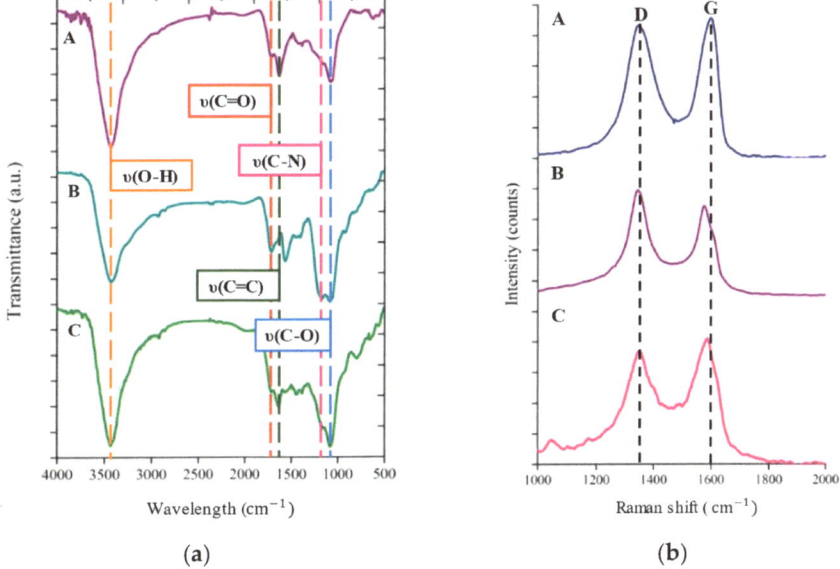

Figure 2. (a) FTIR spectra and (b) Raman spectra of (A) GO, (B) Ppy-rGO prepared in deionized water, and (C) MOFs-W(OCH$_2$CH$_3$)$_6$/Ppy-rGO prepared in DMF.

Meanwhile, the right-shifted D and G peaks of the MOFs-W(OCH$_2$CH$_3$)$_6$/Ppy-rGO Raman spectrum relative to Ppy-rGO were generated upon the addition of tungsten or a new disorder that was attributed to the modification of the fine structure of graphene. The right-shifting peak was related to the increment in stacked layers of graphene, which can be seen in color intensity in EDX images (Figure S2), multiple peaks in 2D peaks of the Raman spectrum (Figure S3A) [39], and the increment number of carbon in the CHNS/O database of MOFs-W(OCH$_2$CH$_3$)$_6$/Ppy-rGO (Table S2). The structure of MOFs-W(OCH$_2$CH$_3$)$_6$/Ppy-rGO nanocrystal was more disordered than the structure of Ppy-rGO. Table 1 shows the Raman spectra of GO, Ppy-rGO, and MOFs-W(OCH$_2$CH$_3$)$_6$/Ppy-rGO.

The degree of the defect increases simultaneously with an increment in the I_D/I_G ratio and intensity in the D peak over the G peak. The intensity ratio between Ppy-rGO and MOFs-W(OCH$_2$CH$_3$)$_6$/Ppy-rGO was decreased due to the electronic interaction between the metal oxide and Ppy-rGO nanomaterials, which forms multiple layers [40]. The decrease in intensity of the D band of the MOFs-W(OCH$_2$CH$_3$)$_6$/Ppy-rGO nanocrystal was attributed to the rearrangement of the sp^2 structure upon the bond interaction between them.

This result was supported by the low-intensity 2D peaks of MOFs-W(OCH$_2$CH$_3$)$_6$/Ppy-rGO nanocrystal, which indicates the low crystallinity of graphene. The low crystallinity of MOFs-W(OCH$_2$CH$_3$)$_6$/Ppy-rGO nanocrystals was related to the low intensity of the D peak over the G peak, or low I_D/I_G ratio. XRD spectra confirm the characteristic peaks of the as-prepared MOFs-W(OCH$_2$CH$_3$)$_6$/Ppy-rGO samples [41–44], as shown in Figure S5 and Table S2. Briefly, the I_D/I_G ratio will start to decrease as an increasing defect density results in a more amorphous carbon structure [45]. Peaks located at 2738.8 cm^{-1} in the GO spectrum, 2714.1 cm^{-1} in the Ppy-rGO spectrum, and 2736.0 cm^{-1} in the GO spectrum were 2D lines representing the overtone of the D line, which results from an inelastic scattering from a second phonon, as shown in Figure S3A.

The FTIR spectra of GO, Ppy-rGO, and MOFs-W(OCH$_2$CH$_3$)$_6$/Ppy-rGO nanocrystals were recorded in the region of 500–4000 cm^{-1}. More oxygen-functional groups, such as carboxyl, hydroxyl, and carbonyl groups, were observed with respect to the presence of the reducing agent, pyrrole. The presence of oxygenated functional groups decreased as reduction and hydrothermal processes were conducted. In Figure 2a, the intense peaks observed at 3435, 3434, and 3433 cm^{-1} are ascribed to the stretching vibration of O-H for GO, Ppy-rGO, and MOFs-W(OCH$_2$CH$_3$)$_6$/Ppy-rGO nanocomposites, respectively [46,47]. The peaks located at 1714, 1708, and 1675 cm^{-1} illustrated the stretching vibration of the C=O carboxyl group of GO, Ppy-rGO, and MOFs-W(OCH$_2$CH$_3$)$_6$/Ppy-rGO from a conjugated dimer acid [48,49], respectively. Since all the nanocomposites have a sp^2 hybridized conjugated hexagonal structure, it was possible to observe the dimer stretch peaks. Meanwhile, peaks located at 1630, 1659, and 1634 cm^{-1} were attributed to the C=C aromatic stretch peaks of GO, Ppy-rGO, and MOFs-W(OCH$_2$CH$_3$)$_6$/Ppy-rGO, respectively, for cyclic conjugated cyclic alkenes. Additionally, the peaks located at 1565 and 1563 cm^{-1} were ascribed to N-H bending of Ppy-rGO and MOFs-W(OCH$_2$CH$_3$)$_6$/Ppy-rGO, which described the interaction between pyrrole and GO after the reduction process. The peaks found at 1158, 1164, and 1131 cm^{-1} were due to the C-N stretch peaks of GO, Ppy-rGO, and MOFs-W(OCH$_2$CH$_3$)$_6$/Ppy-rGO, respectively. However, the high-intensity ratio for the C-N stretch peak observed in the Ppy-rGO spectrum showed the interaction between nitrogen and carbon atoms upon the reduction process. The peaks located at 1079, 1087, and 1078 cm^{-1} were attributed to C-O bending or GO, Ppy-rGO, and MOFs-W(OCH$_2$CH$_3$)$_6$/Ppy-rGO [50], respectively. Upon reduction of GO, the O-H stretch peak in the Ppy-rGO spectrum was reduced. Two possibilities might have occurred: first, the interaction of oxygen from hydroxyl groups with pyrrole, and second, the reduction of the hydroxyl group into a carbonyl group. In the FTIR spectrum shown in Figure 2, the transmittance of C-O increased as more epoxy groups were produced due to the interaction between tungsten ethoxide and Ppy-rGO. The O-H stretch peak at MOFs-W(OCH$_2$CH$_3$)$_6$/Ppy-rGO spectrum was greatly intense due to the possibility that some of the oxygen in tungsten ethoxide has changed into a hydroxyl group.

Figure 3 illustrates XPS spectra exhibiting the chemical composition in the GO, Ppy-rGO, and MOFs-W(OCH$_2$CH$_3$)$_6$/Ppy-rGO samples. Based on the spectra in Figure 4A, the profound peaks located at 284.3 eV and 284.8 eV represent C sp^2 for aromatic molecules of Ppy-rGO and MOFs-W(OCH$_2$CH$_3$)$_6$/Ppy-rGO nanocrystal, respectively. The peak found at 285.1 eV in the MOFs-W(OCH$_2$CH$_3$)$_6$/Ppy-rGO XPS C 1s spectrum represents C sp^3, caused by the rearrangement of the aromatic structure after the interaction between the metal oxide compound and Ppy-rGO. Further, the peaks found at 285.9, 287.8, and 290.9 eV represent the oxygen-containing functional groups such as (C=O) carbonyl, (C-O) hydroxyl, and (C-OOH) carboxyl [51], respectively. In Figure 3b, the sharp peak found at 532.8 eV, for both the Ppy-rGO and MOFs-W(OCH$_2$CH$_3$)$_6$/Ppy-rGO spectra, stands for the (C=O) carboxyl group and (W-O) [52]. Additionally, two peaks that were found at 533.3 and 535.3 eV both represent (C-O) carbonyl and (C-O) hydroxyl, respectively. As for the N1s spectra (Figure 3c), the strong peak located at 400.2 eV represents pyrrolic N molecules. Meanwhile, peaks located at 397.9, 398.5, 401.7, and 402.7 eV represent peaks for graphene-N, graphene-NH$_2$, graphitic N, and N-O, respectively [53,54]. According to the XPS spectra, the existence of an amide group in the rGO compound due to the pyrrole-reduction of GO and oxygen-containing functional groups was clearly observed. The absence of the N1s spectrum of GO can be proven from the full electron spectroscope (SPCA) energy spectrum in Figure S3C–E. The presence of oxygen-containing functional groups such as carbonyl, carboxyl, and hydroxyl improve the capability of the nanocomposites in the detection of ammonium analytes. All the assigned peaks were referred to the XPS Casa database.

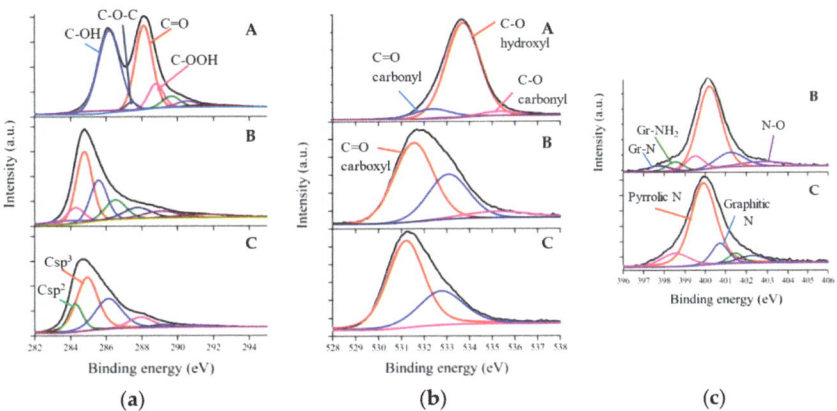

Figure 3. Deconvoluted XPS spectra of the (**a**) C1s (**b**) O1s and (**c**) N1s region of (A) GO, (B) Ppy-rGO, and (C) MOFs-W(OCH$_2$CH$_3$)$_6$/Ppy-rGO, respectively.

Stoichiometry tungsten ethoxide (W(OCH$_2$CH$_3$)$_6$) consists of covalent bonds formed by sharing a total of six electrons from the oxygen of ethoxy groups to fill up six valence electrons of the tungsten to achieve octet rule. The shared electrons are less attracted by the nucleus of W as it increases the Coulombian force of attraction for the remaining core-level electrons, leading to an increment in their binding energy (B.E.). Nevertheless, all electrons surround the nucleus of W metallic alone. As for WO$_3$, it is attracted to electronegative oxygen. The energy needed to eject electrons from the positive center (W or WO$_3$) is higher as compared to the neutral center or W in metal form. Therefore, the B.E. of the tungsten in WO$_6$ is much greater than that of W in metal form. Likewise, the value of B.E. of core level electrons will be observed as being lower if there is oxygen vacancy present in WO$_6$ than in stoichiometric WO$_6$. In Figure S4, WO$_6$ was observed to acquire a lesser B.E. value than those of metallic tungsten and lower than those of stoichiometric WO$_6$, affirming the existence of sub-stoichiometric WO$_6$. Tungsten was presented in two chemical states, namely W^{6+} and W^{5+}. The deconvolution of the 4f core level into a doublet was due to spin–orbit coupling of electrons. A spectrum split into a doublet proves there were two chemical states of tungsten [55,56]. An intense doublet at a higher energy level incredibly illustrates the W^{6+} oxidation states with B.E. located at 35.78 and 37.98 eV, which correspond to 4f$_{5/2}$ and 4f$_{7/2}$, respectively. In addition, a doublet with a considerably low intensity represents the W^{5+} oxidation state. The existence of W in the W^{5+} oxidation state verifies the oxygen deficiency that is responsible for the lower electrical conductivity of the W(OCH$_2$CH$_3$)$_6$ nanocrystal.

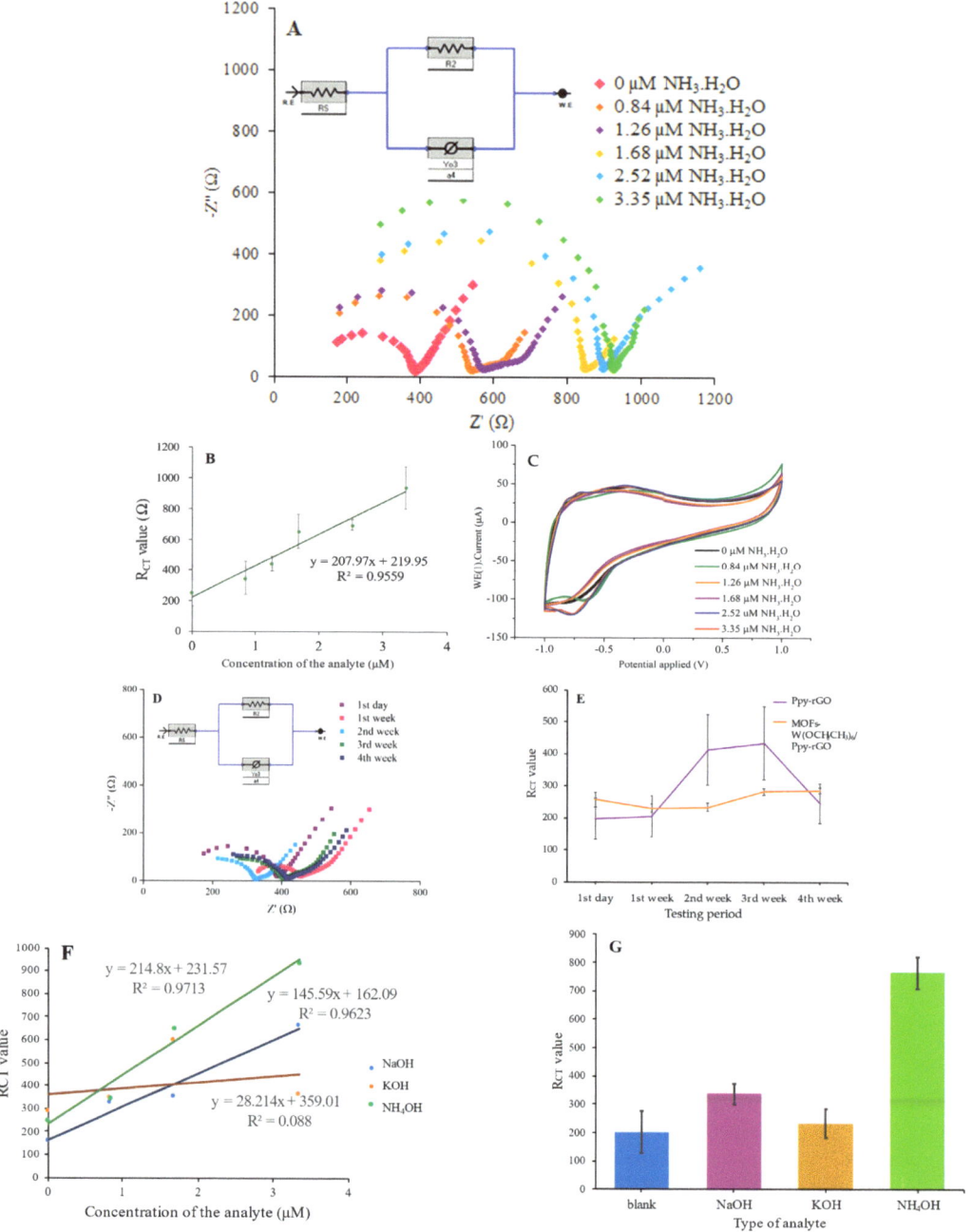

Figure 4. (**A**) EIS curve and the fitted equivalent circuit; (**B**) regression linear plot; (**C**) CV curve at different concentrations in an OPW between −1.0 to 1.0 V; (**D**) EIS curve of the stability test for MOFs-W(OCH$_2$CH$_3$)$_6$/Ppy-rGO; (**E**) the stability performance comparison of MOFs-W(OCH$_2$CH$_3$)$_6$/Ppy-rGO and Ppy-rGO; (**F**) regression linear plot of MOFs-W(OCH$_2$CH$_3$)$_6$/Ppy-rGO drop-casted by NaOH, KOH and NH$_4$OH analytes; and (**G**) the selectivity performance of MOFs-W(OCH$_2$CH$_3$)$_6$/Ppy-rGO upon detection of NH$_4^+$, K$^+$, and Na$^+$ ions.

EIS is a useful tool that can be used to find out about internal resistance and make electrochemical label-free detection easier. The resulting Nyquist plots state that a high-frequency range corresponds to the equivalent series resistance (R_s), and the medium- to high-frequency range corresponds to the charge transfer resistance (R_{ct}) of the cells. MOFs-W(OCH$_2$CH$_3$)$_6$/Ppy-rGO is an n-type semiconductor sensor. For an n-type sensor, the surface absorbs oxygen molecules when exposed to air. Eventually, the oxygen molecules absorbed on the surface absorb electrons from MOFs-W(OCH$_2$CH$_3$)$_6$/Ppy-rGO, which decreases the concentration of electrons and increases the resistance of the material. The chemical reaction can be described as follows [57,58]:

$$O_2 \text{ (g)} + e^- \rightarrow O_2^- \text{ (aq)} \tag{1}$$

$$O_2 \text{ (g)} + e^- \rightarrow 2O^{2-} \text{ (aq)} \tag{2}$$

$$2O^{2-} \text{ (aq)} \rightarrow 2O_2^- \text{ (aq)}^- \tag{3}$$

When surface-depleted metal oxides come into contact with the reducing ammonium ion, electrons will be transported back to the material to reduce its resistance. However, the resistance of the material increases as the concentration of ammonium solution increases. Presumably, when the ammonium hydroxide concentration is diluted to a certain lower concentration, two ions will be produced, specifically NH$_4^+$ and OH$^-$, as shown in Equation (4). The ammonium ions can dissociate into ammonia gas (NH$_3$) and hydrogen ions (H$^+$) as written in Equation (5).

$$NH_3 \cdot H_2O \text{ (l)} \rightarrow NH_4^+ \text{ (aq)} + OH^- \text{ (aq)} \tag{4}$$

$$NH_4^+ \text{ (aq)} \leftrightarrow NH_3 \text{ (g)} + H^+ \text{ (aq)} \tag{5}$$

$$WO_3 \text{ (aq)} + NH_3 \text{ (g)} \leftrightarrow [WO_3(NH_3)]^{5+} \tag{6}$$

$$[WO_3(NH_3)]^{5+} \text{ (aq)} + O_2 \text{ (g)} + 4H^+ \text{ (aq)} \leftrightarrow [WO_3(NH_3)(H_2O)_2]^{6+} \text{ (aq)} \tag{7}$$

$$[WO_3(NH_3)(H_2O)_2]^{6+} \text{ (aq)} \leftrightarrow [WO_3(NH_3)_2]^{5+} \text{ (aq)} + 2H_2O \text{ (l)} \tag{8}$$

The addition of ammonium ions caused the formation of a complex between W(VI) and neutral ammonia, which is easily oxidized by dissolved oxygen to W(VII) (Equation (6)) [59], followed by electrochemical reduction back to W(VI), which can be observed as a cathodic current increase (Equation (7)). The formation of tungsten and water is the product of a reduction reaction, in which water is the by-product (Equation (8)). That reduction reaction may cause a slight increment of resistance. As the concentration is higher, there will be an increment in the number of ions, which increases the resistance value as observed in Figure 4A.

The Nyquist plots portrayed a depressed semi-circle and a spike in the high- and low-frequency ranges. EIS was carried out to illustrate and explore the sensing performance of MOFs-W(OCH$_2$CH$_3$)$_6$/Ppy-rGO nanocomposite. The semi-circle diameter on the abscissa of the Nyquist plot in the high-frequency range of the electrochemical systems was attributed to the charge-transfer resistance (R_{ct}). Meanwhile, the lower value of R_{ct} indicated a higher conductivity with a high ion transfer or diffusion rate into the electrodes. Impedance measurements usually generate Nyquist plots that can be categorized into three types. (i) a depressed semi-circle, (ii) a tilted spike, or (iii) a depressed semi-circle Nyquist plot with a tilted spike. A depressed semi-circle was formed when a constant phase element with a resistor connected in parallel was present in an equivalent circuit, as shown in Figure 4A,D. The constant phase element (CPE) or "leaky capacitor," which is connected in series with a resistor, can be observed by the plot as a shape of a spike. Thus, the Nyquist plot that appears with a depressed semi-circle and a tilted spike corresponds to a combination of a parallel connection of a resistor and CPE that are connected in series with another resistor or CPE. The depressed semi-circle corresponds to the bulk material, while the tilted spike represents the electrical double layer. The electrochemical sensor showed remarkable stability with repetitive detection of NH_4^+ ions for four weeks consecutively at RT and a subsequent washing process with PBS solution (Table S5 and S6). The improved stability of MOFs-W(OCH$_2$CH$_3$)$_6$/Ppy-rGO as compared to Ppy-rGO is due to the presence of tungsten, as discussed earlier in Raman and XPS analyses. The redox peak current has increased correspondingly to an increment of scan rate from 20 mVs^{-1} to 100 mVs^{-1}, as confirmed by linear regression with an R^2 of 0.9559 as shown in Figure S6A.

The LOD and LOQ were used to measure the sensitivity of the sensor based on the concentration of ammonium solution. LOD is the minimum number of analytes a sensor can detect. Meanwhile, LOQ is the smallest amount of the minimum concentration of a substance that is possible to determine with acceptable precision, accuracy, and uncertainty. Both LOD and LOQ values were analyzed using regression analysis and calculated using the formula equation as followed:

$$\text{LOD} = 3.3 \times \frac{\text{SD}}{\text{m}} \quad (9)$$

$$\text{LOQ} = 10 \times \frac{\text{SD}}{\text{m}} \quad (10)$$

where SD is the standard deviation (μM), and m is the gradient of the linear equation of bulk resistance measured (Ω) versus the concentration of the analytes (μM). The standard deviation (SD) is calculated using the following formula:

$$\text{SD} = \frac{\text{SE}}{\sqrt{\text{N}}} \quad (11)$$

where SE represents the standard error after regression analysis. Meanwhile, N represents four different concentrations of analytes: 0 μM, 0.85 μM, 1.26 μM, 1.68 μM, 2.52 μM, and 3.35 μM. The MOFs-W(OCH$_2$CH$_3$)$_6$/Ppy-rGO sample was placed on SPE device and the analytes were drop cast on the sample accordingly starting from 0 μM analytes to the highest concentrated analytes.

In this regression analysis, the dependable variable is the concentration of the targeted analytes, whereas the independent variable is the R_{ct} values, which were tabulated in Table S4. The linear graph of R_{ct} against concentration is plotted as in Figure 4B to obtain the linear equation, which is Y = mX + C, where m is the gradient of the linear plot, C is the y-intercept, Y is the y-axis and X is the x-axis. The regression coefficient (R^2) can be obtained in two ways: through regression analysis and a linear plot. The R^2 value is a measure of how well a linear regression model fits the data. In other words, the closer the R^2 value is to 1, the more variability the model explains. The standard error (SE) obtained from the regression analysis was 37.565 μM. Hence, by using Equation (11), the SD was calculated to be 15.336 μM. Based on the calculation formulas in Equations (9) and (10), the LOD and LOQ obtained were 0.278 and 0.843 μM, respectively. The fitting curve of EIS in the electrolytes is shown in Figure 4A. The electrochemical performance of the MOFs-W(OCH$_2$CH$_3$)$_6$/Ppy-rGO SPE was evaluated every after drop-casted with a constant volume of ammonium solution electrolyte at 5 concentrations.

CV measurement is an electrochemical method that is widely used to understand the electrochemical behavior of materials. The CV curves were recorded at similar scan rates and operating potential windows (OPW) from −1.0 to 1.0 V, as shown in Figure 5c. CV measurements of the W-MOF/Ppy-rGO were performed at a scan rate of 100 mVs^{-1}. The CV curves portrayed a quasi–rectangular shape with no faradic peaks representing double–layer capacitance. Double–layer capacitance (C_{dl}) can also be expressed as a constant–phase element (CPE). The C_{dl} or CPE are obtained from the impedance if and only if the polarizable electrode is illustrated as a connection of the R_s and C_{dl}. The coupling of the resistance with the surface capacitance, the adsorption of ions, and the chemical inhomogeneities of the surface are the results of the polarizable electrode. The CV curve of every concentration exhibited a similar pattern at the same scan rate without any apparent distortion, which contributes to great electronic and ionic transport inside the supercapacitor electrode material [60].

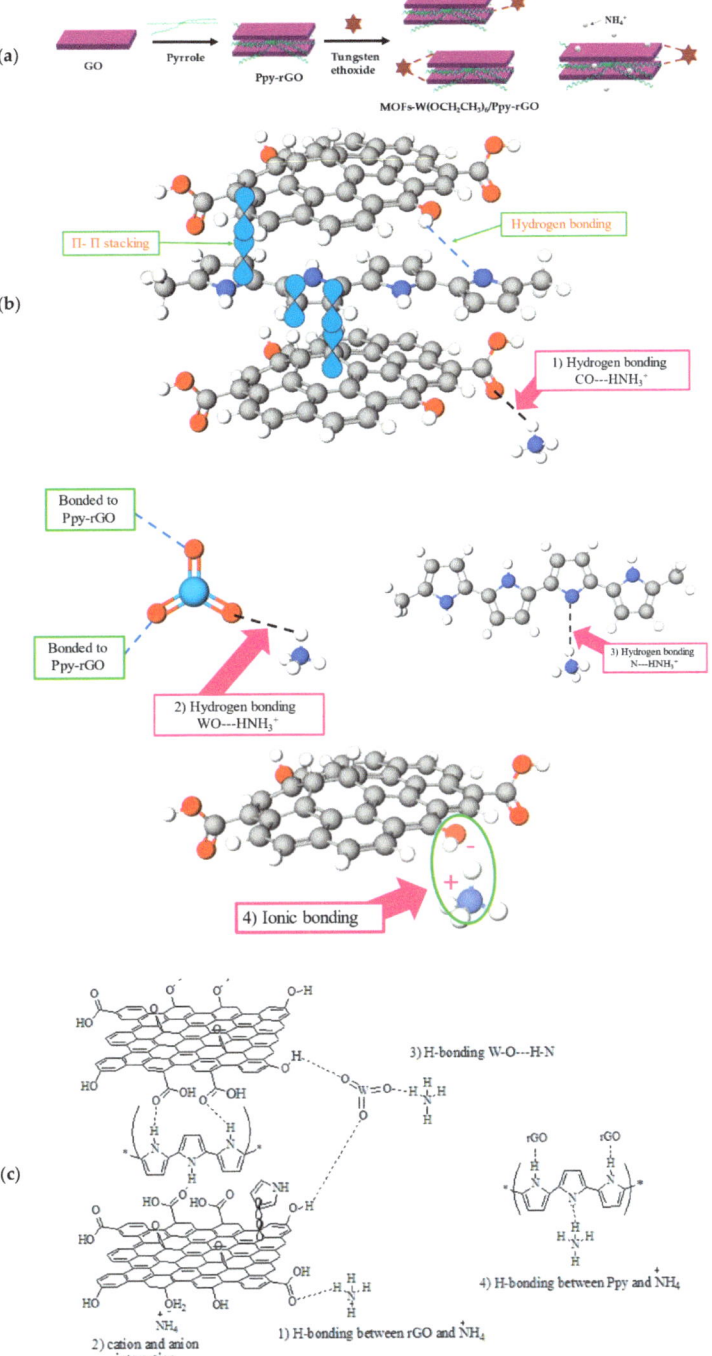

Figure 5. Schematic illustration of the (**a**) preparation of MOFs-W(OCH$_2$CH$_3$)$_6$/Ppy-rGO nanocrystal and detection of ammonium ions, (**b**,**c**) multiple ways of ion sensing of MOFs-W(OCH$_2$CH$_3$)$_6$/Ppy-rGO nanocrystal upon detection of ammonium ions.

Ammonium ion solution is a weak base. It dissociates partially in the aqueous solution, which gives insufficient NH_4^+ and OH^- ions and is thus less conductive. As the concentration increases, there will be an increase in the number of ions, which in turn increases the resistance value. Whereas the NaOH and KOH solutions are strong bases. They would dissociate completely into Na^+, K^+, and OH^- ions in an aqueous solution. In addition, they are very conductive and corrosive, which reduces the thickness of the MOFs-$W(OCH_2CH_3)_6$/Ppy-rGO nanocomposite. As a result, the R_s value shifted inconsistently, as observed in Figure S6B,C, which is correlated to the CV curve in Figure S6D,E. The R_s value is linearly proportional to the thickness of the drop-casted nanocomposites [61]. However, due to their high conductivity, the resistivity value of each concentration of Na^+ and K^+ ion solution is low. R_{ct} value are tabulated in the Table S7. In fact, resistivity is inversely proportional to conductivity values. Hence, as expected, the EIS-fitted curves of the strong base solution showed a low resistivity value.

PBS was dropped on the material to help the device maintain n-type behavior such that electrons are the majority carriers with a larger electron concentration as compared to hole concentration. This solution has critical effects on the sensing performance of MOFs-$W(OCH_2CH_3)_6$/Ppy-rGO. It consists of ions that influence the surface charges that exist on the nanocomposites' surface. After a quick wash with PBS solution, the ammonium solution was dropped onto the material, and EIS and CV analyses were performed. The synergistic combination of the n-type properties of tungsten and the intrinsic characteristics of rGO led to an increase in the number of active sites and in the electrical conductivity needed to detect ammonium ions spiked in distilled water. Polypyrrole was anchored on the surface of the rGO nanosheets by two types of interaction: hydrogen bonding and π-π stacking, as shown in Figure 5b. The combination of these compounds has resulted in a large surface area for the fast transport of charge carriers and high availability of active surface sites for NH_4^+ ions. The addition of tungsten ethoxide to Ppy-rGO has enhanced the sensitivity due to the availability of multiple pathways for the detection of NH_4^+. The performance of MOFs-$W(OCH_2CH_3)_6$/Ppy-rGO is attributed to the hydrogen bonding between $W(OCH_3CH_5)_6$ and Ppy-rGO, which led to the p-n junction structure [62] as shown in Figure 5. As the NH_4^+ ions are adsorbed onto the sensor, the resistance increases due to the presence of electrons from the conduction band, which were transferred via efficient pathways provided by rGO. Particularly, there are four possible ways of NH_4^+ ion detection: hydrogen bonding between pyrrole and analytes; phenoxy groups from rGO and analytes; tungsten ethoxy and the analytes; and ion interaction between oxygen anion and NH_4^+ ion (Figure 5).

There are many transduction techniques for NH_4^+ ion detection, such as potentiometry, differential pulse voltammetry, square-wave voltammetry, colorimetry, amperometry, conductometry, impedance spectrometry, spectrophotometry, field-effect transistors (FET), and CV. Common reagents used to detect NH_4^+ ions are Nessler, indophenol blue (IPB), Berthelot, and Riegler to be used in the spectrophotometry technique. This technique provides highly accurate data and presents obvious color changes with the presence of NH_4^+ ions, as shown in Table S8 [7–9,59,63–69]. However, the reagents used have a negative effect on the environment, as most of them are toxic and corrosive [70]. This optical approach also requires complicated sample pretreatment and experimental operations [71]. Another approach will be using a FET-based sensor, which is very convenient due to its fast measurement capabilities, is portable, and allows operation under harsh and corrosive environmental conditions [72]. However, the consistency and reliability of sensing performance will be poor due to the corrosion of electrodes.

4. Conclusions

In summary, facile synthesis of MOFs-$W(OCH_2CH_3)_6$/Ppy-rGO nanocrystals has been demonstrated through chemical reduction and hydrothermal techniques. The combination of graphene in tungsten-MOF provides high current mobility for improved NH_4^+ ion detection. The developed electrochemical sensor demonstrates high sensitivity as observed from the linear increment in the resistivity of the MOFs-W(OCH$_2$CH$_3$)$_6$/Ppy-rGO electrode against increasing concentrations of NH$_4^+$ ions. The MOFs-W(OCH$_2$CH$_3$)$_6$/Ppy-rGO electrode exhibited excellent stability towards NH$_4^+$ ion detection and good recovery for a four-week testing period at RT. The cyclic voltammetry showed a large redox peak current at a scan rate of 100 mVs^{-1}. The LOD and LOQ of the MOFs-W(OCH$_2$CH$_3$)$_6$/Ppy-rGO upon detection of NH$_4^+$ ions reached ppb levels at 0.278 µM (9.74 ppb) and 0.843 µM (29.54 ppb), respectively. The CV curves have remained consistent for all concentrations, showing excellent electronic and ionic transport in MOFs-W(OCH$_2$CH$_3$)$_6$/Ppy-rGO nanocomposites. The improved electrochemical performance of the MOFs-W(OCH$_2$CH$_3$)$_6$/Ppy-rGO after the drop-casted ammonium ions solution was attributed primarily to the synergistic effects of MOFs-W(OCH$_2$CH$_3$)$_6$/Ppy-rGO, which provide outstanding electrochemical performance and, in turn, high charge capacity. The mobility properties of Ppy-rGO were enhanced with the presence of tungsten ethoxide, resulting in better ion detection. The outcome of this work based on MOFs-W(OCH$_2$CH$_3$)$_6$/Ppy-rGO is expected to be of interest to sensing in agricultural areas.

Supplementary Materials: The following supporting information can be downloaded at: https://www.mdpi.com/article/10.3390/chemosensors11030201/s1. Figure S1: Schematic illustration of (A) preparation of MOFs-W(OCH$_2$CH$_3$)$_6$/Ppy-rGO-based electrochemical sensor device and analysis method, and (B) the schematic illustration of SPE device preparation and detection of NH$_4^+$ ions; Figure S2: Compilation of EDX image of (A) GO, (B) Ppy-rGO, and (C) MOFs-W(OCH$_2$CH$_3$)$_6$/Ppy-rGO, and elemental mapping of carbon for (D) GO, (E) Ppy-rGO, and (F) MOFs-W(OCH$_2$CH$_3$)$_6$/Ppy-rGO; nitrogen for (G) Ppy-rGO, and (H) MOFs-W(OCH$_2$CH$_3$)$_6$/Ppy-rGO; oxygen for (I) GO, (J) Ppy-rGO, and (K) MOFs-W(OCH$_2$CH$_3$)$_6$/Ppy-rGO; sulfur for (L) GO, (M) Ppy-rGO, and (N) MOFs-W(OCH$_2$CH$_3$)$_6$/Ppy-rGO; and tungsten for (O) MOFs-W(OCH$_2$CH$_3$)$_6$/Ppy-rGO, respectively; Figure S3: Full Raman spectra of (A) GO, Ppy-rGO, and MOFs-W(OCH$_2$CH$_3$)$_6$/Ppy-rGO, and (B) MOFs-W(OCH$_2$CH$_3$)$_6$/Ppy-rGO at 30 wt%, 60 wt%, and 90 wt% of Ppy-rGO added to tungsten ethoxide, and the full electron spectroscope for chemical analysis (SPCA) energy spectrum for (C) GO, (D) Ppy-rGO, and (E) MOFs-W(OCH$_2$CH$_3$)$_6$/Ppy-rGO; Table S1: Raman shifts and intensity ratios of GO, Ppy-rGO, and MOFs-W(OCH$_2$CH$_3$)$_6$/Ppy-rGO; Figure S4: Deconvoluted narrow scan XPS spectrum of the W4f of MOFs-W(OCH$_2$CH$_3$)$_6$/Ppy-rGO; Figure S5: XRD spectra of (a) GO, Ppy-rGO, and MOFs-W(OCH$_2$CH$_3$)$_6$/Ppy-rGO, and (b) W(OCH$_2$CH$_3$)$_6$; Table S2: Crystallite size and XRD data of the triclinic phase of the nanocomposites; Table S3: Empirical formula calculation based on CHNS/O database for GO, Ppy-rGO, and MOFs-W(OCH$_2$CH$_3$)$_6$/Ppy-rGO; Table S4: Concentrations of the targeted analytes and the recorded charge transfer resistance; Table S5: Weekly R$_{CT}$ value for four weeks of the testing period; Figure S6: (A) CV curve of W-MOF/Ppy-rGO at different scan rates, EIS curve of MOFs-W(OCH$_2$CH$_3$)$_6$/Ppy-rGO drop-casted by (B) NaOH and (C) KOH analytes, and CV curve of (D) NaOH and (E) KOH analytes; Table S6: Reproducibility test with concentrations of the targeted analytes and the recorded charge transfer resistance; Table S7: Selectivity test with concentrations of NH$_4^+$, Na$^+$, and K$^+$ analytes solution and recorded charge transfer resistance; Figure S7: FESEM images of (A) Ppy-rGO and (B) MOFs-W(OCH$_2$CH$_3$)$_6$/Ppy-rGO after degradation; Table S8: The performance comparison of the W-MOF/Ppy-rGO-SPE device as an ammonium ions sensor with the developed ammonium ions sensor. References [7–9,39,41–44,59,60,63–69] are cited in the supplementary materials.

Author Contributions: Writing–reviewing, methodology, and original draft preparation: S.M.M.H., Conceptualization, editing–reviewing supervision, project administration: M.S.M.S. All authors have read and agreed to the published version of the manuscript.

Funding: This research was funded by the Lloyd's Register Foundation International Consortium of Nanotechnologies (LRF-ICON) grant number 015ME0-117.

Institutional Review Board Statement: Not applicable.

Informed Consent Statement: Not applicable.

Data Availability Statement: Not applicable.

Acknowledgments: The authors extend their appreciation to Universiti Teknologi PETRONAS and the Center of Innovative Nanostructures and Nanodevices (COINN) for providing technical support and research facilities.

Conflicts of Interest: The authors declare no conflict of interest.

References

1. Vitousek, P.M.; Aber, J.D.; Howarth, R.W.; Likens, G.E.; Matson, P.A.; Schindler, D.W.; Schlesinger, W.H.; Tilman, D.G. Human alteration of the global nitrogen cycle: Sources and consequences. *Ecol. Appl.* **1997**, *7*, 737–750. [CrossRef]
2. Esteban, R.; Ariz, I.; Cruz, C.; Moran, J.F. Mechanisms of ammonium toxicity and the quest for tolerance. *Plant Sci.* **2016**, *248*, 92–101. [CrossRef] [PubMed]
3. Ouyang, Y.; Norton, J.M.; Stark, J.M.; Reeve, J.R.; Habteselassie, M.Y. Ammonia-oxidizing bacteria are more responsive than archaea to nitrogen source in an agricultural soil. *Soil Biol. Biochem.* **2016**, *96*, 4–15. [CrossRef]
4. Vega-Mas, I.; Marino, D.; Sanchez-Zabala, J.; Gonzalez-Murua, C.; Estavillo, J.M.; González-Moro, M.B. CO_2 enrichment modulates ammonium nutrition in tomato adjusting carbon and nitrogen metabolism to stomatal conductance. *Plant Sci.* **2015**, *241*, 32–44. [CrossRef] [PubMed]
5. Zheng, M.; Chen, J.; Zhang, L.; Cheng, Y.; Lu, C.; Liu, Y.; Singh, A.; Trivedi, M.; Kumar, A.; Liu, J. Metal Organic Framework as an Efficient Adsorbent for Drugs from Wastewater. *Mater. Today Commun.* **2022**, *31*, 103514. [CrossRef]
6. Moorcroft, M.J.; Davis, J.; Compton, R.G. Detection and determination of nitrate and nitrite: A review. *Talanta* **2001**, *54*, 785–803. [CrossRef] [PubMed]
7. Amirjani, A.; Fatmehsari, D.H. Colorimetric detection of ammonia using smartphones based on localized surface plasmon resonance of silver nanoparticles. *Talanta* **2018**, *176*, 242–246. [CrossRef]
8. Ling, T.L.; Ahmad, M.; Heng, L.Y. UV-vis spectrophotometric and artificial neural network for estimation of ammonia in aqueous environment using cobalt (II) ions. *Anal. Methods* **2013**, *5*, 6709–6714. [CrossRef]
9. Azmi, N.E.; Ahmad, M.; Abdullah, J.; Sidek, H.; Heng, L.Y.; Karuppiah, N. Biosensor based on glutamate dehydrogenase immobilized in chitosan for the determination of ammonium in water samples. *Anal. Biochem.* **2009**, *388*, 28–32. [CrossRef]
10. Jackson, P.E. Ion chromatography in environmental analysis. In *Encyclopedia of Analytical Chemistry*; John Wiley & Sons Ltd.: Chichester, UK, 2000; pp. 2779–2801.
11. Kazanskaya, N.; Kukhtin, A.; Manenkova, M.; Reshetilov, N.; Yarysheva, L.; Arzhakova, O.; Volynskii, A.; Bakeyev, N. FET-based sensors with robust photosensitive polymer membranes for detection of ammonium ions and urea. *Biosens. Bioelectron.* **1996**, *11*, 253–261. [CrossRef]
12. Bhardwaj, S.K.; Mohanta, G.C.; Sharma, A.L.; Kim, K.-H.; Deep, A. A three-phase copper MOF-graphene-polyaniline composite for effective sensing of ammonia. *Anal. Chim. Acta* **2018**, *1043*, 89–97. [CrossRef] [PubMed]
13. Shustova, N.B.; Cozzolino, A.F.; Reineke, S.; Baldo, M.; Dincă, M. Selective turn-on ammonia sensing enabled by high-temperature fluorescence in metal–organic frameworks with open metal sites. *J. Am. Chem. Soc.* **2013**, *135*, 13326–13329. [CrossRef]
14. Li, Y.-P.; Li, S.-N.; Jiang, Y.-C.; Hu, M.-C.; Zhai, Q.-G. A semiconductor and fluorescence dual-mode room-temperature ammonia sensor achieved by decorating hydroquinone into a metal–organic framework. *Chem. Commun.* **2018**, *54*, 9789–9792. [CrossRef]
15. Liu, R.; Zhao, L.; Yu, S.; Liang, X.; Li, Z.; Li, G. Enhancing Proton Conductivity of a 3D Metal–Organic Framework by Attaching Guest NH3 Molecules. *Inorg. Chem.* **2018**, *57*, 11560–11568. [CrossRef] [PubMed]
16. Katz, M.J.; Howarth, A.J.; Moghadam, P.Z.; DeCoste, J.B.; Snurr, R.Q.; Hupp, J.T.; Farha, O.K. High volumetric uptake of ammonia using Cu-MOF-74/Cu-CPO-27. *Dalton Trans.* **2016**, *45*, 4150–4153. [CrossRef] [PubMed]
17. Nair, R.R.; Blake, P.; Grigorenko, A.N.; Novoselov, K.S.; Booth, T.J.; Stauber, T.; Peres, N.M.; Geim, A.K. Fine structure constant defines visual transparency of graphene. *Science* **2008**, *320*, 1308. [CrossRef] [PubMed]
18. Lee, C.; Wei, X.; Kysar, J.W.; Hone, J. Measurement of the elastic properties and intrinsic strength of monolayer graphene. *Science* **2008**, *321*, 385–388. [CrossRef]
19. Balandin, A.A.; Ghosh, S.; Bao, W.; Calizo, I.; Teweldebrhan, D.; Miao, F.; Lau, C.N. Superior thermal conductivity of single-layer graphene. *Nano Lett.* **2008**, *8*, 902–907. [CrossRef]
20. Demon, S.Z.N.; Kamisan, A.I.; Abdullah, N.; Noor, S.A.M.; Khim, O.K.; Kasim, N.A.M.; Yahya, M.Z.A.; Manaf, N.A.A.; Azmi, A.F.M.; Halim, N.A. Graphene-based Materials in Gas Sensor Applications: A Review. *Sens. Mater.* **2020**, *32*, 759–777. [CrossRef]
21. Singh, E.; Meyyappan, M.; Nalwa, H.S. Flexible graphene-based wearable gas and chemical sensors. *ACS Appl. Mater. Interfaces* **2017**, *9*, 34544–34586. [CrossRef]
22. Gong, X.; Liu, G.; Li, Y.; Yu, D.Y.W.; Teoh, W.Y. Functionalized-graphene composites: Fabrication and applications in sustainable energy and environment. *Chem. Mater.* **2016**, *28*, 8082–8118. [CrossRef]
23. Zheng, H.; Tachibana, Y.; Kalantar-Zadeh, K. Dye-sensitized solar cells based on WO_3. *Langmuir* **2010**, *26*, 19148–19152. [CrossRef]
24. Cao, J.; Zhang, D.; Yue, Y.; Wang, X.; Srikhaow, A.; Sriprachuabwong, C.; Tuantranont, A.; Zhang, X.; Wu, Z.-S.; Qin, J. Strongly coupled tungsten oxide/carbide heterogeneous hybrid for ultrastable aqueous rocking-chair zinc-ion batteries. *Chem. Eng. J.* **2021**, *426*, 131893. [CrossRef]
25. Yang, G.; Zhu, X.; Cheng, G.; Chen, R.; Xiong, J.; Li, W.; Wei, Y. Engineered tungsten oxide-based photocatalysts for CO_2 reduction: Categories and roles. *J. Mater. Chem. A* **2021**, *9*, 22781–22809. [CrossRef]

26. Han, W.; Shi, Q.; Hu, R. Advances in electrochemical energy devices constructed with tungsten oxide-based nanomaterials. *Nanomaterials* **2021**, *11*, 692. [CrossRef] [PubMed]
27. Zhang, Y.; Yun, S.; Wang, C.; Wang, Z.; Han, F.; Si, Y. Bio-based carbon-enhanced tungsten-based bimetal oxides as counter electrodes for dye-sensitized solar cells. *J. Power Sources* **2019**, *423*, 339–348. [CrossRef]
28. Jeevitha, G.; Mangalaraj, D. Ammonia sensing at ambient temperature using tungsten oxide (WO_3) nanoparticles. *Mater. Today Proc.* **2019**, *18*, 1602–1609. [CrossRef]
29. Gupta, S.P.; More, M.A.; Late, D.J.; Walke, P.S. High-rate quasi-solid-state hybrid supercapacitor of hierarchical flowers of hydrated tungsten oxide nanosheets. *Electrochim. Acta* **2021**, *366*, 137389. [CrossRef]
30. Jaroenapibal, P.; Boonma, P.; Saksilaporn, N.; Horprathum, M.; Amornkitbamrung, V.; Triroj, N. Improved NO_2 sensing performance of electrospun WO_3 nanofibers with silver doping. *Sens. Actuators B Chem.* **2018**, *255*, 1831–1840. [CrossRef]
31. Ma, J.; Ren, Y.; Zhou, X.; Liu, L.; Zhu, Y.; Cheng, X.; Xu, P.; Li, X.; Deng, Y.; Zhao, D. Pt nanoparticles sensitized ordered mesoporous WO_3 semiconductor: Gas sensing performance and mechanism study. *Adv. Funct. Mater.* **2018**, *28*, 1705268. [CrossRef]
32. Ban, F.; Majid, S.R.; Huang, N.M.; Lim, H. Graphene oxide and its electrochemical performance. *Int. J. Electrochem. Sci.* **2012**, *7*, 4345–4351.
33. Hizam, S.M.M.; Soaid, N.I.; Saheed, M.S.M.; Mohamed, N.M.; Kait, C.F. Study of Electrical Conductivity of Pyrrole-Reduced Graphene Oxide Pellet. In Proceedings of the 2021 IEEE International Conference on Sensors and Nanotechnology (SENNANO), Port Dickson, Malaysia, 22–24 September 2021; pp. 150–154.
34. Wang, Y.; Zhang, L.; Hu, N.; Wang, Y.; Zhang, Y.; Zhou, Z.; Liu, Y.; Shen, S.; Peng, C. Ammonia gas sensors based on chemically reduced graphene oxide sheets self-assembled on Au electrodes. *Nanoscale Res. Lett.* **2014**, *9*, 251. [CrossRef] [PubMed]
35. Esfandiar, A.; Irajizad, A.; Akhavan, O.; Ghasemi, S.; Gholami, M.R. Pd–WO_3/reduced graphene oxide hierarchical nanostructures as efficient hydrogen gas sensors. *Int. J. Hydrogen Energy* **2014**, *39*, 8169–8179. [CrossRef]
36. Lee, A.Y.; Yang, K.; Anh, N.D.; Park, C.; Lee, S.M.; Lee, T.G.; Jeong, M.S. Raman study of D* band in graphene oxide and its correlation with reduction. *Appl. Surf. Sci.* **2021**, *536*, 147990. [CrossRef]
37. Xu, H.; Wu, X.; Li, X.; Luo, C.; Liang, F.; Orignac, E.; Zhang, J.; Chu, J. Properties of graphene-metal contacts probed by Raman spectroscopy. *Carbon* **2018**, *127*, 491–497. [CrossRef]
38. Vidano, R.; Fischbach, D.; Willis, L.; Loehr, T. Observation of Raman band shifting with excitation wavelength for carbons and graphites. *Solid State Commun.* **1981**, *39*, 341–344. [CrossRef]
39. Saheed, M.S.M.; Mohamed, N.M.; Singh, B.S.M.; Saheed, M.S.M. Precursor and pressure dependent 3D graphene: A study on layer formation and type of carbon material. *Diamond Relat. Mater.* **2017**, *79*, 93–101. [CrossRef]
40. Peng, F.; Wang, S.; Yu, W.; Huang, T.; Sun, Y.; Cheng, C.; Chen, X.; Hao, J.; Dai, N. Ultrasensitive ppb-level H_2S gas sensor at room temperature based on WO_3/rGO hybrids. *J. Mater. Sci. Mater. Electron.* **2020**, *31*, 5008–5016. [CrossRef]
41. Wojtoniszak, M.; Chen, X.; Kalenczuk, R.J.; Wajda, A.; Łapczuk, J.; Kurzewski, M.; Drozdzik, M.; Chu, P.K.; Borowiak-Palen, E. Synthesis, dispersion, and cytocompatibility of graphene oxide and reduced graphene oxide. *Colloids Surf. B* **2012**, *89*, 79–85. [CrossRef]
42. Zhu, Y.; Murali, S.; Cai, W.; Li, X.; Suk, J.W.; Potts, J.R.; Ruoff, R.S. Graphene and graphene oxide: Synthesis, properties, and applications. *Adv. Mater.* **2010**, *22*, 3906–3924. [CrossRef]
43. Zhang, Y.; Park, S.-J. In situ shear-induced mercapto group-activated graphite nanoplatelets for fabricating mechanically strong and thermally conductive elastomer composites for thermal management applications. *Compos. Part A Appl. Sci. Manuf.* **2018**, *112*, 40–48. [CrossRef]
44. Ansari, S.M.; Khan, M.Z.; Anwar, H.; Ikram, M.; Sarfraz, Z.; Alam, N.; Khan, Y. Tungsten Oxide–reduced Graphene Oxide Composites for Photoelectrochemical Water Splitting. *Arab. J. Sci. Eng.* **2021**, *46*, 813–825. [CrossRef]
45. Childres, I.; Jauregui, L.A.; Park, W.; Cao, H.; Chen, Y.P. Raman spectroscopy of graphene and related materials. In *New Developments in Photon and Materials Research*; Nova Science: Hauppauge, NY, USA, 2013; Volume 1, pp. 1–20.
46. Hu, N.; Yang, Z.; Wang, Y.; Zhang, L.; Wang, Y.; Huang, X.; Wei, H.; Wei, L.; Zhang, Y. Ultrafast and sensitive room temperature NH_3 gas sensors based on chemically reduced graphene oxide. *Nanotechnology* **2013**, *25*, 025502. [CrossRef] [PubMed]
47. Jeevitha, G.; Abhinayaa, R.; Mangalaraj, D.; Ponpandian, N. Tungsten oxide-graphene oxide (WO_3-GO) nanocomposite as an efficient photocatalyst, antibacterial and anticancer agent. *J. Phys. Chem. Solids* **2018**, *116*, 137–147. [CrossRef]
48. Qin, L.; Liang, F.; Li, Y.; Wu, J.; Guan, S.; Wu, M.; Xie, S.; Luo, M.; Ma, D. A 2D porous zinc-organic framework platform for loading of 5-fluorouracil. *Inorganics* **2022**, *10*, 202. [CrossRef]
49. Qin, L.; Li, Y.; Liang, F.; Li, L.; Lan, Y.; Li, Z.; Lu, X.; Yang, M.; Ma, D. A microporous 2D cobalt-based MOF with pyridyl sites and open metal sites for selective adsorption of CO_2. *Microporous Mesoporous Mater.* **2022**, *341*, 112098. [CrossRef]
50. Tiwari, D.C.; Atri, P.; Sharma, R. Sensitive detection of ammonia by reduced graphene oxide/polypyrrole nanocomposites. *Synth. Met.* **2015**, *203*, 228–234. [CrossRef]
51. Kang, M.-A.; Ji, S.; Kim, S.; Park, C.-Y.; Myung, S.; Song, W.; Lee, S.S.; Lim, J.; An, K.-S. Highly sensitive and wearable gas sensors consisting of chemically functionalized graphene oxide assembled on cotton yarn. *RSC Adv.* **2018**, *8*, 11991–11996. [CrossRef]
52. Vasilopoulou, M.; Soultati, A.; Georgiadou, D.; Stergiopoulos, T.; Palilis, L.; Kennou, S.; Stathopoulos, N.; Davazoglou, D.; Argitis, P. Correction: Hydrogenated under-stoichiometric tungsten oxide anode interlayers for efficient and stable organic photovoltaics. *J. Mater. Chem. A* **2016**, *4*, 17875. [CrossRef]

53. Han, Y.; Wang, T.; Li, T.; Gao, X.; Li, W.; Zhang, Z.; Wang, Y.; Zhang, X. Preparation and electrochemical performances of graphene/polypyrrole nanocomposite with anthraquinone-graphene oxide as active oxidant. *Carbon* **2017**, *119*, 111–118. [CrossRef]
54. Dai, S.; Liu, Z.; Zhao, B.; Zeng, J.; Hu, H.; Zhang, Q.; Chen, D.; Qu, C.; Dang, D.; Liu, M. A high-performance supercapacitor electrode based on N-doped porous graphene. *J. Power Sources* **2018**, *387*, 43–48. [CrossRef]
55. Ganbavle, V.; Agawane, G.; Moholkar, A.; Kim, J.; Rajpure, K. Structural, optical, electrical, and dielectric properties of the spray-deposited WO3 thin films. *J. Mater. Eng. Perform.* **2014**, *23*, 1204–1213. [CrossRef]
56. Kalanur, S.S. Structural, optical, band edge and enhanced photoelectrochemical water splitting properties of tin-doped WO_3. *Catalysts* **2019**, *9*, 456. [CrossRef]
57. Yuan, Z.; Zhang, J.; Meng, F.; Li, Y.; Li, R.; Chang, Y.; Zhao, J.; Han, E.; Wang, S. Highly Sensitive Ammonia Sensors Based on Ag-Decorated WO 3 Nanorods. *IEEE Trans. Nanotechnol.* **2018**, *17*, 1252–1258. [CrossRef]
58. Chen, T.-Y.; Chen, H.-I.; Hsu, C.-S.; Huang, C.-C.; Wu, J.-S.; Chou, P.-C.; Liu, W.-C. Characteristics of ZnO nanorods-based ammonia gas sensors with a cross-linked configuration. *Sens. Actuators B Chem.* **2015**, *221*, 491–498. [CrossRef]
59. Zhybak, M.T.; Vagin, M.Y.; Beni, V.; Liu, X.; Dempsey, E.; Turner, A.P.; Korpan, Y.I. Direct detection of ammonium ion by means of oxygen electrocatalysis at a copper-polyaniline composite on a screen-printed electrode. *Microchim. Acta* **2016**, *183*, 1981–1987. [CrossRef]
60. Biswas, S.; Drzal, L.T. Multilayered nanoarchitecture of graphene nanosheets and polypyrrole nanowires for high performance supercapacitor electrodes. *Chem. Mater.* **2010**, *22*, 5667–5671. [CrossRef]
61. Khataee, A.; Dražević, E.; Catalano, J.; Bentien, A. Performance optimization of differential pH quinone-bromide redox flow battery. *J. Electrochem. Soc.* **2018**, *165*, A3918. [CrossRef]
62. Hashemi, S.A.; Mousavi, S.M.; Naderi, H.R.; Bahrani, S.; Arjmand, M.; Hagfeldt, A.; Chiang, W.-H.; Ramakrishna, S. Reinforced polypyrrole with 2D graphene flakes decorated with interconnected nickel-tungsten metal oxide complex toward superiorly stable supercapacitor. *Chem. Eng. J.* **2021**, *418*, 129396. [CrossRef]
63. Ahmad, R.; Tripathy, N.; Khan, M.Y.; Bhat, K.S.; Ahn, M.-s.; Hahn, Y.-B. Ammonium ion detection in solution using vertically grown ZnO nanorod based field-effect transistor. *RSC Adv.* **2016**, *6*, 54836–54840. [CrossRef]
64. Ribeiro, A.; Silva, F.; Pereira, C.M. Electrochemical sensing of ammonium ion at the water/1, 6-dichlorohexane interface. *Talanta* **2012**, *88*, 54–60. [CrossRef] [PubMed]
65. Zazoua, A.; Kazane, I.; Khedimallah, N.; Dernane, C.; Errachid, A.; Jaffrezic-Renault, N. Evidence of ammonium ion-exchange properties of natural bentonite and application to ammonium detection. *Mater. Sci. Eng. C* **2013**, *33*, 5084–5089. [CrossRef] [PubMed]
66. Heng, L.Y.; Alva, S.; Ahmad, M. Ammonium ion sensor based on photocured and self-plasticising acrylic films for the analysis of sewage. *Sens. Actuators B* **2004**, *98*, 160–165. [CrossRef]
67. Ling, T.L.; Ahmad, M.; Heng, L.Y.; Seng, T.C. The effect of multilayer gold nanoparticles on the electrochemical response of ammonium ion biosensor based on alanine dehydrogenase enzyme. *J. Sens.* **2011**, *2011*, 754171. [CrossRef]
68. Tan, L.L.; Musa, A.; Lee, Y.H. Reflectance based optical fibre sensor for ammonium ion using solid-state Riegler's reagent. *Sens. Actuators B* **2012**, *173*, 614–619. [CrossRef]
69. Tan, L.L.; Ahmad, M.; Lee, Y.H. A novel optical ammonia sensor based on reflectance measurements for highly polluted and coloured water. *Sens. Actuators B* **2012**, *171*, 994–1000. [CrossRef]
70. Li, D.; Xu, X.; Li, Z.; Wang, T.; Wang, C. Detection methods of ammonia nitrogen in water: A review. *TrAC Trends Anal. Chem.* **2020**, *127*, 115890. [CrossRef]
71. Giakisikli, G.; Anthemidis, A.N. Automatic pressure-assisted dual-headspace gas-liquid microextraction. Lab-in-syringe platform for membraneless gas separation of ammonia coupled with fluorimetric sequential injection analysis. *Anal. Chim. Acta* **2018**, *1033*, 73–80. [CrossRef]
72. Bushra, K.A.; Prasad, K.S. Based field-effect transistor sensors. *Talanta* **2022**, *239*, 123085. [CrossRef]

Disclaimer/Publisher's Note: The statements, opinions and data contained in all publications are solely those of the individual author(s) and contributor(s) and not of MDPI and/or the editor(s). MDPI and/or the editor(s) disclaim responsibility for any injury to people or property resulting from any ideas, methods, instructions or products referred to in the content.

Communication

Dual-Response Photofunctional Covalent Organic Framework for Acid Detection in Various Solutions

Wenyue Ma [1], Zijian Gu [1], Guocui Pan [1], Chunjuan Li [2], Yu Zhu [2], Zhaoyang Liu [1], Leijing Liu [1], Yupeng Guo [2], Bin Xu [1] and Wenjing Tian [1,*]

[1] State Key Laboratory of Supramolecular Structure and Materials, College of Chemistry, Jilin University, Changchun 130012, China
[2] National Chemistry Experimental Teaching Demonstration Center, Jilin University, Changchun 130012, China
* Correspondence: wjtian@jlu.edu.cn

Abstract: The detection of acid in different solution environments plays a significant role in chemical, environmental and biological fields. However, reducing the constraints of detecting environment, such as aqueous, organic solvents and mixed phases of aqueous and organic phases, remains a challenge. Herein, by combining N, N, N′, N′-tetrakis(4-aminophenyl)-1,4-phenylenediamine (TPBD) and terephthalaldehyde (TA) via Shiff-base condensation, we constructed a covalent organic framework (COF) TPBD-TA COF. The COF exhibits color change from red to dark red as well as fluorescence quenching with the increase of acid contents in either aqueous or organic solvents, or a mixture of aqueous and organic solvents, due to the weak donor-acceptor interactions within the COF as well as the weak proton ionization ability of the solutions. Therefore, regardless of the detection environment, TPBD-TA COF can realize color and fluorescence dual-response to acid with the detection limit as low as 0.4 μmol/L and 58 nmol/L, respectively, due to the protonation of the nitrogen atoms on imine bonds of the COF.

Keywords: covalent organic framework; color change; fluorescence quenching; acid detection

1. Introduction

Covalent organic frameworks (COFs), two-dimensional [1–7] or three-dimensional [8–11] crystalline porous organic polymers, constructed by organic building blocks through covalent bonds [12–17] have been applied in energy storage [18,19], gas adsorption and separation [20,21], catalysis [22], chemical separation [23,24] and optoelectronics [25,26] due to their controllable structure, easy functionalization, excellent crystallinity and porosity [27,28]. The ordered channels formed by periodic pores of COFs provide more space for host-guest molecules to interact with each other, which is beneficial for sensing and detecting. Acid detection in solution plays an important part in environmental surveillance, water quality, chemical reaction control and medical diagnosis [29]. Photofunctional materials, especially COFs, play a significant role in the field of detection [30–32]. Using COFs as chemosensors to detect acid in solution via colorimetric and fluorescence methods with the advantage of the high sensitivity, fast response and easy operation has already achieved significant progress. For example, Liu et al. constructed the first fluorescent pH sensor COF-JLU4 [33] to detect pH in aqueous solution. Wang and his co-workers prepared COF-HQ, whose emission intensity and colour were sensitive to pH in aqueous solution [34]. Auras et al. synthesized star-shaped dual-pore perylene-based 2D COFs and realized acid vapor detection in nonaqueous solution [35]. Gao et al. achieved the reversible color and fluorescent response of VCOF-PyrBpy to acids/pH in water [36]. The above examples show color switching or/and fluorescence decreasing response to acid in either aqueous or nonaqueous solutions. However, it is generally difficult to realize acid detection in different kinds of solutions (aqueous solution, organic solvents and mixture of aqueous and organic solvents) simultaneously. This is because H^+ from different solvents will interact with the frameworks weakly or strongly, resulting in more or less protonation to change the intramolecular electronic transitions of the frameworks, and further causing the change of photophysical properties of the COFs [29,36].

In fact, the protonation of the linkage or linker of COFs will cause the intraframework charge transfer or change the π-electron conjugation of the framework, thereby altering the photophysical properties of the COFs; as such, the COFs can be used as chemosensors for acid detecting. In order to realize acid detection in different kinds of solutions, selecting the structure of COFs with weak donor-acceptor charge transfer interactions is anticipated. In this case, different solvents will not induce the obvious charge transfer effect of the COFs, thereby breaking the detection environmental limitations for acid detection.

Herein, we use N, N, N′, N′-tetrakis(4-aminophenyl)-1,4-phenylenediamine (TPBD) with rich nitrogen atoms and terephthalaldehyde (TA) to prepare photofunctional TPBD-TA COF through Schiff-base reaction. The COF exhibits obvious absorption red-shifts in UV-vis spectra and the fluorescence intensity of the COF decreases with increasing acid concentration in different kinds of solution (aqueous solutions, organic solvents and mixed solution of aqueous and organic phases) due to the protonation of nitrogen atoms on imine bonds. Thus, TPBD-TA COF, which exhibits weak donor-acceptor interactions, can serve as a versatile chemosensor for detecting acid in different solutions.

2. Materials and Methods

2.1. Materials and Instrumentations

N, N, N′, N′-tetrakis(4-aminophenyl)-1,4-phenylenediamine (TPBD) and terephthalaldehyde (TA) were purchased from Leyan. o-Dichlorobenzene, n-BuOH, acetone, tetrahydrofuran (THF), ethanol (EtOH), acetonitrile (MeCN), methanol (MeOH), toluene, HCl, trifluoroacetic acid (TFA), trifluoromethanesulfonic acid (TFMS), 4-methylbenzenesulfonic acid (TsOH) and benzaldehyde were purchased from Titan.

The solid-state ^{13}C cross-polarization/magic-angle spinning (CP/ MAS) NMR spectra were collected using a Bruker AVANCE III 400 WB spectrometer. The time-of-flight mass spectra were recorded using a Kratos MALDI-TOF mass system (Bruker, Billerica, MA, USA). Fourier transform infrared (FT-IR) spectra were recorded on a Vertex 80 V spectrometer (Bruker, Rosenheim, Germany). The sample was grinded into powder and dried, then mixed with dried KBr (powder) and pressed into piece. UV-Vis spectra were recorded with a Shimadzu UV-2550 spectrophotometer (Tokyo, Japan). Fluorescence spectroscopy was taken using a Shimadzu RF-5301 PC spectrometer. SEM images were recorded using scanning electron microscopy (Hitachi Regulus 8100, Hitachi, Japan). A drop of the aqueous solution was dripped directly onto a silicon wafer and air-dried. The TEM image was recorded using a JEM-2100F instrument (JEOL Ltd., Beijing, China) with an accelerating voltage of 200 kV. The sample was prepared by placing a drop of the stock solution on a 300-mesh, carbon-coated copper grid and air-dried before measurement. The N_2 adsorption–desorption isotherms were measured using a Quantachrome Autosorb-iQ2 analyzer (Boynton Beach, FL, USA). Powder X-ray diffraction (PXRD) was performed with a Riguku D/MAX2550 diffractometer (Rigaku, Tokyo, Japan) using Cu-Kα radiation, 40 kV, 200 mA at room temperature.

2.2. The Solid-State UV-Vis Absorption Test of TPBD-TA COF on Detecting Acid

TPBD-TA COF was dispersed in aqueous solution, organic solvents and mixed solutions of aqueous and organic solvents with different amounts of acid, then the COF powder was filtered. The filtered powder was used to perform solid-state absorption spectrum tests.

2.3. The Fluorescence Test of TPBD-TA COF on Detecting Acid

TPBD-TA COF was dispersed in aqueous solution, organic solvents and mixed solutions of aqueous and organic solvents, then certain concentrations of the acid were gradually added to the COF suspension. The COF suspension (0.5 mg mL^{-1}) with different concentrations of acid was used to perform fluorescence tests.

2.4. The Recyclability Test of the COF

The PL spectra of TPBD-TA COF in ethanol solution and after adding TFA were recorded. Then, we washed the COF with water until the pH of the solvent was 7 and then dried the COF at 120 °C under vacuum for 4 h for the measurement of PXRD of the COF. The COF was used to detect acid for next cycle by repeating the above processes.

2.5. The Preparation of COF Filter Paper

Polystyrene (PS) (300 mg) was dissolved in toluene (2 mL) by stirring overnight. Then, TPBD-TA COF (10 mg) suspension (2 mL) was added into the toluene solution of PS and stirred at room temperature for several hours. The mixed solution was filtrated on the filter paper, thereby forming the COF filter paper.

3. Results and Discussion

3.1. Synthesis and Characterization

TPBD-TA COF (Figure S1) was synthesized by using N, N, N′, N′- tetrakis (4-aminophenyl)-1,4-phenylenediamine (TPBD) as the knot and terephthalaldehyde (TA) as the linker through Schiff-base reaction [37]. The COF were characterized by powder X-ray diffraction (PXRD), Fourier transform infrared spectroscopy (FT-IR), solid state ^{13}C CP/ MAS NMR, thermogravimetric analysis (TGA), field transmission electron microscopy (TEM) and scanning electron microscopy (SEM).

FT-IR spectra (Figure S3a) and solid-state ^{13}C CPMAS NMR spectra (Figure S3b) of TPBD-TA COF confirmed the formation of imine bonds (at 1621 cm^{-1} and 156 ppm, respectively). PXRD measurement (Figure S3c) indicated the crystallinity of the COF. As shown in Figure S3c, the PXRD profile of TPBD-TA COF displays a series of reflection peaks at 2.47°, 4.45°, 4.49°, 6.49°, 8.93° and 21.13°, which can be assigned to (100), (110), (200), (210), (310) and (001) facets of TPBD-TA COF, respectively. The (001) plane at 21.13° is attributed to π-π stacking between the adjacent layers of the COF. In order to investigate the structure and stacking mode of the COF, theoretical simulations were carried out using Materials Studio software packages. Pawley refinements with the unit cell (P-3, a = b = 43.35 Å and c = 4.46 Å) producing the Pawley-refined pattern of TPBD-TA COF matched well with the experimental results (Rwp = 7.7%). By comparing the PXRD patterns of the simulated TPBD-TA COF with the experimental pattern (red curve), it is indicated that the COFs adopt the eclipsed AA-stacking mode instead of the staggered AB-stacking mode. The porosity of the COF was proved by nitrogen adsorption measurements at 77 K. In Figure S3d, a steep N_2 adsorption at low relative pressure indicates that the COF has micropores; in addition, there was a hysteresis loop in the degassing process, showing that the COF has mesopores. The Brunauer−Emmett−Teller (BET) surface area was calculated to be 510.2 m^{2}g^{-1}. The pore size distribution was at 1.73 and 2.94 nm, calculated by using nonlocal density functional theory (NLDFT). The TEM image (Figure S4a) showed that TPBD-TA COF manifested a layered stacking structure. The SEM image (Figure S4b) revealed the COF was microparticles with sheet-like morphologies. In the TGA experiment (Figure S5), the COF exhibited slight weight loss at about 450 °C under an N_2 atmosphere, suggesting high thermostability. To investigate the chemical stability of the COF, we measured XRD (Figure S6) of the COF immersed in different solvents for 2 days at room temperature. There was a slight change of the PXRD curves of the COF immersed in different solvents for 48 h when compared with the original (black curve); however, the crystallinity of the COF immersed in different solvents was still retained, indicating the excellent chemical stability of TPBD-TA COF.

We investigated the photophysical properties of TPBD-TA COF by recording UV-vis absorption spectra and fluorescence emission spectra. The COF was dispersed in various solvents, such as acetone, tetrahydrofuran (THF), ethanol (EtOH), acetonitrile (MeCN), methanol (MeOH) and H_2O; we then filtered the powders to perform solid-state UV-vis absorption tests. As shown in Figure S7, the absorption bands of the COF showed slight red- or blue-shift with the increase of the solvent polarity, and the fluorescence spectra of the suspension in acetone, THF, EtOH, MeCN, MeOH and H$_2$O featured emission peaks at 623, 633, 635, 624, 637 and 629 nm, respectively (Figure S8); this means that there is no obvious solvatochromic behavior for the COF in solvents with different polarity since the weak donor-acceptor structure of TPBD-TA COF will not induce apparent intraframework charge transfer by increasing the polarity of solvents.

3.2. Acid Sensing

Supported by the chemical stability and photophysical properties of TPBD-TA COF, we investigated the acid-response properties of the COF in various solutions via UV−Vis absorption and fluorescence spectra.

As shown in Figure 1b, the absorption spectra of TPBD-TA COF in HCl aqueous solution displayed a gradual increase in the new absorption bands at the range from 600 nm to 850 nm when the concentration of HCl in water increased from 0 μM to 16.67 μM. Similarly, the new absorption band emerged with the increasing concentration of TFA in water (Figure S9a). The response of TPBD-TA COF to inorganic acid (HCl) and organic acid (TFA) in water solution illustrates that COF can detect different kinds of acid in aqueous solution. When TsOH, an organic acid, was dispersed in MeCN, the absorption bands of TPBD-TA COF gradually shifted towards longer wavelengths as the concentration of TsOH increased (Figure S9b). When the COF was scattered in MeOH and the concentration of TFA increased from 0 mM to 0.5 mM, the additional absorption band of the COF gradually emerged in the range from 600 nm to 840 nm (Figure S9c). The response of TPBD-TA COF to TsOH and TFA in organic solvents demonstrates that COF can detect organic acid in different organic solvents. The reason for the response of TPBD-TA COF to inorganic and organic acids is that H$^+$ from acid can protonate nitrogen atoms on imine bonds of the COF (Figure 1a); this will increase the donor-acceptor interaction between TPBD and TA moiety of the protonated COF, leading to the red-shifted absorption upon successive adding of H$^+$. Therefore, TPBD-TA COF can be used as an acidochromism sensor to detect both inorganic and organic acid in aqueous solutions as well as organic acid in organic solvents.

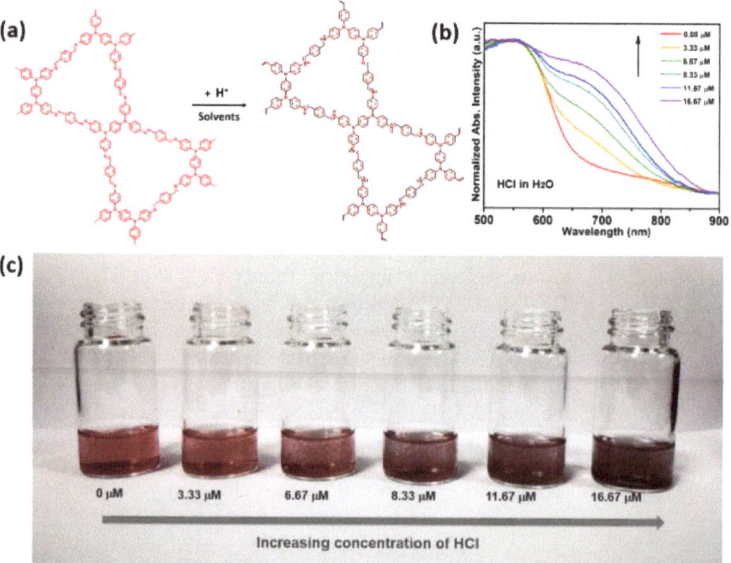

Figure 1. (a) Protonation process of TPBD-TA COF in acidic solutions, (b) solid-state absorption spectra of TPBD-TA COF in HCl aqueous solution, (c) color change of TPBD-TA COF in HCl aqueous solution with increased HCl concentration.

A pronounced color change of the COF from red to dark red (Figure 1c and Figure S10) was observed when the COF powders were immersed in various solutions with different contents of acid.

We attempted to calculate the detection limit of the acid-responsive COF. The relationship between the absorbance change or fluorescence sensing response and acid concentration was explored using the following equations [38,39]: $\Delta A = 1 + K[C]$ (1), $I_0/I = 1 + K[C]$ (2), where ΔA was the absorbance change of the COF before and after acid addition; I_0 was the initial luminescent intensity of the COF suspensions without acid; I was the luminescent intensity of the COF suspensions with different contents of acid; K was the absorbance change constant or the quenching constant; and [C] was the concentration of acid. Then we calculated the limit of detection (LOD) using the following formula [40,41]: $LOD = 3S/K$ (3), where S was the standard deviation of the absorption or fluorescence response for TPBD-TA COF without adding acid and K was the absorbance change constant or the quenching constant. Therefore, the LOD values for the COF detecting HCl in H_2O, TFA in H_2O, TsOH in MeCN and TFA in MeOH were 0.4 μmol L^{-1}, 14.7 μmol L^{-1}, 3.26 μmol L^{-1} and 26.7 μmol L^{-1}, respectively.

Figure 2 illustrates the change in fluorescence intensity of TPBD-TA COF in response to different acid solutions with varying acid concentrations, as well as the corresponding LOD values. As shown in Figure 2a, with the addition of HCl in aqueous solution, the fluorescence intensity of the COF decreases gradually. The change in the fluorescence intensity maintains a good linear fit with the acid concentrations (Figure 2b), to give an accurate LOD value. Meanwhile, the fluorescence of the COF can be also quenched by TFA in aqueous solution (Figure S11a), indicating that the COF can detect acid in water. As revealed in Figure S11b, when TPBD-TA COF was dispersed in an MeCN solution of TsOH, the fluorescence intensity of the COF can be obviously attenuated with the increase of acid concentration from 0 μM to 60 μM. Similarly, the fluorescence intensity of TPBD-TA COF suspension was quenched by increasing the concentration of TFA in MeOH (Figure S11c). This fluorescence quenching is attributed to the protonated nitrogen atoms on imine bonds, which destroy the π-electron-conjugated system of TPBD-TA COF upon acid addition [42].

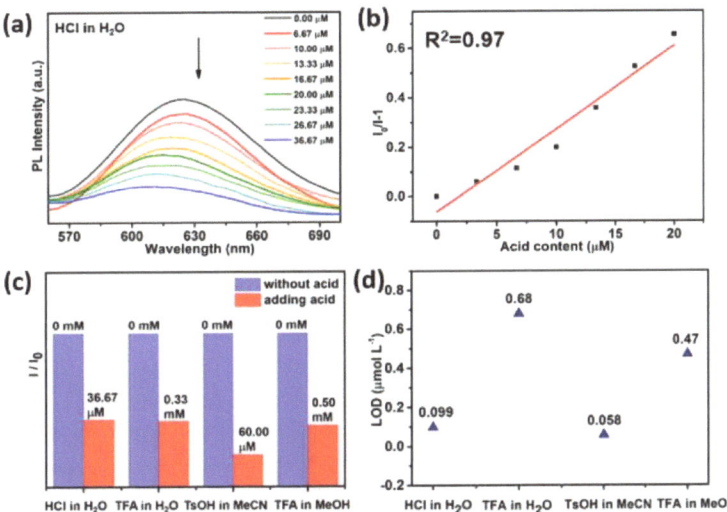

Figure 2. (a) Fluorescence spectra of TPBD-TA COF in HCl aqueous solution with different acid concentration (λex = 365 nm), (b) the linear relationship between acid content and emission ratios, (c) emission ratio before and after adding acid, (d) the LOD values of detecting different acid in different solutions.

It can be seen from Figure 2c that the COF can detect inorganic acid in aqueous solutions and organic acid in either aqueous solutions or organic solvents via fluorescence approaches with the detection limit of 0.099 µmol L^{-1}, 0.68 µmol L^{-1}, 0.058 µmol L^{-1} and 0.47 µmol L^{-1} for detecting HCl in H$_2$O, TFA in H$_2$O, TsOH in MeCN and TFA in MeOH (Figure 2d), respectively.

The detection ability of the COF toward acid in the mixture of aqueous solution and organic solvents was further explored by UV−Vis absorption and fluorescence spectra (Figure 3). As shown in Figure 3a, when the COF was immersed in EtOH solution of TFA with 5% water, an additional absorption band in the range of 600~800 nm and a red-shift of the absorption band was observed as the acid concentration increased. When the water content increased to 10%, 20%, 40%, 60% and 80%, the absorption bands of the COF were red-shifted gradually to 800 nm (Figures S12a–d and 3b). The red-shifted absorption of the COF is attributed to the interaction between H$^+$ from acid and nitrogen atoms of imine bonds of the COF. Of course, different kinds of solvents also released H$^+$, but it is difficult to induce apparent intraframework charge transfer effects within the COF with weak donor−acceptor interaction because the concentration of H$^+$ ionized from solvents is very low. The COF exhibits excellent linear relationships between the absorption change and the acid content (Figure 3e). The detection limit of the COF to detect TFA in mixed solutions of EtOH and H$_2$O was in the range of 0.059 mmol L^{-1} ~ 0.24 mmol L^{-1} (Figure 3f).

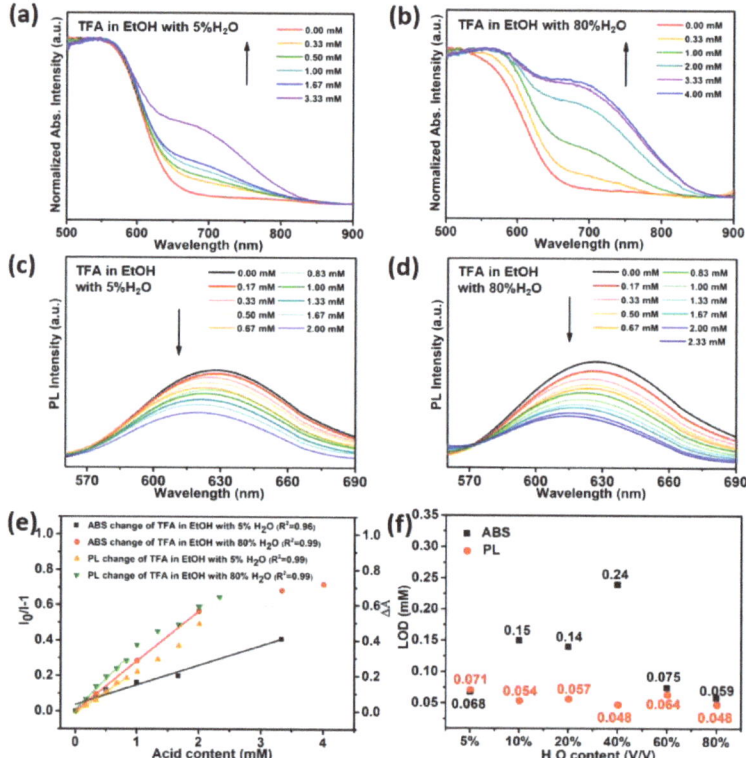

Figure 3. Solid-state absorption spectra of TPBD-TA COF in EtOH solutions of TFA with (**a**) 5% water, (**b**) 80% water (*V*/*V*) and different acid concentration, fluorescence spectra of TPBD-TA COF in TFA EtOH solution with (**c**) 5% water, (**d**) 80% water (*V*/*V*) and different acid concentration, (**e**) the linear relationship between acid content and emission ratios/or absorption change, (**f**) the LOD values of detecting TFA in EtOH solutions with different amount of water.

The acid response of the COF to TFA in acetone aqueous solution (Figure S13), TsOH in MeCN aqueous solution (Figure S14) and trifluoromethanesulfonic acid (TFMS) in EtOH aqueous solution (Figure S15) indicates that TPBD-TA COF can be used as a colorimetric sensor to detect acid in different solutions with the LOD ranging from 0.9 µmol L^{-1} to 0.22 mmol L^{-1}.

The fluorescence spectra of TPBD-TA COF in EtOH solution of TFA with different water (V/V; 5%, 10%, 20%, 40%, 60% and 80%) and different acid concentrations are shown in Figures 3c,d and S16. These spectra indicate that no matter the amount of water in ethanol, the fluorescence intensity of the COF is significantly correlated with the acid content, i.e., the fluorescence intensity gradually decreases with the increase of the acid concentration. In addition, the COF exhibits a good linear relationship between the emission ratio and the acid content (Figure 3e). The LOD value was as low as 0.048 mmol L^{-1} (Figure 3f). The fluorescence quenching effect of the COF was demonstrated in the mixed solutions of water and organic solvents when increasing the acid concentration of TFA, TsOH and TFMS (Figures S17–S19). Because the protonating ability of H$^+$ from solvents to nitrogen atoms on imine bonds is weaker than that of H$^+$ from acid, the solvents hardly lead to the change of π-electron-conjugated system of the COF, thereby facilitating the detection of acid in different kinds of solution by the COF.

To evaluate the recyclability of TPBD-TA COF as a fluorescent sensor to probe acid, we recorded acid-dependent fluorescence spectra for five continuous cycles (Figure S20). The fluorescence was quenched after acid was added, and the color of the COF quickly turned from red to dark red. The emission can return to similar intensities and the color of the COF can convert back to red after removing acid from the suspension of the COF through water washing. Moreover, the crystallinity of the COF remained after five cycles (Figure S21), demonstrating the responsive reproducibility of TPBD-TA COF to acid.

3.3. Acid Sensing Mechanism

In order to prove the protonated site of the COF, we synthesized the model compound (1) and tested absorption and fluorescence responses of TPBD and the model compound to acid, respectively. As shown in Figure S22a, the absorption intensity of TPBD gradually increases and the absorption bands display an obvious blue-shift with increasing acid concentration; this is different from the appearance of a new absorption peak of the COF with the increased acid concentration, illustrating that the acidochromism of the COF does not originate from the protonation of the free amine. In Figure S22b, a new absorption band of the model compound gradually emerged from 450 nm to 600 nm as a result of increasing acid concentration; this is consistent with the acidochromism of the COF, which proves the acidochromism of the COF is from the protonation of the imine nitrogen of the COF. In Figure S22c, the fluorescence intensity of TPBD first decreased and then increased with the increase of acid concentration; this is different from the gradually decreased fluorescence intensity of the COF with the increase of the acid concentration, indicating the fluorescence quenching of the COF is not attributed to the protonation of the free amine. Furthermore, the fluorescence intensity of the model compound gradually decreases when the acid concentration increases (Figure S22d). This is consistent with the response of the COF to acid, illustrating that the emission quenching comes from protonation of imine nitrogen of the COF.

In order to further investigate the acid sensing mechanism of the COF, we calculated the electronic cloud distribution of the COF before and after protonation by using Density functional theory (DFT). Before the addition of acid, partly separation of the electronic cloud density between the highest occupied molecular orbital (HOMO) and the lowest unoccupied molecular orbital (LUMO) can be observed, where the electronic cloud of the HOMO was distributed on TPBD and benzene of TA (Figure S23a); meanwhile, the electron cloud of the LUMO was primarily distributed on TA (Figure S23b). It means that the COF had a weak intramolecular charge transfer (ICT) property before protonation, which is consistent with the photophysical property (Figures S7 and S8). After the addition of acid, the nitrogen atoms in imine bonds of the COF were protonated, leading to well-separated electronic cloud density between HOMO and LUMO. As revealed in Figure S23c,d, the electronic cloud distribution on HOMO was dominated by TPBD units, while that on LUMO was dominated by TA units, indicating the formation of charge transfer (CT) state from TPBD moiety to TA moiety with the increase of acid concentration. Compared with the non-protonated imine, the protonated imine of the COF is a stronger acceptor, leading to a lower bandgap from 1.27 eV to 0.70 eV, as well as red-shifted absorption. That is why we can observe color change of the COF from red to dark red when increasing the acidity.

As shown in Figure S23, when compared with the deprotonated COF, the electron cloud density distribution of the protonated COF has smaller π-electron-conjugated range. The damaged π-electron-conjugated system of the COF by the protonation of acid is the reason for the fluorescence quenching. Meanwhile, the weak H^+ ionization ability of solvent can hardly change the π-electron-conjugated system of the COF and the dipole strength of donor-acceptor pair neither. This is to say that the solvent has little effect on acid-response fluorescence of the COF. Therefore, TPBD-TA COF can be applied as a fluorescent acid sensor to detect acid in different solvents via fluorescence method.

In order to further investigate the acid sensing mechanism of the COF, we calculated the natural transition orbitals (NTOs) of the deprotonated COF and the protonated COF (Figure S23e). The holes of the deprotonated COF were distributed on TPBD and TA moieties, while the electrons were distributed on TA moiety; this indicates that the deprotonated COF had some charge transfer property in the excited state. However, when the COF was protonated, the holes were distributed on TPBD unit and the electrons were primarily distributed on the TA moiety, predicting the obvious charge transfer property of the protonated COF in the excited state. However, the weak transition oscillator strength (f = 0.0870) led to a low probability of the ICT process, resulting in fluorescence quenching of the protonated COF.

The electronic cloud distributions of the COF in different solvents (acetone, ethanol and H_2O) were calculated using DFT to prove that different solvents had little effect on them. As shown in Figure S24a, the electronic cloud between HOMO and LUMO of the COF in acetone was partly overlapped, illustrating the weak intramolecular charge transfer interaction within the COF. Meanwhile, the electronic cloud distributions of the COF in ethanol (Figure S24b) and H_2O (Figure S24c) were the same as that in acetone. Moreover, the electronic cloud distributions of the COF in different kinds of solvents with different polarity were similar to that of the COF without solvent (Figure S23a), providing more evidence to prove that the solvents had almost no influence on photophysical properties of the COF. It endows good properties of TPBD-TA COF in detecting acid in various solutions.

3.4. Acid Sensing Application

The acid detection test paper was prepared by using TPBD-TA COF to detect different kinds of acids in varied solutions. As revealed in Figure 4a, when different solvents were dropped on the papers, there was no color change. However, when different solvents with acid were dropped on the test paper, the color of the paper immediately change from red to dark-red (Figure 4a).

The acid-responsive properties of TPBD-TA COF endow it with significant application potential in information encryption. When we wrote a Chinese knot pattern on the COF filter paper using TFA in EtOH solution, the dark-red Chinese knot pattern was observed, as shown in Figure 4b. Then, the COF filter paper was exposed to NH_3 (20–28%) vapor and the pattern vanished (encrypt information). After the filter paper was heated for about two minutes, the dark-red Chinese knot pattern appeared on the paper again. This process could be repeated at least four times, demonstrating the potential application of the COF in the field of information encryption.

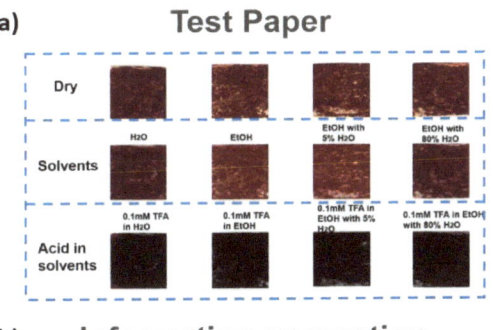

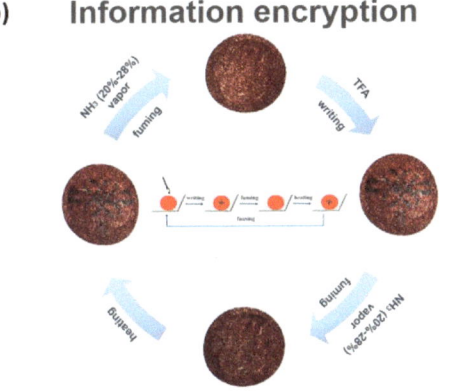

Figure 4. (**a**) Changes in color of TPBD-TA COF-based test paper before and after dropping solvents and acidic solutions, (**b**) visible picture of TPBD-TA COF filter paper through 0.1 mM TFA writing and NH_3 (20–28% aqueous solution) vapor erasing.

4. Conclusions

In summary, we have synthesized a dual-mode acid sensing photofunctional COF through Schiff-base reaction that can detect acid in aqueous solutions, organic solvents and mixed solutions of aqueous and organic solvents due to the weak donor-acceptor interactions within the COF and the weak proton ionization ability of the solvents. The COF shows obvious color change from red to dark red as acid concentration increases in different solutions with a good detecting level (LOD = 0.4 $\mu mol\ L^{-1}$). In addition, different concentrations of acid in various kinds of solvents can also cause the decreased fluorescence intensity of TPBD-TA COF; the LOD value is as low as 0.058 $\mu mol\ L^{-1}$. TPBD-TA COF breaks the constraints of the detecting environment; this endows it with significant prospects in future applications for detecting acid in various solutions.

Supplementary Materials: The following supporting information [37,43,44] can be downloaded at: https://www.mdpi.com/article/10.3390/chemosensors11040214/s1.

Author Contributions: Conceptualization, methodology, software, validation and writing—original draft preparation, W.M.; data curation, Z.G.; theoretical calculation, G.P.; experimental treatment, C.L. and Y.Z.; polish manuscript, Z.L. and L.L.; instrument support, Y.G.; funding acquisition, B.X. and W.T. All authors have read and agreed to the published version of the manuscript.

Funding: This work was supported by the National Natural Science Foundation of China (21835001, 52073116, 21674041, 51773080) and the JLU Science and Technology Innovative Research Team (2021TD-03).

Institutional Review Board Statement: Not applicable.

Informed Consent Statement: Not applicable.

Data Availability Statement: Not available.

Conflicts of Interest: The authors declare no conflict of interest.

References

1. Jiang, S.; Meng, L.C.; Ma, W.Y.; Pan, G.C.; Zhang, W.; Zou, Y.C.; Liu, L.J.; Xu, B.; Tian, W.J. Dual-functional two-dimensional covalent organic frameworks for water sensing and harvesting. *Mater. Chem. Front.* **2021**, *5*, 4193–4201. [CrossRef]
2. Wang, P.; Xu, Q.; Li, Z.P.; Jiang, W.M.; Jiang, Q.H.; Jiang, D.L. Exceptional Iodine Capture in 2D Covalent Organic Frameworks. *Adv. Mater.* **2018**, *30*, 1801991. [CrossRef] [PubMed]
3. Spitler, E.L.; Koo, B.T.; Novotney, J.L.; Colson, J.W.; Uribe-Romo, F.J.; Gutierrez, G.D.; Clancy, P.; Dichtel, W.R. A 2D Covalent Organic Framework with 4.7-nm Pores and Insight into Its Interlayer Stacking. *J. Am. Chem. Soc.* **2011**, *133*, 19416–19421. [CrossRef] [PubMed]
4. Colson, J.W.; Woll, A.R.; Mukherjee, A.; Levendorf, M.P.; Spitler, E.L.; Shields, V.B.; Spencer, M.G.; Park, J.; Dichtel, W.R. Oriented 2D Covalent Organic Framework Thin Films on Single-Layer Graphene. *Science* **2011**, *332*, 228–231. [CrossRef]
5. Evans, A.M.; Parent, L.R.; Flanders, N.C.; Bisbey, R.P.; Vitaku, E.; Kirschner, M.S.; Schaller, R.D.; Chen, L.X.; Gianneschi, N.C.; Dichtel, W.R. Seeded growth of single-crystal two-dimensional covalent organic frameworks. *Science* **2018**, *361*, 53–57. [CrossRef]
6. Mohamed, M.G.; Lee, C.C.; El-Mahdy, A.F.M.; Luder, J.; Yu, M.H.; Li, Z.; Zhu, Z.L.; Chueh, C.C.; Kuo, S.W. Exploitation of two-dimensional conjugated covalent organic frameworks based on tetraphenylethylene with bicarbazole and pyrene units and applications in perovskite solar cells. *J. Mater. Chem. A* **2020**, *8*, 11448–11459. [CrossRef]
7. Lv, Y.K.; Li, Y.S.; Zhang, G.; Peng, Z.X.; Ye, L.; Chen, Y.; Zhang, T.; Xing, G.L.; Chen, L. An In Situ Film to Film Transformation Approach toward Highly Crystalline Covalent Organic Framework Films. *Ccs Chem.* **2022**, *4*, 1519–1525. [CrossRef]
8. Fang, Q.R.; Gu, S.; Zheng, J.; Zhuang, Z.B.; Qiu, S.L.; Yan, Y.S. 3D Microporous Base-Functionalized Covalent Organic Frameworks for Size-Selective Catalysis. *Angew. Chem. Int. Edit.* **2014**, *53*, 2878–2882. [CrossRef]
9. Lin, G.Q.; Ding, H.M.; Chen, R.F.; Peng, Z.K.; Wang, B.S.; Wang, C. 3D Porphyrin-Based Covalent Organic Frameworks. *J. Am. Chem. Soc.* **2017**, *139*, 8705–8709. [CrossRef]
10. El-Kaderi, H.M.; Hunt, J.R.; Mendoza-Cortes, J.L.; Cote, A.P.; Taylor, R.E.; O'Keeffe, M.; Yaghi, O.M. Designed synthesis of 3D covalent organic frameworks. *Science* **2007**, *316*, 268–272. [CrossRef]
11. Zhu, Q.; Wang, X.; Clowes, R.; Cui, P.; Chen, L.J.; Little, M.A.; Cooper, A.I. 3D Cage COFs: A Dynamic Three-Dimensional Covalent Organic Framework with High-Connectivity Organic Cage Nodes. *J. Am. Chem. Soc.* **2020**, *142*, 16842–16848. [CrossRef]
12. Cote, A.P.; Benin, A.I.; Ockwig, N.W.; O'Keeffe, M.; Matzger, A.J.; Yaghi, O.M. Porous, crystalline, covalent organic frameworks. *Science* **2005**, *310*, 1166–1170. [CrossRef] [PubMed]
13. Hunt, J.R.; Doonan, C.J.; LeVangie, J.D.; Cote, A.P.; Yaghi, O.M. Reticular synthesis of covalent organic borosilicate frameworks. *J. Am. Chem. Soc.* **2008**, *130*, 11872–11873. [CrossRef] [PubMed]
14. Li, Y.S.; Chen, W.B.; Xing, G.L.; Jiang, D.L.; Chen, L. New synthetic strategies toward covalent organic frameworks. *Chem. Soc. Rev.* **2020**, *49*, 2852–2868. [CrossRef] [PubMed]
15. Rodriguez-San-Miguel, D.; Montoro, C.; Zamora, F. Covalent organic framework nanosheets: Preparation, properties and applications. *Chem. Soc. Rev.* **2020**, *49*, 2291–2302. [CrossRef] [PubMed]
16. Segura, J.L.; Mancheno, M.J.; Zamora, F. Covalent organic frameworks based on Schiff-base chemistry: Synthesis, properties and potential applications. *Chem. Soc. Rev.* **2016**, *45*, 5635–5671. [CrossRef]
17. Huang, N.; Wang, P.; Jiang, D.L. Covalent organic frameworks: A materials platform for structural and functional designs. *Nat. Rev. Mater.* **2016**, *1*, 1–19. [CrossRef]
18. Hu, Y.M.; Dunlap, N.; Wan, S.; Lu, S.L.; Huang, S.F.; Sellinger, I.; Ortiz, M.; Jin, Y.H.; Lee, S.H.; Zhang, W. Crystalline Lithium Imidazolate Covalent Organic Frameworks with High Li-Ion Conductivity. *J. Am. Chem. Soc.* **2019**, *141*, 7518–7525. [CrossRef]
19. Mohamed, M.G.; Atayde, E.C.; Matsagar, B.M.; Na, J.; Yamauchi, Y.; Wu, K.C.W.; Kuo, S.W. Construction Hierarchically Mesoporous/Microporous Materials Based on Block Copolymer and Covalent Organic Framework. *J. Taiwan Inst. Chem. Eng.* **2020**, *112*, 180–192. [CrossRef]
20. Shan, M.X.; Liu, X.L.; Wang, X.R.; Yarulina, I.; Seoane, B.; Kapteijn, F.; Gascon, J. Facile manufacture of porous organic framework membranes for precombustion CO2 capture. *Sci. Adv.* **2018**, *4*, eaau1698. [CrossRef]
21. Khan, N.A.; Zhang, R.N.; Wu, H.; Shen, J.L.; Yuan, J.Q.; Fan, C.Y.; Cao, L.; Olson, M.A.; Jiang, Z.Y. Solid-Vapor Interface Engineered Covalent Organic Framework Membranes for Molecular Separation. *J. Am. Chem. Soc.* **2020**, *142*, 13450–13458. [CrossRef] [PubMed]
22. Ren, X.M.; Li, C.Z.; Kang, W.C.; Li, H.; Ta, N.; Ye, S.; Hu, L.Y.; Wang, X.L.; Li, C.; Yang, Q.H. Enormous Promotion of Photocatalytic Activity through the Use of Near-Single Layer Covalent Organic Frameworks. *CCS Chem.* **2022**, *4*, 2429–2439. [CrossRef]
23. Kang, Z.X.; Peng, Y.W.; Qian, Y.H.; Yuan, D.Q.; Addicoat, M.A.; Heine, T.; Hu, Z.G.; Tee, L.; Guo, Z.G.; Zhao, D. Mixed Matrix Membranes (MMMs) Comprising Exfoliated 2D Covalent Organic Frameworks (COFs) for Efficient CO2 Separation. *Chem. Mater.* **2016**, *28*, 1277–1285. [CrossRef]
24. Fan, H.W.; Mundstock, A.; Feldhoff, A.; Knebel, A.; Gu, J.H.; Meng, H.; Caro, J. Covalent Organic Framework-Covalent Organic Framework Bilayer Membranes for Highly Selective Gas Separation. *J. Am. Chem. Soc.* **2018**, *140*, 10094–10098. [CrossRef] [PubMed]

25. Rice, A.M.; Dolgopolova, E.A.; Yarbrough, B.J.; Leith, G.A.; Martin, C.R.; Stephenson, K.S.; Heugh, R.A.; Brandt, A.J.; Chen, D.A.; Karakalos, S.G.; et al. Stack the Bowls: Tailoring the Electronic Structure of Corannulene-Integrated Crystalline Materials. *Angew. Chem. Int. Edit.* **2018**, *57*, 11310–11315. [CrossRef]
26. Zhang, L.; Yi, L.; Sun, Z.J.; Deng, H.X. Covalent organic frameworks for optical applications. *Aggregate* **2021**, *2*, 13011. [CrossRef]
27. Diercks, C.S.; Yaghi, O.M. The atom, the molecule, and the covalent organic framework. *Science* **2017**, *355*, eaal1585. [CrossRef]
28. Feng, L.; Wang, K.Y.; Day, G.S.; Zhou, H.C. The chemistry of multi-component and hierarchical framework compounds. *Chem. Soc. Rev.* **2019**, *48*, 4823–4853. [CrossRef]
29. Gilmanova, L.; Bon, V.; Shupletsov, L.; Pohl, D.; Rauche, M.; Brunner, E.; Kaskel, S. Chemically Stable Carbazole-Based Imine Covalent Organic Frameworks with Acidochromic Response for Humidity Control Applications. *J. Am. Chem. Soc.* **2021**, *143*, 18368–18373. [CrossRef]
30. Luo, Y.C.; Zhang, S.; Wang, H.; Luo, Q.; Xie, Z.G.; Xu, B.; Tian, W.J. Precise Detection and Visualization of Cyclooxygenase-2 for Golgi Imaging by a Light-Up Aggregation-Induced Emission Based Probe. *CCS Chem.* **2022**, *4*, 456–463. [CrossRef]
31. Meng, L.C.; Ma, X.B.; Jiang, S.; Zhang, S.; Wu, Z.Y.; Xu, B.; Lei, Z.; Liu, L.J.; Tian, W.J. Twisted Intramolecular Charge Transfer-Aggregation-Induced Emission Fluorogen with Polymer Encapsulation-Enhanced Near-Infrared Emission for Bioimaging. *CCS Chem.* **2021**, *3*, 2084–2094. [CrossRef]
32. Wu, Z.Y.; Zhang, C.Y.; Zhu, Y.L.; Lu, Z.Y.; Liu, H.; Xu, B.; Zhang, X.Q.; Tian, W.J. Visualization of Macrophase Separation and Transformation in Immiscible Polymer Blends. *Ccs Chem.* **2022**, *2022*, 1–11. [CrossRef]
33. Zhang, Y.W.; Shen, X.C.; Feng, X.; Xia, H.; Mu, Y.; Liu, X.M. Covalent organic frameworks as pH responsive signaling scaffolds. *Chem. Commun.* **2016**, *52*, 11088–11091. [CrossRef] [PubMed]
34. Long, C.; He, L.W.; Ma, F.Y.; Wei, L.; Wang, Y.X.; Silver, M.A.; Chen, L.H.; Lin, Z.; Gui, D.X.; Juan, D.W.; et al. Covalent Organic Framework Functionalized with 8-Hydroxyquinoline as a Dual-Mode Fluorescent and Colorimetric pH Sensor. *ACS Appl. Mater. Inter.* **2018**, *10*, 15364–15368.
35. Ascherl, L.; Evans, E.W.; Gorman, J.; Orsborne, S.; Bessinger, D.; Bein, T.; Friend, R.H.; Auras, F. Perylene-Based Covalent Organic Frameworks for Acid Vapor Sensing. *J. Am. Chem. Soc.* **2019**, *141*, 15693–15699. [CrossRef]
36. Bu, R.; Zhang, L.; Liu, X.Y.; Yang, S.L.; Li, G.; Gao, E.Q. Synthesis and Acid-Responsive Properties of a Highly Porous Vinylene-Linked Covalent Organic Framework. *ACS Appl. Mater. Inter.* **2021**, *13*, 26431–26440. [CrossRef]
37. Hao, Q.; Li, Z.J.; Bai, B.; Zhang, X.; Zhong, Y.W.; Wan, L.J.; Wang, D. A Covalent Organic Framework Film for Three-State Near-Infrared Electrochromism and a Molecular Logic Gate. *Angew. Chem. Int. Edit.* **2021**, *60*, 12498–12503. [CrossRef]
38. Sung, T.W.; Lo, Y.L. Ammonia vapor sensor based on CdSe/SiO2 core-shell nanoparticles embedded in sol-gel matrix. *Sens. Actuat. B Chem.* **2013**, *188*, 702–708. [CrossRef]
39. Wang, Y.F.; Liu, X.; Wang, M.K.; Wang, X.X.; Ma, W.Y.; Li, J.Y. Facile synthesis of CDs@ZIF-8 nanocomposites as excellent peroxidase mimics for colorimetric detection of H2O2 and glutathione. *Sens. Actuat. B Chem.* **2021**, *329*, 129115. [CrossRef]
40. Balaji, T.; Sasidharan, M.; Matsunaga, H. Optical sensor for the visual detection of mercury using mesoporous silica anchoring porphyrin moiety. *Analyst* **2005**, *130*, 1162–1167. [CrossRef]
41. Richardson, T.H.; Dooling, C.M.; Jones, L.T.; Brook, R.A. Development and optimization of porphyrin gas sensing LB films. *Adv. Colloid Interface Sci.* **2005**, *116*, 81–96. [CrossRef] [PubMed]
42. Jiang, H.L.; Feng, D.W.; Wang, K.C.; Gu, Z.Y.; Wei, Z.W.; Chen, Y.P.; Zhou, H.C. An Exceptionally Stable, Porphyrinic Zr Metal-Organic Framework Exhibiting pH-Dependent Fluorescence. *J. Am. Chem. Soc.* **2013**, *135*, 13934–13938. [CrossRef] [PubMed]
43. Grigoras, M.; Stafie, L. Synthesis and Characterization of Linear, Branched and Hyperbranched Triphenylamine-Based Polyazomethines. *Des. Monomers Polym.* **2009**, *12*, 177–196. [CrossRef]
44. Frisch, M.J.; Trucks, G.W.; Schlegel, H.B.; Scuseria, G.E.; Robb, M.A.; Cheeseman, J.R.; Scalmani, G.; Barone, V.; Mennucci, B.; Petersson, G.A.; et al. *Gaussian 09, Revision D.01*; Gaussian, Inc.: Wallingford, CT, USA, 2009.

Disclaimer/Publisher's Note: The statements, opinions and data contained in all publications are solely those of the individual author(s) and contributor(s) and not of MDPI and/or the editor(s). MDPI and/or the editor(s) disclaim responsibility for any injury to people or property resulting from any ideas, methods, instructions or products referred to in the content.

Communication

Preparation and Enhanced Acetone-Sensing Properties of ZIF-8-Derived Co$_3$O$_4$@ZnO Microspheres

Xiangxiang Fan [1,2,*], Susu Yang [1], Chun Huang [1], Yujie Lu [1] and Pan Dai [1,2]

[1] School of Information Engineering, Huzhou University, Huzhou 313000, China; yangsusu2739@126.com (S.Y.)
[2] Zhejiang Province Key Laboratory of Smart Management & Application of Modern Agricultural Resources, Huzhou University, Huzhou 313000, China
* Correspondence: fanxiangxiang@zjhu.edu.cn

Abstract: In this work, ZIF-8-derived Co$_3$O$_4$@ZnO microspheres were prepared by a liquid-phase concentration-controlled nucleation strategy. The results of the material characterization showed that Co$_3$O$_4$@ZnO microspheres were obtained, and the surface structure could be controlled with the concentration of the ligand. Compared with pure Co$_3$O$_4$ microspheres, the operating temperature of optimized Co$_3$O$_4$@ZnO microspheres increased by 90 °C after the gas-sensing test. The response to 50 ppm acetone of Co$_3$O$_4$@ZnO microspheres was 4.5 times higher than that of pure Co$_3$O$_4$, and the detection limit reached 0.5 ppm. Meanwhile, Co$_3$O$_4$@ZnO microspheres showed a shorter response-recovery time and better selectivity. The enhanced-sensing mechanism of the ZIF-8-derived Co$_3$O$_4$@ZnO microspheres was also analyzed.

Keywords: gas sensor; microspheres; MOF; ZIF-8

Citation: Fan, X.; Yang, S.; Huang, C.; Lu, Y.; Dai, P. Preparation and Enhanced Acetone-Sensing Properties of ZIF-8-Derived Co$_3$O$_4$@ZnO Microspheres. *Chemosensors* **2023**, *11*, 376. https://doi.org/10.3390/chemosensors11070376

Academic Editor: Manuel Aleixandre

Received: 9 June 2023
Revised: 30 June 2023
Accepted: 3 July 2023
Published: 5 July 2023

Copyright: © 2023 by the authors. Licensee MDPI, Basel, Switzerland. This article is an open access article distributed under the terms and conditions of the Creative Commons Attribution (CC BY) license (https:// creativecommons.org/licenses/by/ 4.0/).

1. Introduction

Acetone is not only an important organic solvent but also a respiratory marker of diabetes. The concentration of acetone in the exhaled breath of a diabetic patient is 1.8 ppm, while the concentration for a healthy person is less than 0.9 ppm. In order to diagnose diabetes, it is of great significance to prepare an acetone sensor with a detection limit of the ppb level [1]. Due to the advantages of their small size, low cost, and easy preparation [2,3], metal-oxide semiconductor (MOS) gas sensors have been widely used for acetone detection, such as In$_2$O$_3$, SnO$_2$, and Co$_3$O$_4$ [4–6]. Compared with the n-type MOS, previous studies have shown that Co$_3$O$_4$-based sensors exhibited the features of lower working temperature and higher selectivity [7,8]. Co$_3$O$_4$ is considered a promising MOS material for acetone-sensing with low power consumption. The response to 200 ppm acetone of Co$_3$O$_4$ nanoparticles prepared by Zhang et al. was 2.6 at 190 °C [9]. Park et al. synthesized Co$_3$O$_4$ hollow microspheres, and the response to 1000 ppm acetone was 7.5 at 100 °C [10]. However, the application of pure Co$_3$O$_4$-based sensors is still limited by their low sensitivities.

To obtain improved gas-sensing properties of Co$_3$O$_4$-based sensors, a general method is to prepare heterojunctions. The response to 50 ppm acetone of Co$_3$O$_4$/ZnO heterogeneous material prepared by Zhao's team was approximately 5.1 times higher than the value of pure Co$_3$O$_4$ [11]. The response to 5 ppm TMA of Co$_3$O$_4$/SnO$_2$ heterogeneous material prepared by Meng's team was 7.2 times higher than the value of pure Co$_3$O$_4$ [12]. Therefore, enhanced sensing properties can be obtained by constructing a heterojunction. The structures of heterogeneous materials include a hybrid heterostructure, a decorated heterostructure, and a multilayer heterostructure [13–16]. Among them, the decorated heterostructure is beneficial for gas sensing due to its good synergistic effect between the core and shell components and larger heterointerface [17,18]. The response to 100 ppm formaldehyde of In$_2$O$_3$@SnO$_2$ heterostructure Nanofibers prepared by Wan et al. was higher than that of pure In$_2$O$_3$ Nanofibers and pure SnO$_2$ nanosheets [19]. A variety of decorated heterostructures have been reported as sensing materials, such as In_2O_3@NiO, CeO_2@$ZnSnO_3$, and Co_3O_4@WO_3 [20–22]. All the reported decorated heterostructures have shown improved gas-sensing properties.

Metal-organic framework (MOF) materials, with their large specific surface area and high porosity, have shown great value for application in the fields of gas storage, energy storage, adsorption, and catalysis [23–25]. Since the porous MOS structure can be obtained via heating treatment, MOF materials have received extensive attention for the preparation of sensors [26–28]. More gas adsorption and reaction sites in porous MOS could significantly enhance the gas-sensing properties [29]. Therefore, more attention has been paid to the decorated heterostructures prepared by MOF. As a representative MOF material, ZIF-8 is bridged by 2-methylimidazole and Zn^{2+}. ZnO could be obtained by annealing treatment of ZIF-8. The material of ZnO exhibits high charge-carrier mobility and high sensitivity [30,31], and ZIF-8-derived ZnO obtained extensive attention as a gas sensor. The response to 500 ppb acetone of ZIF-8-derived ZnO@MoS_2 heterostructures prepared by Chang's group reached 1.5 at 350 °C [32]. Zhan et al. prepared ZIF-8-derived SnO_2@ZnO microspheres, which showed a response of 164 to 70 ppm NO_2 at 300 °C [33]. The excellent gas-sensing performances benefit from the heterojunction and porous structure. However, these ZIF-8-derived n-type ZnO heterogeneous materials were always operated at a high temperature. As a sensing material with low power consumption, Co_3O_4 with ZIF-8-derived ZnO coating was rarely reported.

In this work, ZIF-8-derived Co_3O_4@ZnO microspheres were prepared using the liquid-phase concentration-controlled nucleation strategy. The surface structure was controlled with the concentration of the ligand. The acetone-sensing properties and mechanism of the heterostructure were analyzed.

2. Materials and Methods

2.1. Materials

2-methylimidazole (2-MeIM, AR), zinc acetate ($Zn(COOH_3)_2 \cdot 2H_2O$, AR), cobalt nitrate hexahydrate ($Co(NO_3)_2 \cdot 6H_2O$, AR), sodium hydroxide (NaOH, AR), ethanol (CH_3CH_2OH, AR), and methanol (CH_3OH, AR) were all bought from Aladdin (Shanghai, China). All the reagents were utilized without pretreatment.

2.2. Synthesis of Co_3O_4 Microspheres

First, 58.21 g $Co(NO_3)_2 \cdot 6H_2O$ was added to 40 mL deionized water to obtain a purple solution, then 2 g NaOH was added and stirred for 10 min. Second, the supernatant of the solution was transferred into a 50 mL Teflon-lined autoclave and kept at 180 °C for 5 h. After the reaction, the precipitate was obtained by centrifugation. The precipitate was washed with ethanol and water three times, and the Co_3O_4 precursor was obtained after being dried at 70 °C for 3 h. Finally, Co_3O_4 microspheres were obtained by calcining the precursor at 500 °C for 2 h.

2.3. Synthesis of Co_3O_4@ZnO Microspheres

Solution A was obtained by dissolving 0.0714 g $Zn(COOH_3)_2 \cdot 2H_2O$ in 10 mL of methanol. Solution B was obtained by dissolving a certain amount of 2-methylimidazole and 50 mg of the Co_3O_4 precursor in 10 mL of methanol, followed by sonication for 1 h. Solutions A and B were mixed and stirred for 30 min. After standing for 24 h, the precipitate was obtained by centrifugation. The precipitate was washed with ethanol and water three times, and the precursor of Co_3O_4@ZnO microspheres was obtained after being dried at 60 °C for 12 h. After being annealed at 500 °C for 3 h, Co_3O_4@ZnO microspheres were obtained.

The amount of 2-MeIM added was set as 3.7 times, 37 times, and 110 times that of the molar amount of Zn^{2+}, and the corresponding materials were named CZ-1, CZ-2, and CZ-3.

2.4. Material Characterization

Scanning electron microscopy (SEM, Apreo 2S, ThermoFischer, Waltham, MA, USA) and transmission electron microscopy (TEM, Talos F200S G2, ThermoFischer) were utilized to characterize the morphology and surface structure of the materials. The element compositions were analyzed by energy-dispersive X-ray spectroscopy (EDS, QUANTAX EDS, Bruker, Billerica, MA, USA). The crystal composition and crystallinity were analyzed by X-ray diffraction spectroscopy (XRD, Ultima IV, Rigaku, Tokyo, Japan). X-ray photoelectron spectroscopy (XPS, ESCALAB Xi+, ThermoFischer) was utilized to characterize the chemical ingredients and states of the samples.

2.5. Preparation and Measurement of Sensors

The fabrication method of the sensors is consistent with our previous report [34]. Specifically, 10 mg of Co_3O_4@ZnO microspheres were mixed with 100 μL of ethanol to obtain the homogeneous paste, and it was evenly brushed onto the outer surface of a ceramic tube equipped with a Cr-Ni heating wire and a pair of gold electrodes. After being annealed at 300 °C for 12 h, the acetone sensor was obtained.

As shown in Figure 1, the test system includes a humidity generator, a test chamber, and a source meter. The humidity generator is used to generate different humidity environments. The air was filtered and passed through a saturated salt solution by an air pump. When the air was introduced into the test chamber, the environment with the required humidity was obtained and the relative humidity was determined by a humidity sensor. The measurement of our sensor was performed in an 18 L test chamber whose size was 300 mm × 300 mm × 200 mm. A source meter was used to record the data.

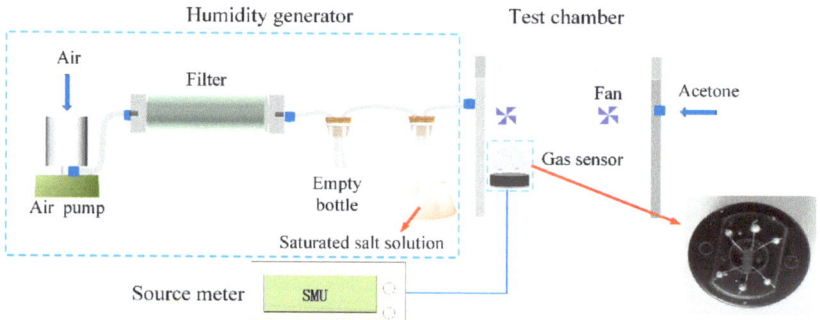

Figure 1. Schematic diagram of gas-sensing test system.

Gas-sensing properties were analyzed by injecting saturated acetone vapor into the test chamber or exhausting acetone from the chamber as per our previous report [35]. When acetone gas was injected, the gas was stirred evenly by the fans. The concentration of acetone is adjusted by changing the volume of saturated steam injected into the chamber [36]. The concentration could be calculated by the formula $C_g = V_g P_g / P_0 V_c$, where V_g is the volume of acetone injected into the chamber, P_g is the saturated vapor pressure of acetone, P_0 is the ambient atmospheric pressure, and V_c is the volume of the test chamber. After the test, the acetone gas in the test chamber was replaced with fresh air through the air pump.

The ratios of $(R_g - R_a)/R_a$ were used to describe the sensitivity of the sensors, where R_g and R_a are the resistances of the sensors in target gas and air. The times required to change the resistance by 90% after injecting or exhausting the target gas were used to define the response and recovery time.

3. Results

3.1. Materials Characterization

Figure 2a presents that the Co_3O_4 particle is a spherical structure with a smooth surface, and the size is approximately 300 nm. As shown in Figure 2b–d, all the samples of CZ-1, CZ-2, and CZ-3 still maintain spherical structures, but the surface characteristics have undergone dramatic changes. With the increase in the 2-methimidazole concentration, Co_3O_4 microspheres are gradually covered by nanoparticles, and the surfaces become rougher when the number of covered grains gradually increased. Therefore, the microspheres with decorated particles can be prepared by this method, and the surface structure can be adjusted by changing the ligand concentration.

Figure 2. SEM images of (**a**) Co_3O_4, (**b**) CZ-1, (**c**) CZ-2, and (**d**) CZ-3.

The XRD spectra of Co_3O_4, CZ-1, CZ-2, and CZ-3 are displayed in Figure 3. The peaks of Co_3O_4 are located at 31.24°, 36.82°, 44.78°, 55.66°, and 59.22°, which correspond to (2 2 0), (3 1 1), (4 0 0), (4 2 2), and (5 1 1) crystal planes of spinel Co_3O_4 (JCPDS No.43-1003). The spectra of CZ-1, CZ-2, and CZ-3 also show identical diffraction peaks of Co_3O_4, indicating that Co_3O_4 crystal is formed in these samples. On the XRD spectrum of CZ-3, three different diffraction peaks are formed from other samples, which are located at 31.6°, 36.24°, and 56.42°. These peaks are indexed to the (1 0 0), (1 0 1), and (1 1 0) crystal planes of ZnO. The results indicate that the CZ-3 contains crystals of Co_3O_4 and ZnO. However, due to the low content of ZnO, the three peaks are not observed on the spectra of CZ-1 and CZ-2.

As shown in Figure 4, all the samples were further analyzed by TEM and EDS to confirm the morphologies and compositions. Figure 4a,b present that pure Co_3O_4 and CZ-1 are both smooth microspheres. Figure 4c,d exhibit that the Co_3O_4 microspheres are coated by ZnO nanoparticles and the amount of ZnO nanoparticles gradually increases with the increase in the 2-methimidazole concentration. In order to verify that CZ-1 is coated by ZnO, EDS was used to analyze the element on the surface. Figure 4e,f show that the element of Zn is evenly distributed on the surface of Co_3O_4 microspheres. The smooth surface of CZ-1 can contribute to the small size of ZnO nanoparticles. Figure 4g,h indicate that the main element in the nanoparticles coated on the microspheres is Zn. Figure 4i shows that the lattice width of the nanoparticles coated on the microspheres is 0.26 nm and is indexed to the (0 0 2) crystal plane of ZnO. The results indicate that Co_3O_4@ZnO microspheres are obtained by the liquid-phase concentration-controlled nucleation strategy.

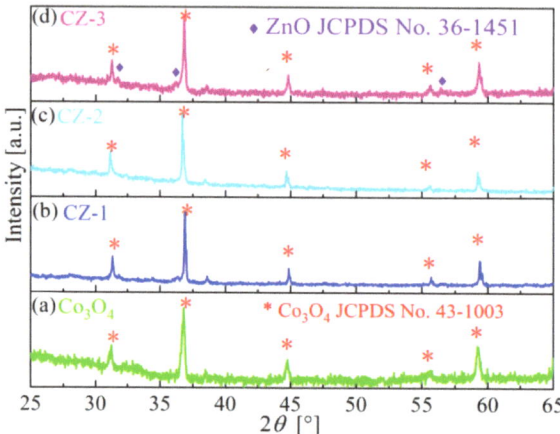

Figure 3. XRD spectra of (**a**) Co$_3$O$_4$, (**b**) CZ-1, (**c**) CZ-2, and (**d**) CZ-3.

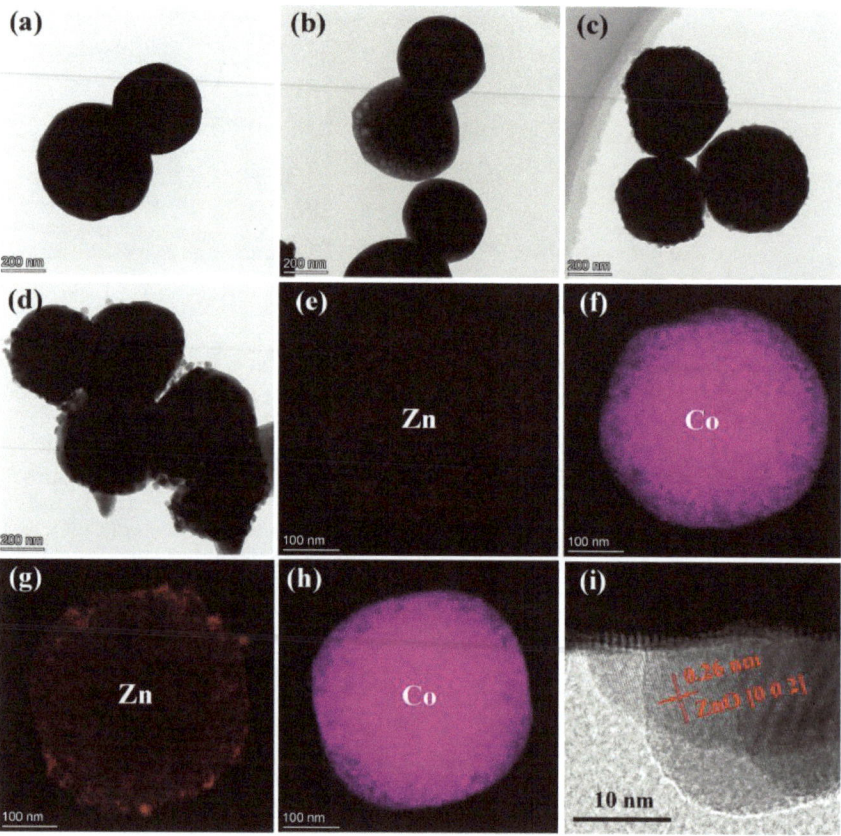

Figure 4. TEM images of (**a**) Co$_3$O$_4$, (**b**) CZ-1, (**c**) CZ-2m and (**d**) CZ-3. EDS mapping images of (**e**,**f**) CZ-1 and (**g**,**h**) CZ-2. (**i**) HRTEM images of CZ-2.

Figure 5a shows that the elements of Zn, O, and Co are presented on the XPS spectrum of CZ-2, while only Co and O appear in Co$_3$O$_4$. Figure 5b shows the high-resolution

XPS spectrum of Zn 2p in CZ-2, which includes two peaks corresponding to Zn^{2+} [37]. Therefore, it is further proved that the nanograins coated on Co_3O_4 microspheres are ZnO. Figure 5c presents the Co 2p spectra of Co_3O_4 and CZ-2. The fitting peaks located at 797 eV and 782.5 eV are associated with Co^{2+}, while the fitting peaks located at 795 eV and 779.5 eV correspond to Co^{3+}. This demonstrates the existence of Co_3O_4 particles in the samples [38]. Figure 5d presents the O 1 s spectra of Co_3O_4 and CZ-2. The three different fitting peaks are related to lattice oxygen (O_L), oxygen vacancies (O_V), and absorbed oxygen (O_C) [39]. Among these, O_V is the key component to affect the gas-sensing performance. The gas-sensing property is attributed to the redox reaction between the target gas with oxygen adsorbed on the surface of the sensing material. As the donor, the oxygen vacancies provide sites for oxygen adsorption and reaction [40]. The area ratio O_V in Co_3O_4 is 29.8%, while that in CZ-2 is 39.1%. The O_V proportion of CZ-2 is higher than that of pure Co_3O_4.

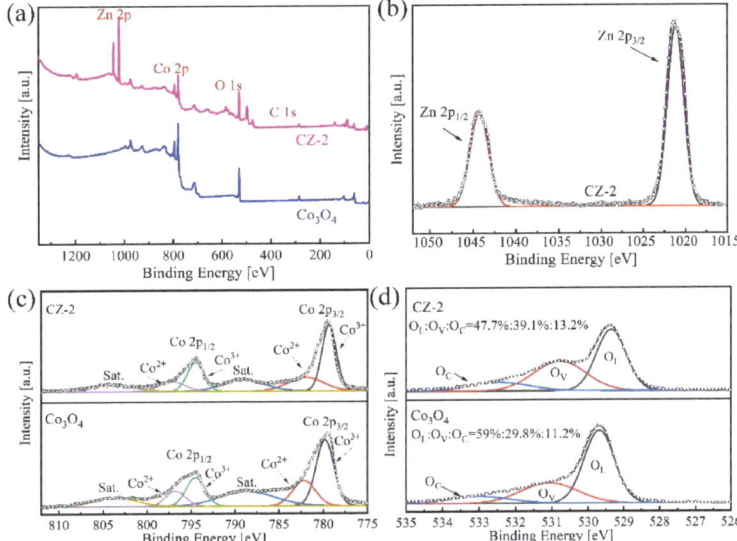

Figure 5. XPS spectra of (**a**) full survey for CZ-2 and Co_3O_4. High-resolution XPS spectra of (**b**) Zn 2p, (**c**) Co 2p, and (**d**) O 1 s.

3.2. Gas-Sensing Properties

The sensing properties of the material are always affected by the operating temperature. Figure 6a shows the resistances of four samples in the air at a temperature ranging from 130 °C to 280 °C. The resistance of all the samples decreases significantly as the temperature increases. Because of the heterojunctions, the resistance of Co_3O_4@ZnO microspheres is higher than that of pure Co_3O_4. The resistance of CZ-2 and CZ-3 even exceeds the range of the source meter at temperatures lower than 190 °C, making it impossible to accurately measure the resistance values. Figure 6b shows the responses of four sensors to 50 ppm acetone, operated at a temperature ranging from 130 °C to 280 °C. Co_3O_4, CZ-1, CZ-2, and CZ-3 reach the maximum response at 160 °C, 190 °C, 250 °C, and 250 °C. The optimal operating temperatures of all the Co_3O_4@ZnO microspheres are higher than the value of pure Co_3O_4 because of the presence of n-type MOS. Among them, CZ-2 shows the highest response at 250 °C, and the resistance in the air is approximately 600 kΩ. The response value of CZ-2 is 4.5 times higher than the value of pure Co_3O_4. Therefore, the acetone sensitivity of Co_3O_4 is enhanced by the ZIF-8-derived ZnO.

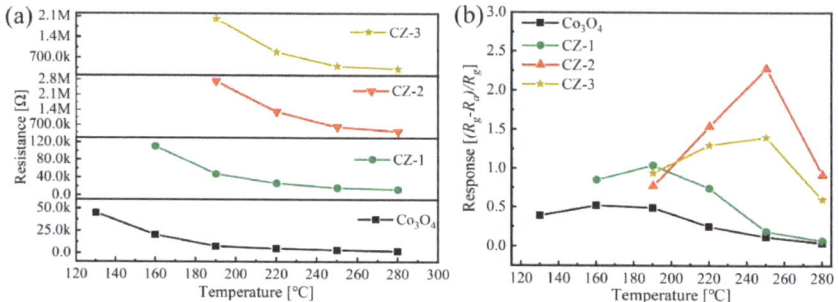

Figure 6. (**a**) Resistances in air and (**b**) responses to 50 ppm acetone of four samples at different operating temperatures.

The responses of pure Co_3O_4 and CZ-2 to 0.5–100 ppm acetone at the optimal operating temperature are displayed in Figure 7a. The response values of CZ-2 are larger than the values of Co_3O_4. Compared with pure Co_3O_4, the detection limit of CZ-2 reaches 500 ppb. As shown in Figure 7b, the response and recovery times of CZ-2 are 32 s and 98 s, which are shorter than that of pure Co_3O_4. Therefore, the sensitivity and response-recovery rates of Co_3O_4 are improved by the ZIF-8-derived ZnO.

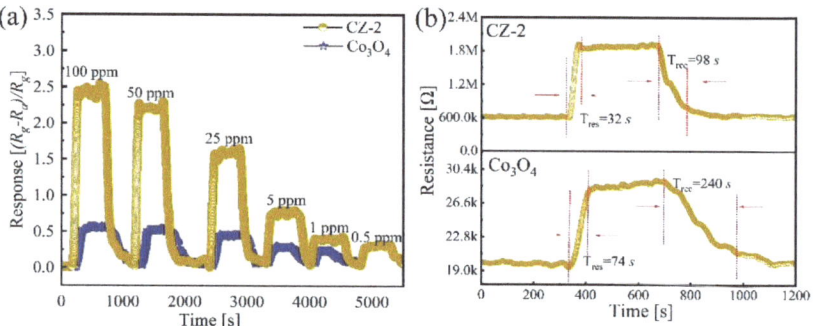

Figure 7. (**a**) Dynamic response curves of pure Co_3O_4 and CZ-2 to 0.5–100 ppm acetone. (**b**) Response-recovery curves of pure Co_3O_4 and CZ-2 to 50 ppm acetone.

As shown in Figure 8a, the responses of pure Co_3O_4 and CZ-2 to 50 ppm of different gases were tested at the optimized temperatures. The two samples show the highest response to acetone, which means that the two sensors have a selective response to acetone. To quantitatively investigate the selectivity of the two sensors, the ratios of $S_{others}/S_{acetone}$ are used and listed in Figure 8b, where $S_{acetone}$ and S_{others} are the responses of the sensor to acetone and other gas. Figure 8b presents that CZ-2 has lower ratios than pure Co_3O_4, which means that CZ-2 has better selectivity. According to the ratios, the selectivity of the sensor depends on the sensitivity of different gases, and the sensitivity of different gases is related to the reaction intensity of the gas on the surface of the sensing material. Due to the reaction intensity being influenced by the amount of adsorbed gas on the surface and the reaction energy, the material composition and working temperature are key factors to determine the selectivity, and the sensor exhibits the highest reaction intensity at the optimal operating temperature [13]. Compared with Co_3O_4, the acetone reaction intensity on the surface of CZ-2 is enhanced more than other gases, resulting in improved selectivity. The results show that the selectivity of Co_3O_4 is enhanced by ZIF-8-derived ZnO.

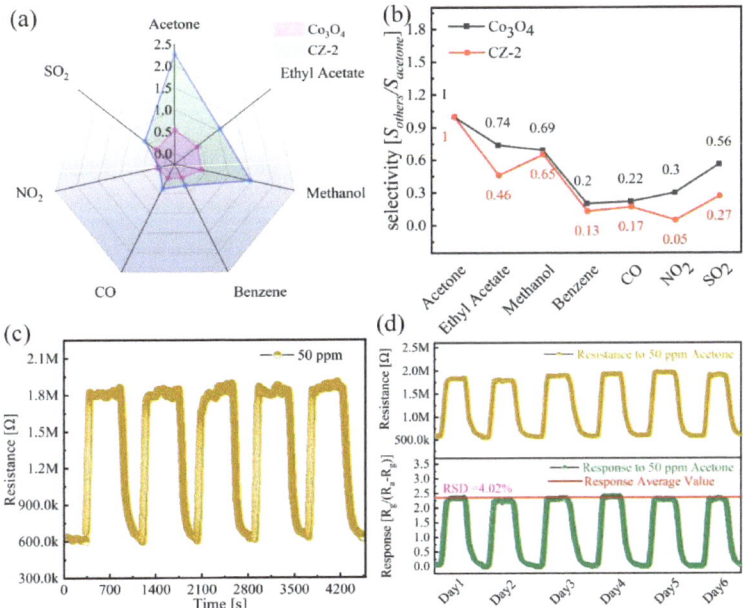

Figure 8. (**a**) Response to 50 ppm different gases and (**b**) ratios ($S_{others}/S_{acetone}$) of pure Co_3O_4 and CZ-2. (**c**) Cycle response curves and (**d**) response curves of CZ-2 during 6 days to 50 ppm acetone.

In order to evaluate the stability of the sensor, the multi-cycle tests of CZ-2 to 50 ppm acetone are shown in Figure 8c. The response curves of CZ-2 remain stable during the multi-cycle tests. The response curves of CZ-2 to 50 ppm acetone over 6 days are presented in Figure 8d. The responses of CZ-2 remain stable, and the relative standard deviation is 4.02%. The results exhibit that Co_3O_4@ZnO microspheres display good stability.

The response to acetone of CZ-2 was also investigated in an environment of relative humidity (RH) ranging from 30% to 80%. In order to quantitatively evaluate the anti-humidity property of CZ-2, the ratios of S_{others}/S_{30} were used as a comparative quantity, where S_{30} and S_{others} are the response in the environment of 30% RH and other RH. Figure 9 shows the ratios of CZ-2 as a function of relative humidity. As the humidity increased from 30% to 80%, the response of the CZ-2 decreased by 31% compared to the original. When the sensor is in a high-humidity environment, water molecules will adsorb on the surface of sensing materials and react with oxygen ions. A hydroxyl group with less activity is generated, and the sensitivity of the sensor decreases [41].

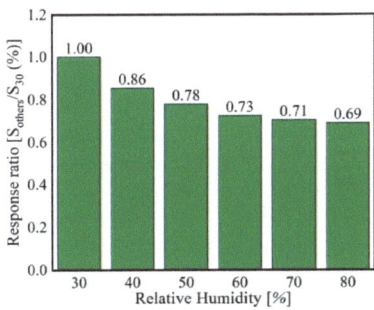

Figure 9. Response ratios (S_{others}/S_{30}) of CZ-2 as a function of relative humidity.

The acetone-sensing properties of different Co_3O_4-based sensors in this work and the previous reports are listed in Table 1 [42–46]. The ZIF-8-derived Co_3O_4@ZnO microspheres exhibit better sensitivity and a lower detection limit compared with the previous reports.

Table 1. Comparison of the gas-sensing performances of Co_3O_4-based sensors to acetone.

Material	Operating Temperature	Response	Detection Limit	References
Co_3O_4 rectangular rods	200 °C	1.94 (50 ppm)	5 ppm	[42]
Co_3O_4/Fe_2O_3 hollow cubes	250 °C	2.27 (100 ppm)	1 ppm	[43]
$Co_3O_4/NiCo_2O_4$ nanocages	238.9 °C	2.09 (100 ppm)	-	[44]
PdO-Co_3O_4 hollow Nanocages	350 °C	1.51 (5 ppm)	0.1 ppm	[45]
$Co_3O_4/ZnSnO_3$ nanorod array	250 °C	2.61 (100 ppm)	-	[46]
Co_3O_4@ZnO microspheres	250 °C	2.3 (50 ppm)	0.5 ppm	This work

4. Discussion

The sensing mechanism of MOS is similar to the description of the previous report [47]. As shown in Figure 10a, when Co_3O_4 is in the air environment, the electrons in the material are captured by the absorbed oxygen molecules to form oxygen ions. A hole-accumulating layer is formed on the surface of the material [48]. The adsorption processes of oxygen are expressed by Formulas (1)–(3). As shown in Figure 10b, when Co_3O_4 is in acetone, the acetone molecules react with the oxygen ions and electrons are released into the sensing material. The reaction processes between oxygen ions with acetone molecules are expressed by Formulas (4) and (5). Due to the reaction, the hole-accumulating layer becomes narrow. For the sensing material of Co_3O_4, the accumulation layer is a conducting channel, and a change in the accumulation layer directly affects the response value.

$$O_2(gas) \rightarrow O_2(ads) \tag{1}$$

$$O_2(ads) + e^- \rightarrow O_2^-(ads) \tag{2}$$

$$O_2^-(ads) + e^- \rightarrow 2O^-(ads) \tag{3}$$

$$CH_3COCH_3(gas) + 4O_2^-(ads) \rightarrow 3H_2O(gas) + 3CO_2(gas) + 4e^- \tag{4}$$

$$CH_3COCH_3(gas) + 8O^-(ads) \rightarrow 3CO_2(gas) + 3H_2O(gas) + 8e^- \tag{5}$$

As the test results show, ZIF-8-derived Co_3O_4@ZnO microspheres present a p-type semiconductor behavior. The improved acetone-sensing properties of CZ-2 contribute to the formed p-n heterojunction between ZnO with Co_3O_4, the increased oxygen vacancy of material, and the porous structure of MOF-derived ZnO.

Firstly, the formed p-n heterojunction between ZnO and Co_3O_4 is the key reason for the enhanced gas sensitivity. As Figure 10c,d show, the band gaps of ZnO and Co_3O_4 are 3.37 eV and 1.6 eV. When they are in contact, the carrier flows between ZnO with Co_3O_4 to make the Fermi level equilibrium. The depletion layers are formed on the interface of two materials, and the resistance primarily depends on the width of the depletion layer. When the sensor is in the air, the reactions of Formulas (1)–(3) occur on the Co_3O_4@ZnO microspheres. The electrons flowing from ZnO to Co_3O_4 decrease, which causes a decrease in the width of the hole-depleting layer on the Co_3O_4 side [49]. When acetone gas is introduced, the reactions of Formulas (4) and (5) occur on the material, leading to the hole-depleting layer increasing. For the Co_3O_4@ZnO microspheres, the resistance change rate caused by the depletion layer is significant, leading to higher sensitivity. Secondly, oxygen vacancy is an active site, which provides a venue for oxygen adsorption and reaction [40]. From XPS characterization, it is clear that the O_V component of CZ-2 is higher than that of pure Co_3O_4. This indicates that more sites of oxygen adsorption and reaction are present on the surface of Co_3O_4@ZnO microspheres. Finally, the porous structure derived by ZIF-8 is also a favorable channel for acetone gas transmission. With the porous structure, the gas reaches the interface of the heterojunction more easily. Therefore, the acetone-sensing properties of Co_3O_4@ZnO microspheres are improved.

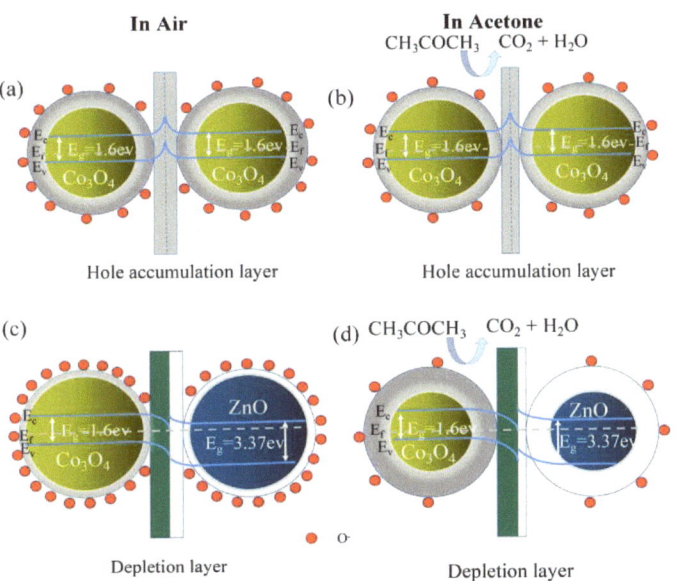

Figure 10. Schematic diagram of energy band structure of (**a**) pure Co_3O_4 in air, (**b**) pure Co_3O_4 in acetone, (**c**) Co_3O_4@ZnO in air and (**d**) Co_3O_4@ZnO in acetone.

5. Conclusions

ZIF-8-derived Co_3O_4@ZnO microspheres were prepared by a liquid-phase concentration-controlled nucleation strategy. This work indicated that the ZIF-8-derived surface structure could be controlled by changing the ligand concentration. Compared with pure Co_3O_4, the Co_3O_4@ZnO microspheres presented improved acetone-sensing properties. The sensor presents higher sensitivity, a faster response-recovery rate, and better selectivity and stability. This work shows that the preparation of an MOF-derived heterostructure is an effective way to enhance the gas-sensing performance of MOS materials.

Author Contributions: Conceptualization, X.F.; methodology, X.F. and P.D.; software, Y.L.; validation, X.F. and S.Y.; formal analysis, X.F. and C.H.; investigation, S.Y.; resources, Y.L.; data curation, S.Y. and C.H.; writing—original draft preparation, X.F. and S.Y.; writing—review and editing, X.F. and Y.L.; visualization, X.F. and S.Y.; supervision, X.F. and P.D.; project administration, X.F. and C.H.; funding acquisition, X.F. All authors have read and agreed to the published version of the manuscript.

Funding: This work was supported by the National Nature Science Foundation of China (No. 62201211).

Institutional Review Board Statement: Not applicable.

Informed Consent Statement: Not applicable.

Data Availability Statement: Not applicable.

Conflicts of Interest: The authors declare no conflict of interest.

References

1. Hanh, N.H.; Van Duy, L.; Hung, C.M.; Xuan, C.T.; Van Duy, N.; Hoa, N.D. High-Performance Acetone Gas Sensor Based on Pt–Zn$_2$SnO$_4$ Hollow Octahedra for Diabetic Diagnosis. *J. Alloys Compd.* **2021**, *886*, 161284. [CrossRef]
2. Isaac, N.; Pikaar, I.; Biskos, G. Metal Oxide Semiconducting Nanomaterials for Air Quality Gas Sensors: Operating Principles, Performance, and Synthesis Techniques. *Microchim. Acta* **2022**, *189*, 196. [CrossRef] [PubMed]
3. Yuan, Z.; Zhao, Q.N.; Xie, C.Y.; Liang, J.G.; Duan, X.H.; Duan, Z.H.; Li, S.R.; Jiang, Y.D.; Tai, H.L. Gold-Loaded Tellurium Nanobelts Gas Sensor for ppt-Level NO$_2$ Detection at Room Temperature. *Sens. Actuators B Chem.* **2022**, *355*, 131300. [CrossRef]
4. Kohli, N.; Hastir, A.; Kumari, M.; Singh, R.C. Hydrothermally Synthesized Heterostructures of In$_2$O$_3$/MWCNT as Acetone Gas Sensor. *Sens. Actuators A Phys.* **2020**, *314*, 112240. [CrossRef]
5. Kim, H.; Cai, Z.; Chang, S.-P.; Park, S. Improved Sub-ppm Acetone Sensing Properties of SnO$_2$ Nanowire-Based Sensor by Attachment of Co$_3$O$_4$ Nanoparticles. *J. Mater. Res. Technol.* **2020**, *9*, 1129–1136. [CrossRef]
6. Kumarage, G.W.; Comini, E. Low-Dimensional Nanostructures Based on Cobalt Oxide (Co$_3$O$_4$) in Chemical-Gas Sensing. *Chemosensors* **2021**, *9*, 197. [CrossRef]
7. Yusof, N.M.; Rozali, S.; Ibrahim, S.; Siddick, S.Z. Synthesis of Hybridized Fireworks-like go-Co$_3$O$_4$ Nanorods for Acetone Gas Sensing Applications. *Mater. Today Commun.* **2023**, *35*, 105516. [CrossRef]
8. Ramakrishnan, V.; Unnathpadi, R.; Pullithadathil, B. P-Co$_3$O$_4$ Supported Heterojunction Carbon Nanofibers for Ammonia Gas Sensor Applications. *J. Mater. Sci.* **2022**, *9*, 61–67.
9. Zhang, R.; Zhou, T.-T.; Wang, L.-L.; Zhang, T. Metal–Organic Frameworks-Derived Hierarchical Co$_3$O$_4$ Structures as Efficient Sensing Materials for Acetone Detection. *ACS Appl. Mater. Interfaces* **2018**, *10*, 9765–9773. [CrossRef]
10. Park, J.; Shen, X.P.; Wang, G.X. Solvothermal Synthesis and Gas-Sensing Performance of Co$_3$O$_4$ Hollow Nanospheres. *Sens. Actuators B Chem.* **2009**, *136*, 494–498. [CrossRef]
11. Zhao, L.P.; Jin, R.R.; Wang, C.; Wang, T.S.; Sun, Y.F.; Sun, P.; Lu, G.Y. Flower-like ZnO-Co$_3$O$_4$ Heterojunction Composites for Enhanced Acetone Sensing. *Sens. Actuators B Chem.* **2023**, *390*, 133964. [CrossRef]
12. Meng, D.; Si, J.P.; Wang, M.Y.; Wang, G.S.; Shen, Y.B.; San, X.G.; Meng, F.L. One-Step Synthesis and the Enhanced Trimethylamine Sensing Properties of Co$_3$O$_4$/SnO$_2$ Flower-like Structures. *Vacuum* **2020**, *171*, 108994. [CrossRef]
13. Fan, X.X.; Xu, Y.J.; He, W.M. High Acetone Sensing Properties of In$_2$O$_3$–NiO One-Dimensional Heterogeneous Nanofibers Based on Electrospinning. *RSC Adv.* **2021**, *11*, 11215–11223. [CrossRef]
14. Cai, Z.; Park, S. Highly Selective Acetone Sensor Based on Co$_3$O$_4$-Decorated Porous TiO$_2$ Nanofibers. *J. Alloys Compd.* **2022**, *919*, 165875. [CrossRef]
15. Karnati, P.; Akbar, S.; Morris, P.A. Conduction Mechanisms in One Dimensional Core-Shell Nanostructures for Gas Sensing: A Review. *Sens. Actuators B Chem.* **2019**, *295*, 127–143. [CrossRef]
16. Rehman, B.; Bhalla, N.K.; Vihari, S.; Jain, S.K.; Vashishtha, P.; Gupta, G. SnO$_2$/Au Multilayer Heterostructure for Efficient CO Sensing. *Mater. Chem. Phys.* **2020**, *244*, 122741. [CrossRef]
17. Xu, Y.S.; Zheng, L.L.; Yang, C.; Zheng, W.; Liu, X.H.; Zhang, J. Chemiresistive Sensors Based on Core-Shell ZnO@TiO$_2$ Nanorods Designed by Atomic Layer Deposition for N-Butanol Detection. *Sens. Actuators B Chem.* **2020**, *310*, 127846. [CrossRef]
18. Xu, Y.H.; Ding, L.J.; Wen, Z.R.; Zhang, M.; Jiang, D.; Guo, Y.S.; Ding, C.F.; Wang, K. Core-Shell LaFeO$_3$@G-C$_3$N$_4$ PN Heterostructure with Improved Photoelectrochemical Performance for Fabricating Streptomycin Aptasensor. *Appl. Surf. Sci.* **2020**, *511*, 145571. [CrossRef]
19. Wan, K.; Wang, D.; Wang, F.; Li, H.J.; Xu, J.C.; Wang, X.Y.; Yang, J.H. Hierarchical In$_2$O$_3$@SnO$_2$ Core–Shell Nanofiber for High Efficiency Formaldehyde Detection. *ACS Appl. Mater. Interfaces* **2019**, *11*, 45214–45225. [CrossRef]
20. Yan, S.; Song, W.N.; Wu, D.; Jin, S.C.; Dong, S.W.; Hao, H.S.; Gao, W.Y. Assembly of In$_2$O$_3$ Nanoparticles Decorated NiO Nanosheets Heterostructures and Their Enhanced Gas Sensing Characteristics. *J. Alloys Compd.* **2022**, *896*, 162887. [CrossRef]
21. Yu, S.W.; Jia, X.H.; Yang, J.; Wang, S.Z.; Li, Y.; Song, H.J. Highly Sensitive and Low Detection Limit of Ethanol Gas Sensor Based on CeO$_2$ Nanodot-Decorated ZnSnO$_3$ Hollow Microspheres. *Ceram. Int.* **2022**, *48*, 14865–14875. [CrossRef]
22. Gui, Y.H.; Yang, L.-L.; Tian, K.; Zhang, H.Z.; Fang, S.M. P-Type Co$_3$O$_4$ Nanoarrays Decorated on the Surface of N-Type Flower-like WO$_3$ Nanosheets for High-Performance Gas Sensing. *Sens. Actuators B Chem.* **2019**, *288*, 104–112. [CrossRef]
23. Felix Sahayaraj, A.; Joy Prabu, H.; Maniraj, J.; Kannan, M.; Bharathi, M.; Diwahar, P.; Salamon, J. Metal–Organic Frameworks (MOFs): The Next Generation of Materials for Catalysis, Gas Storage, and Separation. *J. Inorg. Organomet. Polym. Mater.* **2023**. [CrossRef]
24. Kavian, S.; Hajati, S.; Moradi, M. High-Rate Supercapacitor Based on NiCo-MOF-Derived Porous NiCoP for Efficient Energy Storage. *J. Mater. Sci. Mater. Electron.* **2021**, *32*, 13117–13128. [CrossRef]
25. Najafi, M.; Rahimi, R. Synthesis of Novel Zr-MOF/Cloisite-30B Nanocomposite for Anionic and Cationic Dye Adsorption: Optimization by Design-Expert, Kinetic, Thermodynamic, and Adsorption Study. *J. Inorg. Organomet. Polym. Mater.* **2023**, *33*, 138–150. [CrossRef]
26. Nair, S.S.; Illyaskutty, N.; Tam, B.; Yazaydin, A.O.; Emmerich, K.; Steudel, A.; Hashem, T.; Schöttner, L.; Wöll, C.; Kohler, H. ZnO@ZIF-8: Gas Sensitive Core-Shell Hetero-Structures Show Reduced Cross-Sensitivity to Humidity. *Sens. Actuators B Chem.* **2020**, *304*, 127184. [CrossRef]
27. Liu, C.Q.; Li, D.W.; Tang, W. Enhanced Ethanol Sensors Based on MOF-Derived ZnO/Co$_3$O$_4$ Bimetallic Oxides with High Selectivity and Improved Stability. *Vacuum* **2023**, *214*, 112185. [CrossRef]

28. Arul, C.; Moulaee, K.; Donato, N.; Iannazzo, D.; Lavanya, N.; Neri, G.; Sekar, C. Temperature Modulated Cu-MOF Based Gas Sensor with Dual Selectivity to Acetone and NO_2 at Low Operating Temperatures. *Sens. Actuators B Chem.* **2021**, *329*, 129053. [CrossRef]
29. Wei, Q.; Sun, J.; Song, P.; Li, J.; Yang, Z.X.; Wang, Q. Spindle-like Fe_2O_3/$ZnFe_2O_4$ Porous Nanocomposites Derived from Metal-Organic Frameworks with Excellent Sensing Performance Towards Triethylamine. *Sens. Actuators B Chem.* **2020**, *317*, 128205. [CrossRef]
30. Qin, W.B.; Yuan, Z.Y.; Gao, H.L.; Zhang, R.Z.; Meng, F.L. Perovskite-Structured $LaCoO_3$ Modified ZnO Gas Sensor and Investigation on Its Gas Sensing Mechanism by First Principle. *Sens. Actuators B Chem.* **2021**, *341*, 130015. [CrossRef]
31. Zhou, Q.; Zeng, W.; Chen, W.G.; Xu, L.N.; Kumar, R.; Umar, A. High Sensitive and Low-Concentration Sulfur Dioxide (SO_2) Gas Sensor Application of Heterostructure NiO-ZnO Nanodisks. *Sens. Actuators B Chem.* **2019**, *298*, 126870. [CrossRef]
32. Chang, X.; Li, X.F.; Qiao, X.R.; Li, K.; Xiong, Y.; Li, X.; Guo, T.C.; Zhu, L.; Xue, Q.Z. Metal-Organic Frameworks Derived ZnO@MoS_2 Nanosheets Core/Shell Heterojunctions for ppb-Level Acetone Detection: Ultra-Fast Response and Recovery. *Sens. Actuators B Chem.* **2020**, *304*, 127430. [CrossRef]
33. Zhan, M.M.; Ge, C.X.; Hussain, S.; Alkorbi, A.S.; Alsaiari, R.; Alhemiary, N.A.; Qiao, G.J.; Liu, G.W. Enhanced No_2 Gas-Sensing Performance by Core-Shell SnO_2/ZIF-8 Nanospheres. *Chemosphere* **2022**, *291*, 132842. [CrossRef] [PubMed]
34. Fan, X.X.; Xu, Y.J.; Ma, C.Y.; He, W.M. In-Situ Growth of Co_3O_4 Nanoparticles Based on Electrospray for an Acetone Gas Sensor. *J. Alloys Compd.* **2021**, *854*, 157234. [CrossRef]
35. Fan, X.X.; Wang, J.F.; Sun, C.L.; Huang, C.; Lu, Y.J.; Dai, P.; Xu, Y.J.; He, W.M. Effect of Pr/Zn on the Anti-Humidity and Acetone-Sensing Properties of Co_3O_4 Prepared by Electrospray. *RSC Adv.* **2022**, *12*, 19384–19393. [CrossRef]
36. Fan, X.-X.; He, X.-L.; Li, J.-P.; Gao, X.-G.; Jia, J. Ethanol Sensing Properties of Hierarchical SnO_2 Fibers Fabricated with Electrospun Polyvinylpyrrolidone Template. *Vacuum* **2016**, *128*, 112–117. [CrossRef]
37. Xiong, Y.; Xu, W.-W.; Zhu, Z.Y.; Xue, Q.Z.; Lu, W.B.; Ding, D.G.; Zhu, L. ZIF-Derived Porous ZnO-Co_3O_4 Hollow Polyhedrons Heterostructure with Highly Enhanced Ethanol Detection Performance. *Sens. Actuators B Chem.* **2017**, *253*, 523–532. [CrossRef]
38. Qu, F.D.; Thomas, T.; Zhang, B.X.; Zhou, X.-X.; Zhang, S.D.; Ruan, S.P.; Yang, M.H. Self-Sacrificing Templated Formation of Co_3O_4/$ZnCo_2O_4$ Composite Hollow Nanostructures for Highly Sensitive Detecting Acetone Vapor. *Sens. Actuators B Chem.* **2018**, *273*, 1202–1210. [CrossRef]
39. Cheng, P.F.; Lv, L.; Wang, Y.L.; Zhang, B.; Zhang, Y.; Zhang, Y.Q.; Lei, Z.H.; Xu, L.P. SnO_2/$ZnSnO_3$ Double-Shelled Hollow Microspheres Based High-Performance Acetone Gas Sensor. *Sens. Actuators B Chem.* **2021**, *332*, 129212. [CrossRef]
40. Lee, J.; Choi, Y.; Park, B.J.; Han, J.W.; Lee, H.-S.; Park, J.H.; Lee, W. Precise Control of Surface Oxygen Vacancies in ZnO Nanoparticles for Extremely High Acetone Sensing Response. *J. Adv. Ceram.* **2022**, *11*, 769–783. [CrossRef]
41. Reddeppa, M.; Park, B.-G.; Murali, G.; Choi, S.H.; Chinh, N.D.; Kim, D.; Yang, W.; Kim, M.-D. NO_x Gas Sensors Based on Layer-Transferred N-MoS_2/P-GaN Heterojunction at Room Temperature: Study of UV Light Illuminations and Humidity. *Sens. Actuators B Chem.* **2020**, *308*, 127700. [CrossRef]
42. Wang, S.M.; Cao, J.; Cui, W.; Fan, L.L.; Li, X.; Li, D.J.; Zhang, T. One-Dimensional Porous Co_3O_4 Rectangular Rods for Enhanced Acetone Gas Sensing Properties. *Sens. Actuators B Chem.* **2019**, *297*, 126746. [CrossRef]
43. Xu, K.; Zhao, W.; Yu, X.; Duan, S.L.; Zeng, W. MOF-Derived Co_3O_4/Fe_2O_3 PN Hollow Cubes for Improved Acetone Sensing Characteristics. *Phys. E* **2020**, *118*, 113869. [CrossRef]
44. Qu, F.D.; Jiang, H.F.; Yang, M.H. MOF-Derived Co_3O_4/$NiCo_2O_4$ Double-Shelled Nanocages with Excellent Gas Sensing Properties. *Mater. Lett.* **2017**, *190*, 75–78. [CrossRef]
45. Koo, W.-T.; Yu, S.; Choi, S.-J.; Jang, J.-S.; Cheong, J.Y.; Kim, I.-D. Nanoscale PdO Catalyst Functionalized Co_3O_4 Hollow Nanocages Using MOF Templates for Selective Detection of Acetone Molecules in Exhaled Breath. *ACS Appl. Mater. Interfaces* **2017**, *9*, 8201–8210. [CrossRef] [PubMed]
46. Xu, K.; Zhao, W.; Yu, X.; Duan, S.L.; Zeng, W. Enhanced Ethanol Sensing Performance Using Co_3O_4–$ZnSnO_3$ Arrays Prepared on Alumina Substrates. *Phys. E* **2020**, *117*, 113825. [CrossRef]
47. Li, P.P.; Jin, H.Q.; Yu, J.; Chen, W.M.; Zhao, R.Q.; Cao, C.Y.; Song, W.G. NO_2 Sensing with CdS Nanowires at Room Temperature under Green Light Illumination. *Mater. Futures* **2022**, *1*, 025303. [CrossRef]
48. Srirattanapibul, S.; Nakarungsee, P.; Issro, C.; Tang, I.-M.; Thongmee, S. Enhanced Room Temperature NH_3 Sensing of rGO/Co_3O_4 Nanocomposites. *Mater. Chem. Phys.* **2021**, *272*, 125033. [CrossRef]
49. Kim, H.-J.; Lee, J.-H. Highly Sensitive and Selective Gas Sensors Using P-Type Oxide Semiconductors: Overview. *Sens. Actuators B Chem.* **2014**, *192*, 607–627. [CrossRef]

Disclaimer/Publisher's Note: The statements, opinions and data contained in all publications are solely those of the individual author(s) and contributor(s) and not of MDPI and/or the editor(s). MDPI and/or the editor(s) disclaim responsibility for any injury to people or property resulting from any ideas, methods, instructions or products referred to in the content.

Article

Pyrene-Derived Covalent Organic Framework Films: Advancements in Acid Vapor Detection

Shaikha S. AlNeyadi *, Mohammed T. Alhassani, Ali S. Aleissaee and Ibrahim AlMujaini

Department of Chemistry, College of Science, UAE University, Al-Ain 15551, United Arab Emirates; 202010213@uaeu.ac.ae (M.T.A.); 202012257@uaeu.ac.ae (A.S.A.); 201900767@uaeu.ac.ae (I.A.)
* Correspondence: shaikha.alneyadi@uaeu.ac.ae

Abstract: The expansion of global industry results in the release of harmful volatile acid vapors into the environment, posing a threat to various lifeforms. Hence, it is crucial to prioritize the development of swift sensing systems capable of monitoring these volatile acid vapors. This initiative holds great importance in safeguarding a clean and safe environment. This paper presents the synthesis and characterization of pyrene-based covalent organic frameworks (COFs) that exhibit exceptional crystallinity, thermal stability, and intense fluorescence. Three COFs—PP–COF, PT–COF, and PE–COF—were synthesized, demonstrating large surface areas and robust thermal stability up to 400 °C. The fluorescence properties and intramolecular charge transfer within these COFs were significantly influenced by their Schiff base bonding types and π-stacking degrees between COF layers. Notably, PE-COF emerged as the most fluorescent of the three COFs and exhibited exceptional sensitivity and rapid response as a fluorescent chemosensor for detecting HCl in solution. The reversible protonation of imine bonds in these COFs allowed for the creation of highly sensitive acid vapor sensors, showcasing a shift in spectral absorption while maintaining structural integrity. This study highlights the potential of COFs as reliable and reusable sensors for detecting harmful acid vapors and addressing environmental concerns arising from industrial activities.

Keywords: acid vapor; sensor; COF; pyrene; film

Citation: AlNeyadi, S.S.; Alhassani, M.T.; Aleissaee, A.S.; AlMujaini, I. Pyrene-Derived Covalent Organic Framework Films: Advancements in Acid Vapor Detection. *Chemosensors* **2024**, *12*, 37. https://doi.org/10.3390/chemosensors12030037

Academic Editor: Chunsheng Wu

Received: 25 December 2023
Revised: 20 February 2024
Accepted: 22 February 2024
Published: 3 March 2024

Copyright: © 2024 by the authors. Licensee MDPI, Basel, Switzerland. This article is an open access article distributed under the terms and conditions of the Creative Commons Attribution (CC BY) license (https://creativecommons.org/licenses/by/4.0/).

1. Introduction

The rapid evolution of human lifestyles has led to significant environmental changes, particularly concerning air pollution, which poses a serious threat to both environmental and human well-being [1,2]. Among various air pollutants, hydrogen chloride (HCl) gas stands out due to its contribution to acid rain, generation of dioxins, and severe health effects, including respiratory issues and mucosal damage [2–5]. HCl's origins in incineration plants, pharmaceutical, and metallurgical sectors, among others, necessitate effective monitoring to prevent environmental and health hazards [3,5]. Recent incidents of HCl vapor leaks in various locations globally underline the urgency of developing efficient detection methods [6,7]. Thus, the development of a highly sensitive platform for detecting hazardous HCl vapor becomes imperative to monitor and avert accidental threats. In recent years, several approaches like optochemical, optical, gas chromatographic, and electrochemical sensors have been developed for detecting HCl vapor, each with its limitations, such as time-consuming analysis, low sensitivity, and high cost [8]. Presently, there is a growing demand for improved sensing materials capable of detecting HCl with remarkable sensitivity, focusing on materials like metal oxides, lanthanide complexes, and conducting molecules and aiming for lower detection limits [9]. Among these, conducting materials draw significant attention due to their conductivity arising from conjugated π–π bonds, enabling their utilization in electrical and optical sensor devices [9]. Organic π-conjugated small molecules offer distinct advantages, including customizable structures, mechanical flexibility, and solution processability, making them promising candidates for cost-effective, portable gas sensors operable at room temperature. However, challenges persist in fabricating organic semiconductor gas sensors, including issues related to sensitivity, recovery, and response time [9,10]. Over the years, various sensor technologies have been developed to detect gaseous HCl, including conductometric, amperometric, solid electrochemical, and optochemical sensors. However, these methods suffer from drawbacks, such as prolonged response times, limited sensitivity, high operational costs, and safety concerns. In light of these challenges, there is an urgent need for the development of more accessible, efficient, rapid, and safer sensors capable of onsite detection to effectively identify the presence of HCl.

In this context, Within the dynamic sphere of material science, covalent organic frameworks (COFs) have rapidly ascended as pivotal materials, especially in the cutting-edge field of sensor technology [11–13]. These crystalline, porous structures that are ingeniously bonded by strong covalent connections between light elements, span a remarkable range of applications from gas storage and pollutant removal to catalysis, adsorption, and optoelectronics [14]. Their deployment in the specific arena of chemical sensing, with a notable emphasis on HCl gas detection, exemplifies their expansive utility. Distinguished by their extensive surface areas, superior thermal resilience, and the capacity to form highly fluorescent frameworks, COFs transcend the traditional barriers faced by non-fluorescent variants. This advancement not only showcases their adaptable architectures and exceptional material qualities but also positions COFs at the forefront of research in material science, particularly as the exploration of their sensing capabilities continues to grow, heralding a new era of sensor development [15,16].

In this study, we synthesized three highly stable 2D COFs—PP–COF, PT–COF, and PE–COF—showcasing remarkable Brunauer–Emmett–Teller (BET) surface areas of up to 1350 $m^2\ g^{-1}$. These COFs were created through Schiff base formation under solvothermal conditions, employing 4,4′,4″,4‴-(pyrene-1,3,6,8-tetrayl)tetraaniline and three distinct formyl species: 4,4′,4″-(1,3,5-triazine-2,4,6-triyl)tribenzaldehyde, hexakis (4-aldehyde phenoxy) cyclotriphosphazene, and 4,4′,4″,4‴-(ethane-1,1,2,2-tetrayl)tetra benzaldehyde, as illustrated in Scheme 1. Capitalizing on their substantial surface areas, exceptional stability, and strong fluorescence, these COFs demonstrated their capacity as sensitive chemosensors for detecting HCl. Among the COFs studied, PE-COF initially emits a faint yellow fluorescence. However, upon exposure to solid-state HCl gas, it rapidly transforms into a vibrant dark orange emission in less than a second. This swift and notable color shift is easily observable to the naked eye under regular visible light. Our study aims to showcase an innovative system capable of detecting toxic acid vapors. Highlighting its ease of processing, cost-effectiveness, portability, and functionality under ambient conditions, we demonstrate its potential in constructing an electronic prototype for on-field applications. Our goal is to illustrate how this technology revolutionizes real-time monitoring of hazardous substances.

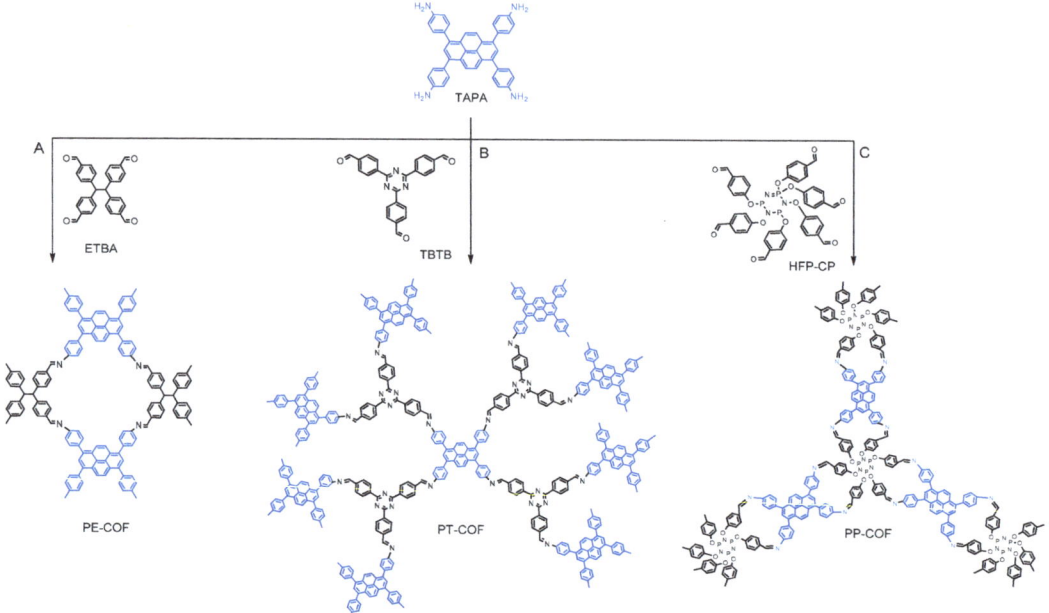

Scheme 1. Synthesis details for PE-COF, PT-COF, and PP-COF, outlining catalysts, solvents, reaction times, and temperatures: (**A**) PE-COF is created with a mesitylene/1,4-dioxane solution (1:1 v/v, 2 mL) and AcOH (6 M, 0.2 mL), and is heated for 3 days at 120 °C. Both (**B**) PT-COF and (**C**) PP-COF are made using a DCB/n-BuOH mix (9:1 v/v, 1.5 mL) with AcOH (6 M, 0.3 mL), and are heated for 5 days at the same temperature.

2. Results and Discussion

In this investigation, we capitalize on the distinctive electron configuration inherent in imine linkages to serve as the defining functional constituent within a series of two-dimensional covalent organic frameworks (COFs). These COFs were meticulously synthesized using 1,3,6,8-tetra(aminophenyl)pyrene (TAPA, as depicted in Scheme 1) as a foundational building block. Upon subjecting the imine bonds within these materials to protonation, our observations reveal a noteworthy redshift in absorption, particularly in the near-infrared region, accompanied by the emergence of absorption bands induced by protonation. Importantly, these alterations manifest without compromising the structural integrity or crystalline attributes of the frameworks. We designed and created three COFs with outstanding stability and porosity using Schiff base reactions, as shown in Scheme 1. In the synthesis of pyrene-based covalent organic frameworks (COFs), the approach centered on employing 1,3,6,8-tetra(aminophenyl)pyrene (TAPA) as a pivotal four-connected building block, serving as an amine linker. This strategic choice aimed to capitalize on TAPA's structural versatility. The coupling of TAPA with diverse compounds, including 4,4′,4″,4‴-(ethane-1,1,2,2-tetrayl)tetra benzaldehyde (ETBA) (Scheme 1A), 4,4′,4″-(1,3,5-Triazine-2,4,6-triyl)tribenzaldehyde (Scheme 1B), and hexa(4-formyl-phenoxy)cyclotriphosphazene (Scheme 1C). The synthesis process yielded three distinct two-dimensional (2D) structures termed PP–COF, PT–COF, and PE–COF. Detailed methods for synthesizing organic linkers, including reagents, reaction conditions, and purification techniques, are provided. The supporting data (SI, Section S2, Schemes S1–S3 and Section S3, Figures S1–S6) encompass comprehensive characterization techniques, such as spectroscopic analysis (e.g., NMR). The creation of the three targeted COFs utilized a solvothermal technique [17]. Acetic acid served as a catalyst, and the building blocks were suspended in a solution composed of a mixture of (mesitylene/1,4-dioxane) or o-(dichlorobenzene/n-butanol). Subsequently, the compounds were subjected to solvothermal conditions at 120 °C for 3–5 days (SI, Section S4). The successful creation of imine bonds was shown by the FT-IR spectroscopy study of PP–COF, PT–COF, and PE–COF, which showed strong stretching vibrations of the C=N unit in the range of 1622–1625 cm^{-1}. Further evidence of the success of the Schiff base condensation process was provided by the spectra, which also revealed the removal of the N–H (3439–3398 cm^{-1}) stretching vibrations from the amino group in the amine linker and the C=O (1648–1660 cm^{-1}) vibration from the aldehyde linker (SI, Section S5, Figure S7). To examine the porous architectures of PP–COF, PT–COF, and PE–COF, nitrogen sorption isotherms at 77 K were employed (Figure 1A). Prior to the nitrogen sorption measurement, the COF samples underwent an overnight pretreatment at 100 °C in a vacuum. The nitrogen absorption of PP–COF, PT–COF, and PE–COF displayed rapid increments at lower pressures (P/P_0 = 0 to 0.1), showcasing their microporous characteristics. The Brunauer–Emmett–Teller (BET) surface area measurements revealed values of 1350 m^2 g^{-1}, 1098 m^2 g^{-1}, and 730 m^2 g^{-1} for PP–COF, PT–COF, and PE–COF, respectively. Evaluation of the total pore volumes at a P/P_0 ratio of 0.99 showcased a notable pore volume of 1.83 cm^3 g^{-1} for PP–COF, surpassing both PT–COF (0.98 cm^3 g^{-1}) and PE–COF (1.23 cm^3 g^{-1}). Furthermore, nonlocal density functional theory (NLDFT) was employed to determine the pore sizes of the three COFs. The results revealed pore sizes of 16 Å and 18 Å for PP–COF and 12 Å and 17 Å for both PT–COF and PE–COF (Figure 1B).

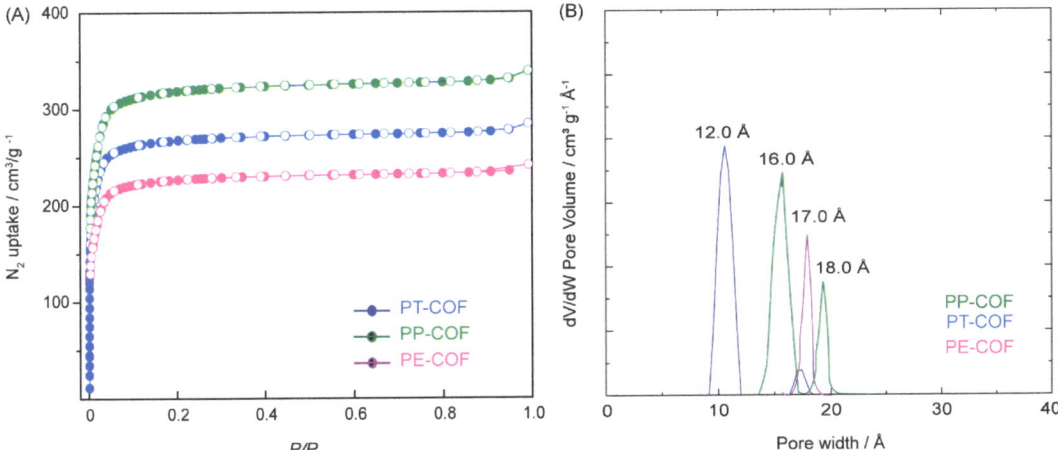

Figure 1. (**A**) Nitrogen sorption isotherms of PP–COF, PT–COF, and PE–COF, at 77 K, with ● representing adsorption and ○ representing desorption; (**B**) Pore size distributions of PP-COF, PT-COF, and PE-COF.

The crystalline characteristics of PP–COF, PT–COF, and PE–COF were investigated through powder X-ray diffraction (PXRD) analysis, as depicted in Figure 2. In the PXRD diffractograms of PT–COF (Figure 2A), distinct peaks corresponding to the (100), (110), (200), (210), (300), and (001) reticular planes were observed at 3.98°, 5.06°, 9.34°, 11.68°, 14.85°, and 24.80°, respectively. Similarly, the PXRD pattern of PP–COF (Figure 2B) exhibited peaks at 2.68°, 5.58°, 9.82°, 12.54°, and 23.35°, which were attributed to the (110), (102), (200), (004), and (204) reticular planes, respectively. In the case of PE–COF (Figure 2C), six distinctive peaks were identified, with the most intense peak at 5.56°, corresponding to the 100 facet, while others appeared at 6.78°, 7.31°, 9.52°, 11.67°, and 12.98°, representing the (110), (210), (220), (300), and (001) reticular planes, respectively. To gain comprehensive insights into their structures, the PXRD patterns of the investigated COFs underwent simulation via the Pawley refinement technique. The simulated PXRD patterns (depicted by the red curve in Figure 2A–C) exhibited close alignment with the experimental PXRD patterns (depicted by the black curve in Figure 2A–C). This agreement is evident from the difference pattern (indicated by the green curve in Figure 2A–C), which confirms the accuracy of the structural analysis. In the case of PT–COF (Figure 2A), the experimental PXRD pattern exhibited strong conformity with the simulated patterns derived from the AA-staggered stacking model, displaying favorable agreement factors (Rp = 1.33% and Rwp = 2.67%) with optimized parameters (a = b = 19.46 Å, c = 3.64 Å, $\alpha = \beta = 90°$, and $\gamma = 120°$, resulting in a unit cell volume of 1279.51 Å^3). For PP–COF, the cell parameters were finetuned through Pawley refinements, showcasing minimal distinction between the experimental curve and simulated profiles produced by the AA-staggered stacking model (a = b = 30.17 Å, c = 3.43 Å, $\alpha = \beta = 90°$, and $\gamma = 120°$, leading to a unit cell volume of 2704.10 Å^3), yielding negligible residuals (Rp = 1.23% and Rwp = 3.27%) (Figure 2B). Moreover, the experimental PXRD patterns of PE–COF (Figure 2C) strongly aligned with the XRD pattern generated from the AA-eclipsed stacking models, indicating excellent agreement. Conversely, significant disparities were observed when comparing the pattern derived from their corresponding AB' staggered stacking model with the experimental results. The refined unit cell parameters for PE–COF were determined as a = 21.23 Å, b = 21.20 Å, c = 4.37 Å, with $\alpha = \beta = \gamma = 90°$, revealing residuals of Rwp = 3.13% and Rp = 4.27% and a unit cell volume of 1964.51 Å^3 (SI, Section S5, Figure S8) and (SI, Section S10, Tables S1–S3).

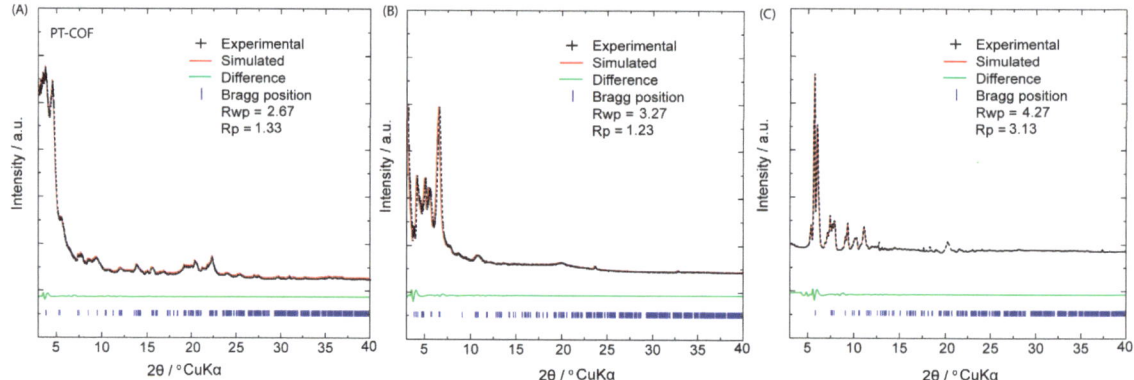

Figure 2. PXRD patterns of (**A**) PT–COF, (**B**) PP–COF, and (**C**) PE–COF.

In our investigation, we aimed to assess the chemical and thermal stability of the synthesized PP–COF, PT–COF, and PE–COF for potential practical applications. Our findings revealed that these COFs exhibited exceptional stability, rendering them highly desirable for applications requiring resistance to chemical reactions and elevated temperatures. Through thermogravimetric analysis (TGA), we observed that these COFs demonstrated outstanding thermal stability. Even at high temperatures, they maintained their structural integrity without significant weight loss or decomposition. Notably, these COFs exhibited no signs of breakdown until reaching 400 °C in a nitrogen environment. Furthermore, we conducted chemical stability tests on the PP–COF, PT–COF, and PE–COF samples in various solvents at room temperature over a 24-h period. These solvents included boiling water, ethanol, N,N-dimethylformamide, dimethylsulfoxide, 3 M HCl at 25 °C, and 3 M NaOH at 25 °C. Surprisingly, post-testing, the PXRD patterns of the COF samples remained robust and unchanged, indicating the preservation of their remarkable crystallinity even under challenging conditions (SI, Section S6, Figures S9–S11). This study confirms the robust chemical stability of PP–COF, PT–COF, and PE–COF, making them ideal candidates for functionalization. The appeal of these COFs as chemically and thermally stable materials is further amplified by their crystalline structure and porous attributes. To facilitate spectroscopic analysis and sensor uses, we grew thin films of PT–COF, PP–COF, and PE–COF on fused silica (SI, Section S4). Despite the susceptibility of imine nitrogen atoms to strong acid protonation, the closely packed 2D COF structures provide stability against hydrolysis, enabling complete and reversible protonation even in humid conditions without notable degradation [18]. To figure out the sensing mechanism of PT–COF, PP–COF, and PE–COF toward HCl, we recorded FT-IR spectrum of the sample of PT–COF, PP–COF, and PE–COF treated with gaseous HCl (termed PT–COF-HCl, PP–COF-HCl, and PE–COF-HCl). Compared with the IR spectra of pristine PE–COF, PT–COF, and PP–COF, the most apparent changes are highlighted in pink where new peaks at approximately 1658, 1660, and 1655 cm^{-1} alongside a reduced intensity in the C=N band (1622–1625 cm^{-1}, Figure 3A–C). These alterations signify the formation of protonated imine bonds (C=NH$^+$), a result of the swift protonation of imine nitrogen atoms by HCl gas [19].

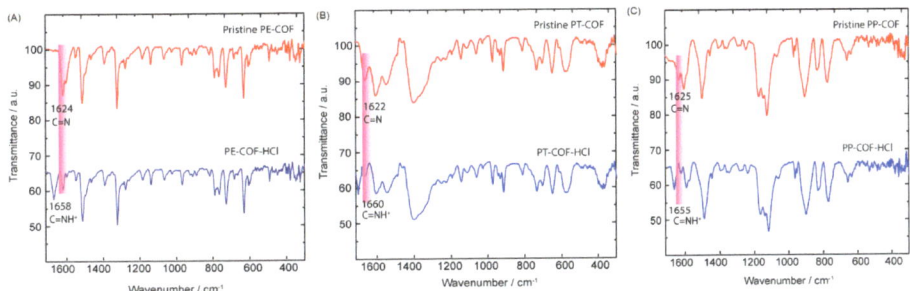

Figure 3. Comparison of the deprotonated and protonated (**A**) PE–COF, (**B**) PT–COF, and (**C**) PP–COF, and powders. The most apparent changes are highlighted in pink. The protonation leads to the appearance of the C=NH$^+$ stretching mode around 1655–1660 cm^{-1}, accompanied by an attenuation of the imine C=N stretching mode around 1622–1625 cm^{-1}.

The photophysical characteristics of our COFs were investigated by analyzing their fluorescence emission spectra in a solution. The COFs were dispersed in 1,4-dioxane at a concentration of 1 mg mL^{-1}, followed by excitation of the suspensions using light at a wavelength of 366 nm. The resulting fluorescence spectra of PT–COF, PE–COF, and PP–COF exhibited emission maxima at 450, 525, and 510 nm, respectively. These distinct emission peaks indicate the unique fluorescence behavior inherent in each COF variant. These features encouraged us to investigate the sensing properties of our COFs toward HCl.

Interestingly, upon exposure to HCl gas, suspensions of PT–COF and PP–COF (1 mg mL^{-1}) in 1,4-dioxane underwent a notable color shift, swiftly transitioning from yellow to dark brown. Subsequent exposure to triethylamine vapor facilitated the rapid restoration of the COF's original yellow color, highlighting the remarkable reversibility inherent in the process of sensing gaseous HCl. Additionally, these color changes were visibly apparent and occurred within response times of less than 1 s—significantly faster than those of previously reported HCl sensors [20]. A video showing the repeated protonation and deprotonation of a pyrene COF film (PP–COF) is supplied in the SI. Similar color transformations were observed in PE–COF from yellow to dark orange upon exposure to HCl vapor demonstrates strong interactions between our synthesized COFs and HCl molecules.

The reversibility of the color change remained uncompromised even after the COFs were subjected to 10 cycles of alternating treatments with HCl and triethylamine (TEA). This endurance underscores the robustness of the COFs in maintaining their reversible properties. Given the distinct and rapid color change that is perceptible without aid, this porous material stands as a commendable candidate for detecting HCl gas. To delve deeper into the structural impact of HCl gas followed by TEA gas exposure, the PT–COF, PP–COF, and PE–COF powders underwent analysis using powder X-ray diffraction

(XRD) and Fourier-transform infrared spectroscopy (FT-IR). As depicted in (SI, Section S7, Figures S12 and S13), minimal changes were observed in the major peaks of both the powder XRD patterns and IR spectra before and after exposure to HCl/TEA gas. These subtle alterations suggest that the fundamental framework of PT–COF, PP–COF, and PE–COF remained largely intact despite exposure to HCl/TEA gas, indicating their robustness in withstanding chemical exposure without significant structural damage.

The influence of varying HCl concentrations on COF fluorescence was studied using a 1,4-dioxane solution of HCl due to challenges in precisely monitoring gaseous HCl concentrations. The fluorescence spectra of the PT–COF, PP–COF, and PE–COF suspensions in 1,4-dioxane revealed distinct responses to different HCl levels (Figure 4). At 1 mmol L^{-1} HCl, the fluorescence emission maxima significantly decreased at 450 nm, 525 nm, and 510 nm for PT–COF, PE–COF, and PP–COF, respectively, while new peaks emerged at 540 nm, 625 nm, and 611 nm. These signals vanished at 5 mmol L^{-1} HCl, with subsequent gradual increases from 1 to 50 mmol L^{-1}, showing no further changes in emission. Calibration curves for HCl concentrations (1 to 50 mmol L^{-1}) exhibited linear correlations at 540 nm, 625 nm, and 611 nm, demonstrating detection limits of approximately 17, 10, and 20 nmol L^{-1} for PT–COF, PE–COF, and PP–COF, respectively (SI, Section S7, Figure S14).

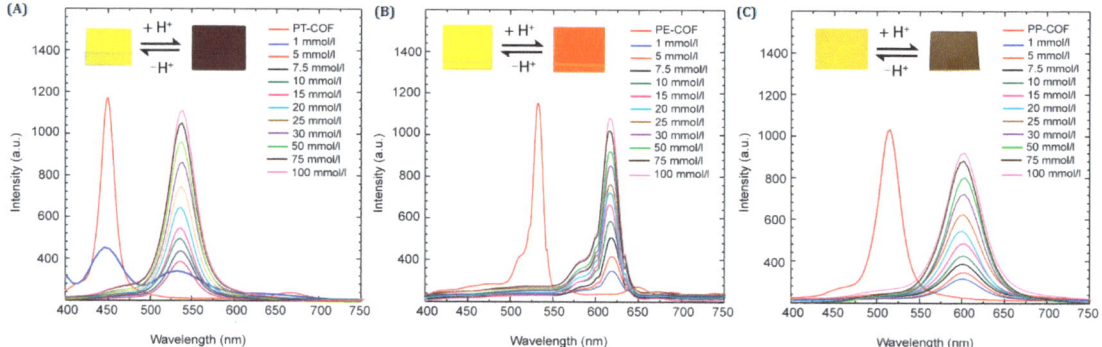

Figure 4. Fluorescence spectra of (**A**) PT–COF, (**B**) PE–COF, and (**C**) PP–COF (1 mg/mL) collected from its 1,4-dioxane suspension with different concentrations of HCl (λ_{ex} = 366 nm). The insets highlight the color transformation of COF films upon protonation: PT–COF from yellow to dark brown with 1 mmol of HCl, PE–COF from yellow to orange, and PP–COF from yellow to brown, visually demonstrating the protonation impact on each COF type.

The sensing mechanism of PT–COF, PE–COF, and PP–COF toward HCl involves the probable protonation of imine nitrogen atoms within the COF skeletons when exposed to an HCl atmosphere. This protonation influences the conjugated structure of COF, leading to significant changes in fluorescence emission and color. Unlike many HCl-responsive materials found in the literature, where protonation sites typically locate at heteroaromatic nitrogen or amino units, in PT–COF, PP–COF, and PE–COF, protonation occurs on abundant and periodically distributed imine nitrogen atoms within the imine-linked COF skeletons. As demonstrated by Ascherl et al. [15], the resulting protonated imine (C=NH$^+$) groups in COF structures act as stronger electron acceptors than their free imine (C=N) counterparts, accelerating charge transfer toward pyrene units, potentially causing a red-shift in transition photoexcitation energies. Specifically, PE–COF rapidly turned dark orange from yellow when exposed to HCl vapor (Figure 5A), showcasing swift nitrogen atom protonation. Its fluorescence peak shifted from 525 nm to 625 nm with HCl and reverted to 525 nm post-TEA treatment (Figure 5B).

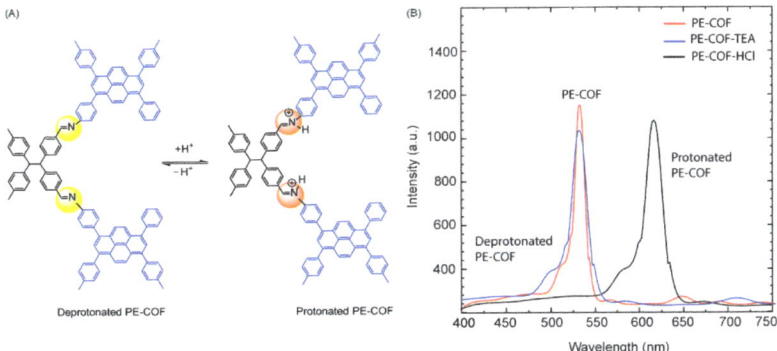

Figure 5. (**A**) Deprotonation, and protonation processes within the PE–COF framework fluorescent responses of PE–COF film; (**B**) Reaction to HCl gas at 75 mmol/L (λmax = 366 nm); Response to TEA gas.

Monitoring pH levels is crucial across various domains like chemistry, environment, and engineering. Two-dimensional COFs containing proton-donating/accepting units, exhibit pH-responsive behaviors driven by nitrogen atom protonation/deprotonation, altering color or fluorescence. These COFs, designed for sensing acid solution pH, also function as effective chemosensors for acidic gasses due to their porous structures that allow gaseous analyte diffusion. The pivotal protonation/deprotonation of imine bonds within COFs, influenced by agents like HCl and TEA, significantly affects their electronic properties, which is crucial for accurate acid sensing in acidic environments.

We delved into the impact of varying humidity levels on the detection capabilities of PT–COF, PP–COF, and PE–COF in the presence of HCl gas. Testing these materials across a spectrum of humidity environments (40%, 50%, 60%, 70%, and 80%) within a controlled constant temperature and humidity chamber (SI, Section S8, Figure S15) revealed an intriguing outcome. Surprisingly, the fluorescence intensities of PT–COF, PP–COF, and PE–COF remained consistently unaffected by shifts in humidity levels while detecting HCl gas. Even under fluctuating humidity conditions, these materials consistently showcased steadfast and robust fluorescence responses. This resilience underscores the stability of their detection capacity, seemingly unswayed by changes in humidity. Further probing into the mechanisms behind this resilience might unveil prospects for deploying these materials in environments characterized by diverse humidity levels, expanding their range of potential applications.

In the exploration of PT–COF, PP–COF, and PE–COF films' reactivity towards co-existing acid gasses, our investigation was extended to evaluate their responses to CO_2, SO_2, H_2S, and Cl_2. As depicted in Figure 6A–C, exposure to CO_2 and SO_2 resulted in no discernible changes in fluorescence intensity within the COF films. However, the presence of H_2S and Cl_2 induced a modest reduction in fluorescence intensity, by approximately ~20%. The discrepancy in fluorescence reduction induced by H_2S and Cl_2 compared to gasses like CO_2 and SO_2 may be attributed to their distinct chemical reactivity and interactions with the COF materials. H_2S and Cl_2 exhibit heightened chemical reactivity, with H_2S being renowned for its reducing properties and Cl_2 being a potent oxidizing agent. These characteristics likely prompt more substantial chemical interactions with the COF, potentially altering fluorescence intensity. Additionally, H_2S and Cl_2 might possess specific binding affinities or interactions with functional groups within the COF structure, potentially influencing electronic states or modifying emission properties, thereby accentuating the observed decrease in fluorescence intensity. Notably, these findings underscore the COFs' limited reactivity towards certain gasses while indicating a moderate interaction with hydrogen sulfide and chlorine gas. Importantly, amidst these varied responses, the COF materials exhibited a distinctive and consistent reactivity to hydrogen chloride (HCl), demonstrating selectivity in detecting this specific acid gas. This observed selectivity emphasizes the potential utility of PT–COF, PP–COF, and PE–COF for targeted and precise detection of hydrogen chloride, elucidating their applicability in scenarios where specific gas identification is paramount.

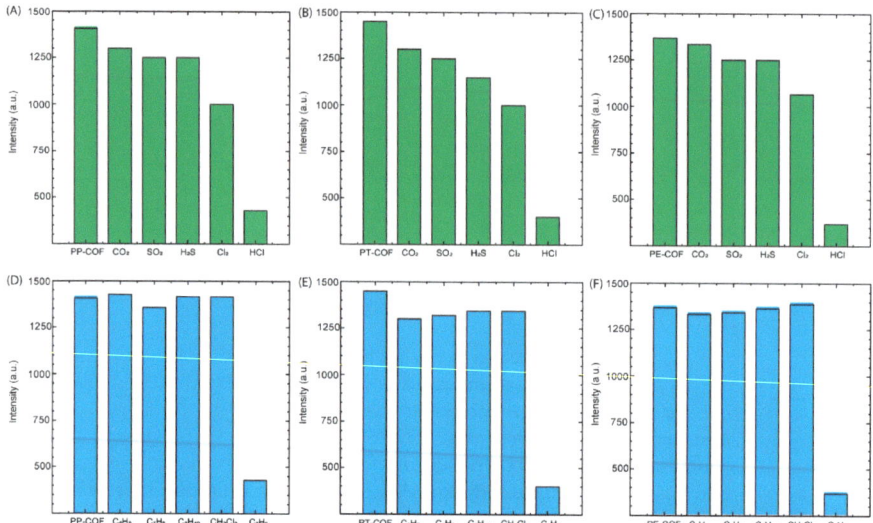

Figure 6. PP–COF, PT–COF, and PE–COF films' responses to other potential coexisting interference gasses: (**A–C**) acid gasses; (**D–F**) volatile organic compounds.

In our investigation, special attention was given to the role of volatile organic compounds (VOCs) as potential interfering agents in the detection of hazardous gasses. The influence of four commonly encountered VOCs—benzene (C_6H_6), toluene (C_7H_8), xylene (C_8H_{10}), and dichloromethane (CH_2Cl_2)—was meticulously evaluated for their impact on the fluorescence-based detection capabilities of three covalent organic frameworks (COFs): PT–COF, PP–COF, and PE–COF. The experimental outcomes, as depicted in Figure 6D–F, reveal a remarkable stability in the fluorescence response of the COFs upon exposure to these VOCs. Notably, fluorescence intensity remained largely unchanged, indicating a negligible interaction between the COFs and the VOCs. This observation is particularly significant, as it suggests that the molecular architecture of PT–COF, PP–COF, and PE–COF provides a selective pathway for hydrogen chloride gas detection, effectively mitigating the potential cross-sensitivity issues commonly associated with the presence of VOCs in environmental and industrial settings. These findings underscore the potential of COFs in advancing gas sensing technology. Their ability to selectively detect target gasses amidst interfering VOCs without losing sensitivity or specificity is crucial for environmental monitoring, industrial safety, and public health. The stability and selectivity demonstrated by PT–COF, PP–COF, and PE–COF pave the way for further research to broaden the application of COF-based sensors in diverse detection environments.

In our study, we assessed the selectivity of PT–COF, PP–COF, and PE–COF for hydrogen chloride (HCl) by exposing their suspensions to different acids at a concentration of 20 mmol L^{-1}. The results, detailed in the Supplementary Materials (SI), Section S9, Figure S16, revealed that only HCl triggered a significant redshift in the fluorescence emission peaks of these COFs to 540 nm for PT–COF, 625 nm for PP–COF, and 611 nm for PE–COF, along with a substantial increase in fluorescence intensity. This indicates that these COFs are highly selective for HCl, showcasing a distinct fluorescence response hierarchy: HCl > HBr > H_2SO_4 > HNO_3 > H_3PO_4 > CH_3COOH.

This specificity sequence suggests that the response of COFs to various acids is influenced by the strength of the acids and their interaction with COFs. It appears that the ability of an acid to protonate the functional groups within the COF structure plays a critical role. This protonation alters the COFs' electronic structure, affecting their fluorescence properties. Strong acids like HCl, which are proficient proton donors, induce notable changes in the electronic configuration of the COFs, resulting in more significant fluorescence shifts. In contrast, weaker acids such as acetic acid (CH_3COOH) have less impact, likely due to their lower proton-donating capacity, leading to minimal changes in the COFs' structure and fluorescence response. This behavior underscores the potential of these COFs as selective sensors for detecting HCl, offering insights into designing advanced materials for gas sensing applications.

3. Conclusions

In conclusion, the synthesis of stable fluorescence COFs—PT–COF, PP–COF, and PE–COF—via distinct polycondensation reactions demonstrates their structural diversity and robust chemical compositions, as verified using FTIR analysis. These COFs exhibit exceptional characteristics: high crystallinity, impressive thermal stability up to 400 °C, and remarkable surface areas spanning 730–1098 $m^2\,g^{-1}$ as determined using PXRD and BET analyses. The developed COF-based chemosensor showcases the rapid and sensitive detection of gaseous HCl, manifesting significant color and fluorescence changes that are reversible upon exposure to TEA vapor. Notably, the sensor's recovery post-exposure to TEA vapor accentuates its remarkable reversibility. This study presents COFs as promising platforms for responsive sensing applications, particularly in detecting toxic gasfses, marking an innovative direction in utilizing COFs for advanced air quality monitoring devices.

Supplementary Materials: The following supporting information can be downloaded at: https://www.mdpi.com/article/10.3390/chemosensors12030037/s1, Scheme S1: Synthesis of 1,3,6,8-tetra(aminophenyl)pyrene; Scheme S2: Synthesis of 4,4′,4″,4‴-(ethane-1,1,2,2-tetrayl) tetrabenzaldehyde; Scheme S3: Synthesis of hexa(4-formyl-phenoxy)cyclotriphosphazene; Figure S1: ^{1}H-NMR spectrum of L1; Figure S2: ^{13}C-NMR spectrum of L1; Figure S3: ^{1}H-NMR spectrum of L2; Figure S4: ^{13}C-NMR spectrum of L2; Figure S5: ^{1}H-NMR spectrum of L3; Figure S6: ^{13}C-NMR spectrum of L3; Figure S7: FTIR spectra of PT-COF (blue), PP-COF (green), and PE-COF (purple), respectively; Figure S8: Comparison of PXRD patterns of (A) PT-COF, (B) PP-COF, and (C) PE-COF; Figure S9: (A) PXRD patterns of PP-COF after treatment in different organic solvents for 24; (B) TGA of PP-COF; Figure S10: (A) PXRD patterns of PE-COF after treatment in different organic solvents for 24; (B) TGA of PE-COF; Figure S11. (A) PXRD patterns of PT-COF after treatment in different organic solvents for 24; (B) TGA of PT-COF; Figure S12: FT-IR spectra of the (A) PP-COF, (B) PT-COF, and (C) PE-COF powders before and after exposure to HCl gas, and recovery with TEA gas; Figure S13: Powder XRD patterns of the (A) PT-COF, (B) PP-COF, and (C) PE-COF powders before and after exposure to HCl gas, and recovery with TEA gas; Figure S14: Calibration curves of the fluorescence intensities of the (A) PT-COF, (B) PP-COF, and (C) PE-COF plotted with respect to the HCl concentration; Figure S15: Detection of HCl gas under different humidity environments (40%, 50%, 60%, 70%, and 80%) with (A) PT-COF, (B) PE-COF, and (C) PP-COF films; Figure S16: Illustrates the fluorescence emission peaks at 540, 625, and 611 nm for (A) PT-COF, (B) PP-COF, and (C) PE-COF when dispersed in 1,4-dioxane (concentration: 1 mg mL^{-1}; excitation wavelength: 366 nm) following the introduction of an acid concentration of 20 mmol L^{-1}; Table S1: Fractional atomic coordinates in the refined unit cell of PT-COF; Table S2: Fractional atomic coordinates in the refined unit cell of PP-COF; Table S3: Fractional atomic coordinates in the refined unit cell of PE-COF.

Author Contributions: Conceptualization, S.S.A.; Methodology, M.T.A., A.S.A. and I.A.; Data Curation, M.T.A., A.S.A. and I.A.; Original Draft Preparation, writing—Review and Editing, S.S.A. All authors have read and agreed to the published version of the manuscript.

Funding: This research was funded by Research Affairs Sector of UAE University (grant number. G00.004310).

Institutional Review Board Statement: Not applicable.

Data Availability Statement: The data supporting the findings of this study are included within the article and its Supplementary Materials.

Acknowledgments: The authors are grateful to the Alfa chemistry Company for the providing with chemicals assistance.

Conflicts of Interest: There are no competing interests declared by the authors.

References

1. Kelly, F.J.; Fussell, J.C. Air pollution and public health: Emerging hazards and improved understanding of risk. *Environ. Geochem. Health* **2015**, *37*, 631–649. [CrossRef] [PubMed]
2. Lelieveld, J.; Evans, J.S.; Fnais, M.; Giannadaki, D.; Pozzer, A. The contribution of outdoor air pollution sources to premature mortality on a global scale. *Nature* **2015**, *525*, 367–371. [CrossRef]
3. Muthukumar, P.; John, S.A. Highly sensitive detection of HCl gas using a thin film of meso-tetra(4-pyridyl)porphyrin coated glass slide by optochemical method. *Sens. Actuators B Chem.* **2011**, *159*, 238–244. [CrossRef]
4. Nakagawa, K.; Kumon, K.; Tsutsumi, C.; Tabuchi, K.; Kitagawa, T.; Sadaoka, Y. HCl gas sensing properties of $TPPH_2$ dispersed in various copolymers. *Sens. Actuators B Chem.* **2000**, *65*, 138–140. [CrossRef]
5. Baron, M.G.; Narayanaswamy, R.; Thorpe, S.C. Hydrophobic membrane sensors for the optical determination of hydrogen chloride gas. *Sens. Actuators B Chem.* **1996**, *34*, 511–515. [CrossRef]
6. Onyancha, R.B.; Ukhurebor, K.E.; Aigbe, U.O.; Osibote, O.A.; Kusuma, H.S.; Darmokoesoemo, H.; Balogun, V.A. A systematic review on the detection and monitoring of toxic gases using carbon nanotube-based biosensors. *Sens. Bio-Sens. Res.* **2021**, *34*, 100463. [CrossRef]
7. Dalaijamts, C.; Cichocki, J.A.; Luo, Y.-S.; Rusyn, I.; Chiu, W.A. Quantitative Characterization of Population-Wide Tissue- and Metabolite-Specific Variability in Perchloroethylene Toxicokinetics in Male Mice. *Toxicol. Sci.* **2021**, *182*, 168–182. [CrossRef] [PubMed]
8. Zhang, C.; Wu, X.; Huang, C.; Peng, J.; Ji, C.; Yang, Y.; Huang, Y.; Guo, Y.; Luo, X. Flexible and Transparent Microwave–Infrared Bistealth Structure. *Adv. Mater. Technol.* **2019**, *4*, 1900063. [CrossRef]
9. Yin, M.; Yu, L.; Liu, S. Synthesis of thickness-controlled cuboid WO_3 nanosheets and their exposed facets-dependent acetone sensing properties. *J. Alloys Compd.* **2017**, *696*, 490–497. [CrossRef]
10. Huang, L.; Wang, Z.; Zhu, X.; Chi, L. Electrical gas sensors based on structured organic ultra-thin films and nanocrystals on solid state substrates. *Nanoscale Horiz.* **2016**, *1*, 383–393. [CrossRef] [PubMed]
11. Ma, T.; Kapustin, E.A.; Yin, S.X.; Liang, L. Single-crystal x-ray diffraction structures of covalent organic frameworks. *Science* **2018**, *361*, 48–52. [CrossRef] [PubMed]
12. Bisbey, R.P.; Dichtel, W.R. Covalent Organic Frameworks as a Platform for Multidimensional Polymerization. *ACS Cent. Sci.* **2017**, *3*, 533–543. [CrossRef] [PubMed]
13. Uribe-Romo, F.J.; Hunt, J.R.; Furukawa, H.; Klöck, C.; O'Keeffe, M.; Yaghi, O.M. A crystalline imine-linked 3-D porous covalent organic framework. *J. Am. Chem. Soc.* **2009**, *131*, 4570–4571. [CrossRef] [PubMed]
14. Lohse, M.S.; Bein, T. Covalent Organic Frameworks: Structures, Synthesis, and Applications. *Adv. Funct. Mater.* **2018**, *28*, 1705553. [CrossRef]
15. Chung, W.-T.; Mekhemer, I.M.A.; Mohamed, M.G.; Elewa, A.M.; El-Mahdy, A.F.M.; Chou, H.-H.; Kuo, S.-W.; Wu, K.C.W. Recent advances in metal/covalent organic frameworks based materials: Their synthesis, structure design and potential applications for hydrogen production. *Coord. Chem. Rev.* **2023**, *483*, 215066. [CrossRef]
16. Cai, L.-Z.; Jiang, X.-M.; Zhang, Z.-J.; Guo, P.-Y.; Jin, A.-P.; Wang, M.-S.; Guo, G.-C. Reversible Single-Crystal-to-Single-Crystal Transformation and Magnetic Change of Nonporous Copper(II) Complexes by the Chemisorption/Desorption of HCl and H_2O. *Inorg. Chem.* **2017**, *56*, 1036–1040. [CrossRef]
17. Xiao, J.; Chen, J.; Liu, J.; Ihara, H.; Qiu, H. Synthesis strategies of covalent organic frameworks: An overview from nonconventional heating methods and reaction media. *Green Energy Environ.* **2023**, *8*, 1596–1618. [CrossRef]
18. Qian, Y.; Li, J.; Ji, M.; Li, J.; Ma, D.; Liu, A.; Zhao, Y.; Yang, C. Fluorescent Covalent Organic Frameworks: A Promising Material Platform for Explosive Sensing. *Front. Chem.* **2022**, *10*, 943813. [CrossRef] [PubMed]
19. Godoy, C.A. New Strategy for the Immobilization of Lipases on Glyoxyl–Agarose Supports: Production of Robust Biocatalysts for Natural Oil Transformation. *Int. J. Mol. Sci.* **2017**, *18*, 2130. [CrossRef] [PubMed]
20. EL-Mahdy, A.F.M.; Elewa, A.M.; Huang, S.-W.; Chou, H.-H.; Kuo, S.-W. Dual-Function Fluorescent Covalent Organic Frameworks: HCl Sensing and Photocatalytic H_2 Evolution from Water. *Adv. Opt. Mater.* **2020**, *8*, 2000641. [CrossRef]

Disclaimer/Publisher's Note: The statements, opinions and data contained in all publications are solely those of the individual author(s) and contributor(s) and not of MDPI and/or the editor(s). MDPI and/or the editor(s) disclaim responsibility for any injury to people or property resulting from any ideas, methods, instructions or products referred to in the content.

Article

Bipyridyl Ruthenium-Decorated Ni-MOFs on Carbon Nanotubes for Electrocatalytic Oxidation and Sensing of Glucose

Yu Zhang, Chang Liu [†], Rongqiu Yan and Chenghong Lei *

College of Chemistry and Bioengineering, Guilin University of Technology, Guilin 541006, China
* Correspondence: clei@glut.edu.cn
† Current address: Shandong Changyi Petrochemical Co., Ltd., 201 Shihua Road, Weifang 261300, China.

Abstract: Bipyridyl Ruthenium-decorated Ni-MOFs on multi-walled carbon nanotubes (MWCNT-RuBpy@Ni-MOF) were synthesized. In an alkaline solution, the glucose was electrocatalytically oxidized at +0.5 V vs. Ag/AgCl at the composite interface of MWCNT-RuBpy@Ni-MOF on a glassy carbon electrode. The Ni^{3+}/Ni^{2+} redox couples in Ni-MOFs played a key role as the active center for the catalytic oxidation of glucose at the electrode, where both RuBpy and MWCNTs enhanced the current responses to glucose. The resulting enzymeless glucose sensor from MWCNT-RuBpy@Ni-MOF exhibited a wide range of linear responses, high sensitivity and selectivity for the determination of glucose.

Keywords: bipyridyl ruthenium; Ni-MOFs; carbon nanotubes; electrocatalytic oxidation; glucose sensing

1. Introduction

The quantitative detection of glucose plays an important role in food science, agricultural science, biological science and other fields [1–6]. Since biological oxidases and dehydrogenase for substrates such as lactate, glucose and alcohols have limitations such as a low stability, difficulty in reuse and high costs, people explore various ways to detect these substrates via their direct oxidation. However, under normal circumstances, these substances are not easily directly oxidized. The catalytic activity of some MOFs enables the electrocatalytic oxidation of the target analytes without the need for the enzymes [7–11]. Moreover, the catalytic performance of MOFs can be greatly enhanced by incorporating other materials. The rational design and engineering of MOFs and their composite materials provide new opportunities for the development of various new catalytic sensing strategies and detection mechanisms [12–18]. On the other hand, the low conductivity and instability of MOFs in aqueous media limit their application in electrochemical processes and sensors [19].

Theoretically, the Ni-H$_2$O system yields a gradual increase in the valence state of Ni from Ni^{2+} to Ni^{4+} with increasing potential values. Experimentally, cyclic voltametric tests have been efficiently implemented to observe Ni^{2+}/Ni^{3+} and Ni^{3+}/Ni^{4+} electron pairs [20,21]. Indeed, the reduction of high valence Ni^{4+} to low valence Ni^{2+} can be achieved by two consecutive single-electron-transfer channels, Ni^{4+}/Ni^{3+} and Ni^{3+}/Ni^{2+} [22,23]. The observation of isolated oxidation peaks in nickel-based electrochemistry is very important since the higher valence and smaller radius of the transition metal cation are more oxidatively active [24,25]. The redox reactions of glucose catalyzed by nickel-based materials are generally interpreted as the deprotonation and isomerization of glucose by Ni^{3+}/Ni^{2+} redox couples. In fact, some glucose sensors have been developed using Ni-based MOFs composites for improving the catalytic performance and catalytic efficiency of the sensors. For example, the Ni$_2$(dihydroxyterephthalic acid) (also known as CPO-27-Ni^{II})-modified

glassy carbon electrode exhibited a wide linear range of glucose detection with high sensitivity (~585 µA mM^{-1} cm^{-2}) and a low detection limit (1.46 µM) [1]. A layer-assembled flower-like Ni-MOF/carbon nanotube composite was used to fabricate the glucose sensor with a sensitivity of 77.7 µA mM^{-1} cm^{-2} and a linear range of 20 µM to 4.4 mM [3]. The glucose sensor employing ultrathin Ni-MOF nanoribbons was reported with a sensitivity of 1.542 µA mM^{-1} cm^{-2} and a linear range of 1–500 µM [26]. However, in most cases, due to low conductivity and the absence of electron transfer mediators, the sensors based on Ni-MOFs usually have limited sensitivity in the detection of glucose. Exceptionally, Lu and Sun et al. reported that a glucose sensor based on a conductive Ni-MOF material Ni$_3$(HHTP)$_2$ displayed a high sensitivity of 21,744 µA mM^{-1} cm^{-2} [27].

Tris(2,2′-bipyridyl) ruthenium dichloride (RuBpy) is an electron transfer mediator with both fluorescent and electrochemical redox properties. Many reports have focused on the immobilization of RuBpy, including the Langmuir–Blodgett technological approach [28–30], the self-assembly technique [31,32] and the sol-gel method [33–36]. The immobilization materials for RuBpy have also been applied to sensing, such as cation-exchange polymers [37–39], SiO$_2$ nanoparticles [40–43], carbon nanotubes [44,45], metal nanoparticles [46] and MOF materials [47]. Among these materials, multi-walled carbon nanotubes (MWCNTs) have excellent electrical conductivity (1.85 × 10^3 S cm^{-1}) as a carrier material [48]. Current densities close to 10^9 A cm^{-2} have also been reported for MWCNTs [49]. Together with the robust adsorption and mechanical properties of MWCNTs, this makes them strong candidates for microelectronic devices and electrode interconnections in numerous applications.

In this work, the Ni-MOF material was synthesized using p-phthalic acid (PTA) as the ligand. RuBpy was spontaneously adsorbed on the Ni-MOF material to form RuBpy@Ni-MOF. Then, RuBpy@Ni-MOF and MWCNTs were co-suspended in a Nafion solution. The resulting MWCNT-RuBpy@Ni-MOF (Scheme 1) was then loaded on electrodes as an electrocatalyst to construct the enzymeless glucose electrochemical sensor. Both RuBpy as an electron transfer mediator and MWCNTs as a carrier enhanced the catalytic efficiency of the Ni-MOF material for glucose oxidation and thus strengthened the sensor performance.

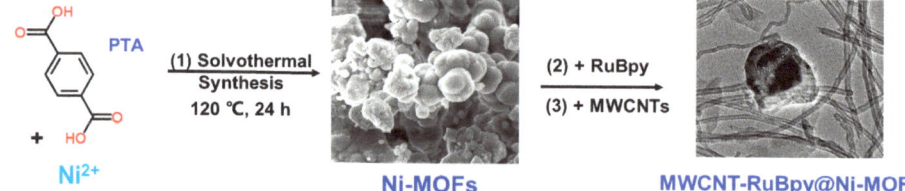

Scheme 1. Synthesis of MWCNT-RuBpy@Ni-MOF.

2. Materials and Methods

All experiments and experimental preparations were carried out at ambient conditions at room temperature (22 ± 1 °C), except for those specified. Ni(NO$_3$)$_2$·6H$_2$O and DMF were purchased from Xilong Scientific (Shantou, China). The Tris(2,2′-bipyridyl)ruthenium dichloride was from D&B Biotech (Shanghai, China). MWCNTs, uric acid and L-Cysteine (L-Cys) were purchased from Aladdin (Shanghai, China). The MWCNTs from Aladdin (Product # C139823) contain 95% of multi-walled carbon nanotubes with an average length of 50 µm, whose interior and outer diameters are in the range of 3–5 nm and 8–15 nm, respectively. P-phthalic acid (PTA), Nafion (5%), glucose, fructose, maltose, D-ribose and sucrose were purchased from Macklin Biochem (Shanghai, China). D-lactose was purchased from Yuanye Biotech (Shanghai, China). L-ascorbic acid and urea were obtained from Rhawn Reagents (Shanghai, China). All reagents were used without further purification.

The solvothermal synthesis of Ni-MOFs in DMF has been reported using either Ni (NO$_3$)$_2$ or NiCl$_2$ as the source of Ni^{2+} with a variety of ligands [1,3,14,50]. In this work, the Ni-MOF material was synthesized using a similar procedure as previously reported [1,3,14,50]. In total, 2.18 g of Ni(NO$_3$)$_2$·6H$_2$O and 0.50 g of PTA were dissolved in 60 mL DMF via stirring, forming into an emerald-green solution. The solution was transferred to a polytetrafluoroethylene-lined reaction kettle and allowed to react at 120 °C for 24 h. Then, it was cooled to room temperature. The reacted mixture was centrifuged at 10,000 rpm for 5 min. The deposit was washed with ethanol and centrifuged five times to remove the excess reactants and residual DMF. The deposit was dried in a vacuum drying oven at 60 °C for 12 h. The dried deposit was ground and, thus, the Ni-MOF material was obtained.

A total of 45 mg of Ni-MOFs was dispersed and suspended in 9 mL of ethanol. A total of 18 mg of RuBpy was dissolved in 9 mL of ethanol. The Ni-MOF suspension and the RuBpy solution were mixed and shaken at 1000 rpm at 4 °C for 10 h to allow RuBpy to be adsorbed on Ni-MOF. The mixture was washed with ethanol and centrifuged at 7000 rpm for 5 min three times. The deposit was dried at room temperature. The dried deposit was ground and, thus, RuBpy@Ni-MOF was obtained.

In total, 0.25 mL of 2 mg/mL of RuBpy@Ni-MOF in 0.05% Nafion aqueous solution was mixed well with 0.25 mL of 1 mg/mL of MWCNTs in 0.05% Nafion aqueous solution until it was well dispersed and ready for use. The glassy carbon electrode (GCE, diameter: 3 mm) was polished until it was subsequently smooth with 1 μm, 0.3 μm and 0.05 μm alumina polishing powder. The GCE surface was then ultrasonically cleaned with 1:3 nitric acid, deionized water and ethanol, respectively. Then, the clean GCE was dried with N$_2$. Then, 5 μL of the prepared suspension of RuBpy@Ni-MOF and MWCNTs in 0.05% Nafion aqueous solution was pipetted onto the GCE surface and allowed to dry at room temperature. Thus, the working electrode MWCNT-RuBpy@Ni-MOF/GCE was obtained. In the absence of MWCNTs, the control electrodes Ni-MOF/GCE and RuBpy@Ni-MOF/GCE were prepared in a similar way.

The three-electrode electrochemical measurements were carried out with a CHI730E electrochemical workstation in 5 mL of 0.1 M NaOH solution, where MWCNT-RuBpy@Ni-MOF/GCE or the control electrode was the working electrode, a Ag/AgCl (3M KCl) electrode was the reference electrode and a platinum wire was the counter electrode. The electrochemical measurements were implemented with cyclic voltammetry and chronoamperometry (current versus time (i-t) curve). For cyclic voltametric tests, the potential range was scanned from 0.0 V to +0.8 V at the scanning rate of 100 mV/s. In i-t tests, the glucose was successively added under constant stirring with the potential set at +0.5 V, except for those specified.

Scanning electron microscopy (SEM) images were acquired by a field emission scanning electron microscope (HT7000, Hitachi SU5000, Tokyo, Japan) operated at 5 kV. Transmission electron microscopy (TEM) images were obtained at a working voltage of 200 kV by a field emission transmission electron microscope (JEM-2100F, Japan Electronics, Tokyo, Japan). X-ray photoelectron spectroscopy (XPS) was measured on an XPS spectrometer (ESCALAB 250Xi, Thermo Electron Corporation, Round Rock, TX, USA).

3. Results and Discussion

The SEM images show the morphologies of Ni-MOFs, RuBpy@Ni-MOF and MWCNT-RuBpy@Ni-MOF (Figure 1). Figure 1a displays that Ni-MOFs had a morphology of aggregated spherical and irregular particles. Figure 1b shows that the morphology of RuBpy@Ni-MOF did not change much but with some fine particles of RuBpy decorated on Ni-MOFs. Figure 1c demonstrates that RuBpy@Ni-MOF particles were embellished and more dispersive on MWCNTs, compared to those aggregates in the absence of MWCNTs (Figure 1a,b). Figure 2a shows the TEM image of Ni-MOFs. The morphology of the Ni-MOF particles shows a spherical shape, and the size is about 100 nm. There are some extra crystals on the surface of RuBpy@Ni-MOF, and thus, it appears much rougher than Ni-MOFs (Figure 2a,b), indicating RuBpy was compounded on Ni-MOFs. Figure 2c shows the TEM image of MWCNT-RuBpy@Ni-MOF. Obviously, MWCNTs were entangled around RuBpy@Ni-MOF.

The image with high-angle annular dark-field transmission electron microscopy (HAADF-TEM) and the EDS elemental distribution mappings of O, Ni, N and Ru of RuBpy@Ni-MOF are displayed in Figure 2d–h. The major and characteristic elements, O, Ni, N and Ru, were uniformly distributed with RuBpy@Ni-MOF (Figure 2d–h). The results demonstrate that RuBpy was uniformly distributed all over the Ni-MOF surface.

Figure 1. SEM images of Ni-MOFs (**a**); RuBpy@Ni-MOF (**b**); MWCNT-RuBpy@Ni-MOF (**c**).

XPS tests of RuBpy@Ni-MOF could reveal the surface electronic state and core energy levels of the elements. Figure 3a shows the full spectra of the representative elements Ni, O, N, C and Ru present in RuBpy@Ni-MOF. The binding energies of Ru and C are located in the same region, and the binding energy of C is large, so the characteristic peak of Ru is partially masked by the characteristic peak of C. Figure 3b–d demonstrate the analytical spectra of Ni 2p, Ru 3d and N 1s. Figure 3b illustrates the core energy level spectra of Ni 2p, showing the presence of two states of Ni. In the fine spectrum of Ni 2p, the two peaks observed at 855.9 eV and 873.7 eV belong to Ni $2p_{3/2}$ and Ni $2p_{1/2}$ of Ni^0, respectively, while the peaks of Ni $2p_{3/2}$ and Ni $2p_{1/2}$ in Ni^{2+} are located at 860.34 eV and 878.0 eV, respectively. In addition, the other two peaks are the satellite peaks of Ni $2p_{3/2}$ and Ni $2p_{1/2}$ of Ni^{2+} with the spin-orbit energy level located at 863.7 eV and 881.2 eV, respectively [51]. Figure 3c demonstrates the analytical spectrum of Ru 3d. The high-resolution XPS spectra of RuBpy@Ni-MOF show Ru 3d has the double peaks Ru $3d_{5/2}$ and Ru $3d_{3/2}$ located at 284.6 eV and 287.9 eV, respectively, with a spin-orbit splitting energy of 3.3 eV. The Coster–Kronig effect is responsible for the $3d_{3/2}$ peak being much wider than the $3d_{5/2}$ peak, which is consistent with the report [52]. Figure 3d demonstrates the analytical spectrum of N 1s. There are two characteristic peaks for N 1s, which are C=N-C and N-Ru, located at 400.0 eV and 406.9 eV, respectively [53]. The results demonstrate that RuBpy@Ni-MOF was successfully synthesized.

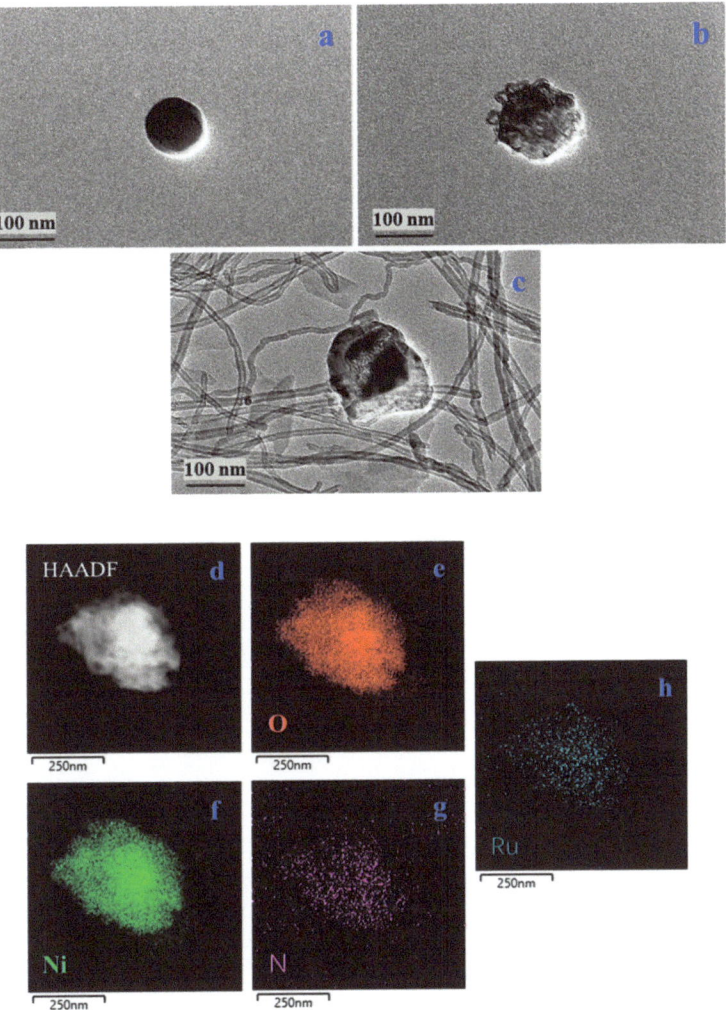

Figure 2. TEM images of Ni-MOFs (**a**); RuBpy@Ni-MOF (**b**); MWCNT-RuBpy@Ni-MOF (**c**); HAADF-TEM image of RuBpy@Ni-MOF (**d**); Single-colored EDS elemental dot mappings of O, Ni, N and Ru, respectively (**e–h**).

Cyclic voltametric experiments were carried out with GCE, Ni-MOF/GCE, RuBpy@Ni-MOF/GCE and MWCNT-RuBpy@Ni-MOF/GCE in the absence and presence of 1 mM glucose in 0.1 M NaOH. Figure 4 shows the well-defined redox peaks resulting from Ni^{3+}/Ni^{2+} redox couples of Ni-MOF/GCE, RuBpy@Ni-MOF/GCE and MWCNT-RuBpy@Ni-MOF/GCE. Compared to the bare GCE, the redox peaks were attributed to the Ni^{2+}/Ni^{3+} redox electron pair [14]. The oxidation peaks of all three electrodes were located around +0.51 V (Figure 4). The reduction peaks of Ni-MOF/GCE and RuBpy@Ni-MOF/GCE were located around +0.33 V, while the reduction peak of MWCNT-RuBpy@Ni-MOF/GCE was located around 0.29 V (Figure 4). Since the baselines of cyclic voltammograms of three modified electrodes were nearly the same as that of the naked GCE during the potential scanning in the direction from the low potential to the high potential prior to the oxidation of Ni^{2+} to Ni^{3+}, in the absence of glucose, the current increase could result from a reduced resistance of the system upon adding RuBpy and MWCNTs. The peak currents of the Ni^{3+}/Ni^{2+}

redox couples were in the range of 0.5–1.0 mA. This is a large current range for GCE with a diameter of 3 mm. In the presence of glucose, the catalytic currents were observed for all three modified electrodes. The catalytic currents, the differences in the oxidation peak currents in the absence and presence of glucose, were actually increased larger and larger when the working electrode switched from Ni-MOF/GCE to RuBpy@Ni-MOF/GCE and then to MWCNT-RuBpy@Ni-MOF/GCE. For example, at +0.55 V, the differences in the oxidation peak currents in the absence and presence of glucose were about 66.0, 71.0 and 100.1 µA for Ni-MOF/GCE, RuBpy@Ni-MOF/GCE and MWCNT-RuBpy@Ni-MOF/GCE, respectively. The results demonstrate that, besides providing greater conductivity and a larger electrode area, MWCNTs could allow the aggregated RuBpy@Ni-MOF particles to be more dispersive on the electrode (Figure 1c), which could thus expose more active sites for the catalytic oxidation of glucose and accordingly enhance the catalytic currents to some extent. Although the differences in the oxidation peak currents in the absence and presence of glucose seemed as a small fraction of the peak currents of three modified electrodes, such catalytic currents over 60.0 µA were still quite large for 3 mm GCE. MWCNT-RuBpy@Ni-MOF/GCE was thus selected as the working electrode for the rest of this work.

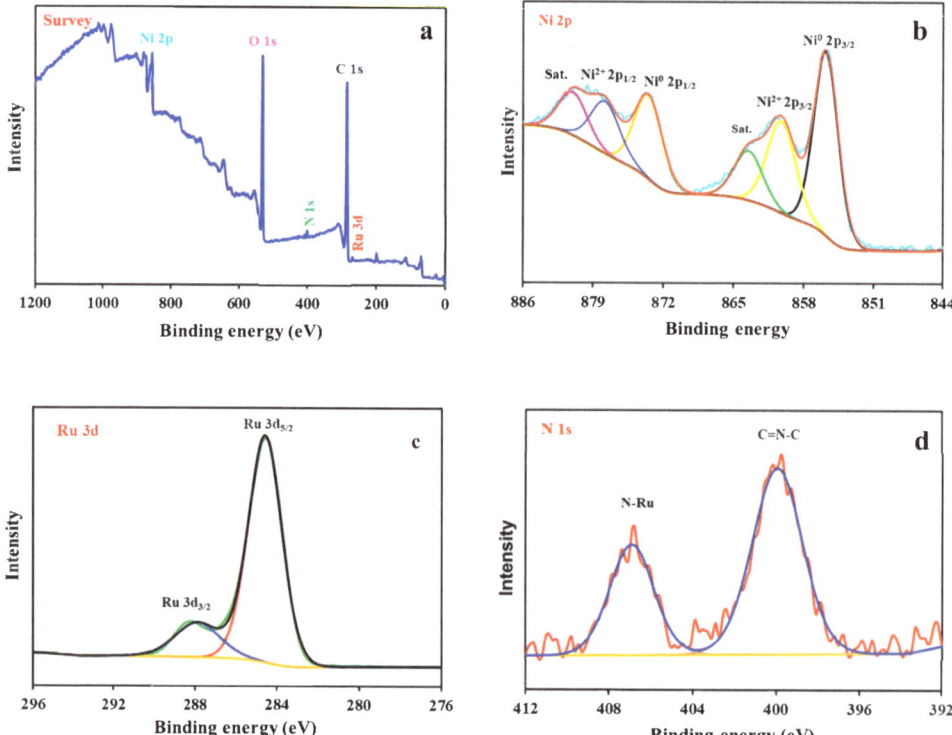

Figure 3. XPS spectra of RuBpy@Ni-MOF (**a**), Ni 2p (**b**), Ru 3d (**c**), and N 1s (**d**).

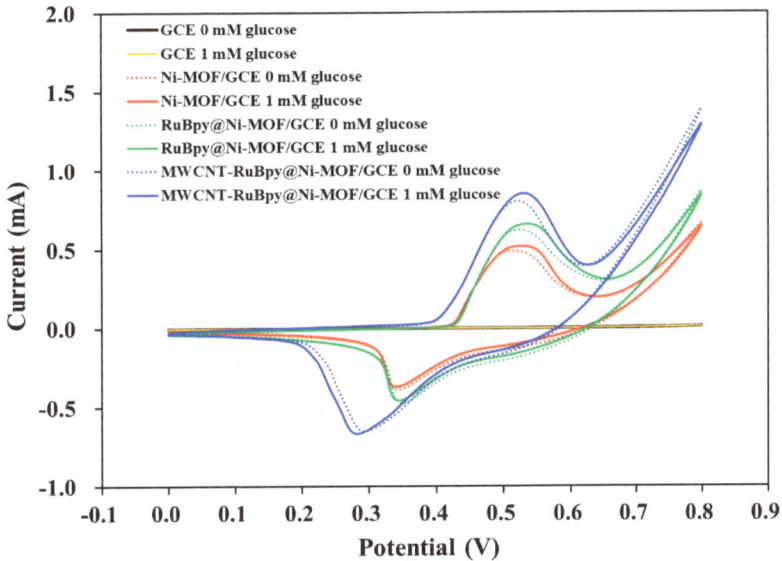

Figure 4. Cyclic voltammograms of GCE, Ni-MOF/GCE, RuBpy@Ni-MOF/GCE and MWCNT-RuBpy@Ni-MOF/GCE in 0.1 M NaOH in the absence and the presence of 1 mM glucose, respectively. The scanning rate: 100 mV/s.

In 0.1 M NaOH, the specific reaction formula was proposed as the following equation [14]:

$$Ni(OH)_2 + OH^- \rightarrow NiOOH + H_2O + e^- \quad (1)$$

$$2NiOOH + glucose \rightarrow 2Ni(OH)_2 + gluconolactone \quad (2)$$

During anodic scanning in 0.1 M NaOH, it was proposed that $Ni(OH)_2$ could be generated at the surface of MWCNT-RuBpy@Ni-MOF/GCE. $Ni(OH)_2$ could lose one electron to generate NiOOH in 0.1 M NaOH, as shown in Equation (1). In the presence of 1 mM glucose, NiOOH could oxidize glucose to generate $Ni(OH)_2$ again, thus increasing the anodic peak current, as shown in Equation (2) [54]. The positive shift of the anodic peak indicates the slower kinetics of the reaction, including the oxidized intermediates and the uptake of glucose at the active site [55].

Figure 5a shows cyclic voltammograms of MWCNT-RuBpy@Ni-MOF/GCE at different scan rates in the presence of 1 mM glucose in 0.1 M NaOH. The redox peak currents increased as the scan rate increased, and the oxidation and reduction peak currents (I_a and I_c) were proportional to the square root of the scan rate (Figure 5b). The results demonstrate that the electrocatalytic glucose oxidation process at MWCNT-RuBpy@Ni-MOF/GCE was a typical diffusion-controlled electrochemical reaction [56].

The i-t curves were tested using MWCNT-RuBpy@Ni-MOF/GCE at different applied potentials with the step concentrations of 0.5 mM, 1 mM, 1.5 mM, 2 mM and 2.5 mM in the testing solution for glucose (Figure 6a). Overall, at +0.45 V and +0.50 V, the current responses were larger than those at +0.55 V and +0.60 V. And the current response at +0.50 V was slightly higher than that at +0.45 V. At +0.60 V, the increase in the current response was not significant, and the baseline current was much larger than that at +0.45, +0.50, and +0.55 V. Considering that the higher the potential, the greater the number of possible interfering substances, +0.50 V was selected as the optimal working potential for subsequent testing. In order to investigate the selectivity of MWCNT-RuBpy@Ni-MOF/GCE, the potential interferences such as AA, UA, L-Cys, Urea, fructose, lactose, maltose, sucrose and D-ribose were tested under the same conditions as for glucose. In human serum, the concentration of glucose is more than 30 times higher than that of these interferences. The experiments were carried out under continuous stirring in the presence of 0.5 mM glucose and 50 µM interfering substances (Figure 6b). The results demonstrate that the responses to all tested substances were minimal. Therefore, the glucose sensor developed in this work could be practically applied for the determination of glucose in some real samples.

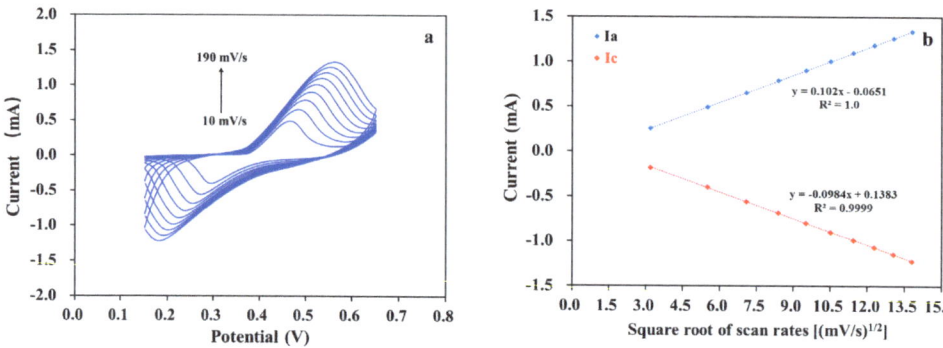

Figure 5. (a) Cyclic voltammograms of MWCNT-RuBpy@Ni-MOF/GCE with scan rates from 10 mV/s to 190 mV/s in 0.1 M NaOH containing 1 mM glucose; (b) Fitting curve of peak currents vs. square root of scan rates.

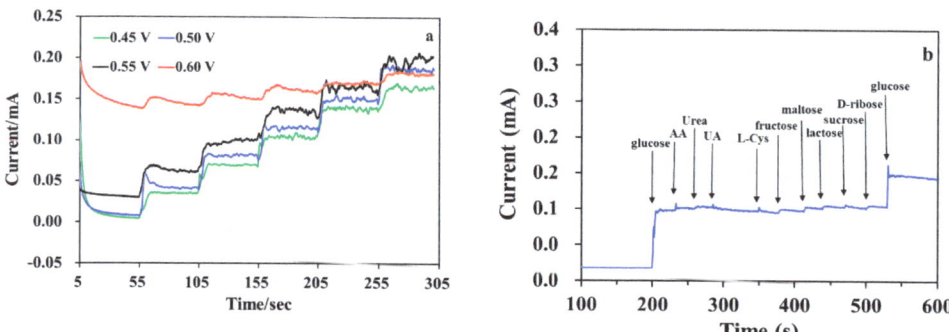

Figure 6. (a) Amperometric responses of MWCNT-RuBpy@Ni-MOF/GCE to successive additions of glucose at the different potentials in 0.1 M NaOH. The concentration of glucose in the testing solution was 0.5 mM, 1 mM, 1.5 mM, 2 mM and 2.5 mM for each step of glucose addition; (b) Current responses (i-t curve) of MWCNT-RuBpy@Ni-MOF/GCE of 0.5 mM glucose and 50 µM AA, Urea, UA, L-Cys, fructose, lactose, maltose, sucrose and D-ribose, respectively, in 0.1 M NaOH at +0.50 V.

For the calibration plot, the i-t curve was obtained with MWCNT-RuBpy@Ni-MOF/GCE at +0.50 V (Figure 7a). The inset represents the i-t curve for low concentrations of glucose with a minimum concentration of 5 µM. The detection limit of the resulting glucose sensor based on MWCNT-RuBpy@Ni-MOF/GCE was as low as 1.5 µM. The fitting curve of response currents vs. glucose concentrations exhibited that the glucose sensor had a wide linear detection range from 5 µM to 3.5 mM, spanning three orders of magnitude (Figure 7b). The R^2 values were 0.9966, 0.9977 and 0.9959 for the linear ranges of 5 µM to 50 µM (upper-left inset of Figure 7b), 50 µM to 0.5 mM (lower-right inset of Figure 7b) and 5 mM to 3.5 mM of glucose concentrations, respectively. When taking the entire linear range from 5 µM to 3.5 mM of glucose into account, its R^2 value was 0.9967 (Figure 7b).

Based on the slope of 0.103 µA·µM^{-1} and the GCE surface area (0.07 cm^2), the sensitivity of the glucose sensor was calculated as 1471.43 µA·mM^{-1}·cm^{-2}.

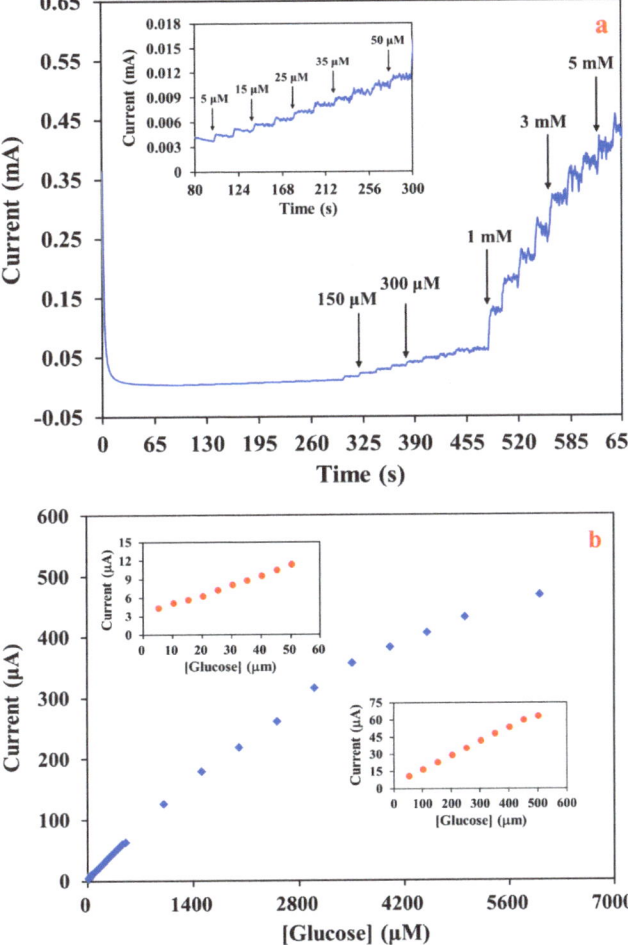

Figure 7. (**a**) Amperometric responses of MWCNT-RuBpy@Ni-MOF/GCE to successive additions of glucose at +0.50 V in a 0.1 M NaOH. Inset: i-t curve of low concentrations of glucose; (**b**) Calibration plot of glucose in a wide linear range.

The response stability of MWCNT-RuBpy@Ni-MOF/GCE was evaluated by continuously recording the i-t curve in the presence of 1 mM glucose (Figure 8a). The current signal was measured over 2500 s. Taking the increment of the current response at the time of addition of glucose at 100 s as 100%, the current signal was retained by 99.65% after 1000 s, 95.24% after a 1500 s scan and 92.51% after 2500 s, respectively. The results show that MWCNT-RuBpy@Ni-MOF/GCE was relatively stable in detecting glucose. The response stability of MWCNT-RuBpy@Ni-MOF in the detection of glucose was also characterized by multiple assays with the same sensor in the presence of 1 mM glucose (Figure 8b). The response stability was retained by 99.52% after 5 assays and by 91.47% after 30 assays, respectively. The data demonstrate that MWCNT-RuBpy@Ni-MOF/GCE had an outstanding response stability.

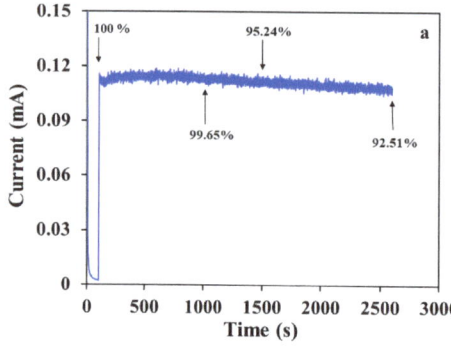 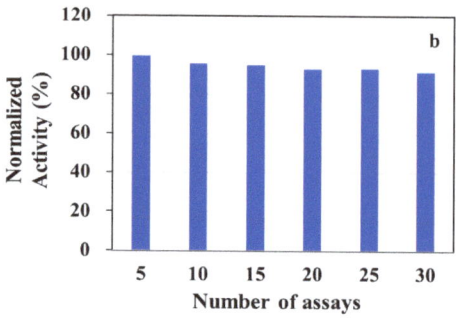

Figure 8. (**a**) Amperometric response stability of MWCNT-RuBpy@Ni-MOF/GCE; (**b**) Stability of MWCNT-RuBpy@Ni-MOF after multiple assays. [Glucose]: 1 mM. Working solution: 0.1 M NaOH. Working potential: +0.50 V.

The reproducibility of multiple sensors was tested with six MWCNT-RuBpy@Ni-MOF/GCEs prepared from the same preparing process (Figure 9). For the response of 0.5 mM glucose, the relative standard deviation (RSD) of 1.11% was acquired with the time window 800 to 1000 s for the six sensors (Figure 9a and the inset). Figure 9b shows the response reproducibility by recording the i-t curves of six sensors under successive additions of glucose. For the glucose concentration steps from 0.1 to 0.8 mM, the RSD of 7.21–15.75% was obtained with the current signal at 10 s of each step plateau for the six sensors (Figure 9b). The results demonstrate that the constructed enzymeless glucose sensors had good performance reproducibility.

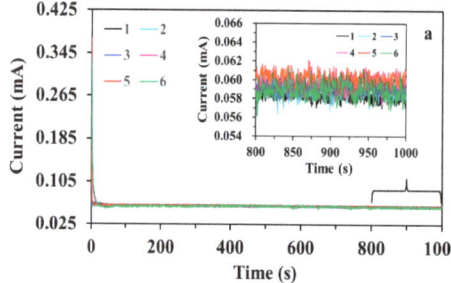 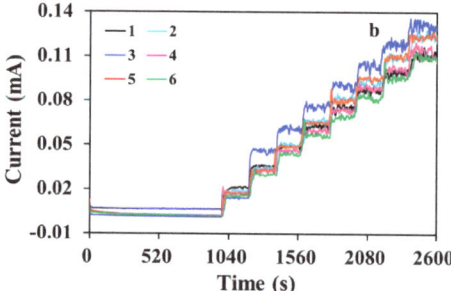

Figure 9. (**a**) Amperometric responses of six MWCNT-RuBpy@Ni-MOF/GCEs in the presence of 0.5 mM glucose at +0.50 V in 0.1 M NaOH; (**b**) i-t curves of six MWCNT-RuBpy@Ni-MOF/GCEs at +0.50 V in 0.1 M NaOH. The glucose concentration in the testing solution for each step was 0.1, 0.2, 0.3, 0.4, 0.5, 0.6, 0.7 and 0.8 mM, respectively.

The chronoamperograms (i-t curves) were measured with different concentrations of glucose at +0.5 V to further evaluate the electrocatalytic kinetics of glucose oxidation at MWCNT-RuBpy@Ni-MOF/GCE (Figure 10a). The electrocatalytic kinetics were studied for the initial rate of current responses in the time window at the very beginning of the development of the i-t curve from 0.4 to 1.3 s (Figure 10a). Figure 10b shows that the linear curve of I_{cat}/I_0 vs. $t^{1/2}$ originated from the chronoamperogram at 0.0 mM and 1.0 mM glucose. Therefore, the catalytic rate constant (k_{cat}) for glucose oxidation can be calculated using the following equation [14]:

$$\frac{I_{cat}}{I_0} = \pi^{1/2}(K_{cat} \cdot C \cdot t)^{1/2} \qquad (3)$$

where I_{cat} and I_0 are the currents in the presence and absence of glucose, respectively; C is the concentration of glucose; and t is the time in s. When [glucose] was 1 mM, based on the slope of the plot of I_{cat}/I_0 vs. $t^{1/2}$, $K_{cat} = 104.1 \times 10^6$ cm^3 M^{-1}s^{-1}, indicating that the composite electrode interface of MWCNT-RuBpy@Ni-MOF/GCE had good electrocatalytic activity for glucose oxidation.

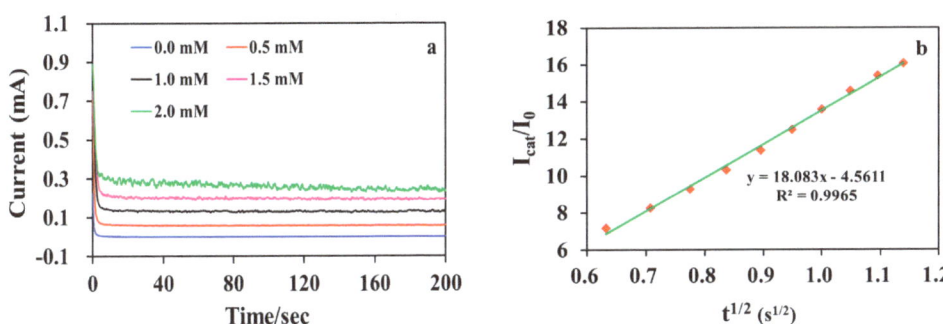

Figure 10. (**a**) Chronoamperograms of MWCNT-RuBpy@Ni-MOF/GCE in the presence of various concentrations of glucose in 0.1 M NaOH; (**b**) The plot of I_{cat}/I_0 vs. $t^{1/2}$, derived from the data of the chronoamperogram for 1 mM glucose.

The comparison of the performances of the enzymeless glucose sensors based on Ni-based materials including Ni-MOFs is in Table 1. Compared to the enzymeless electrochemical glucose sensors based on other Ni-based composite materials, MWCNT-RuBpy@Ni-MOF/GCE exhibited high sensitivity and a wide linear range of responses. However, the catalytic current in the presence of glucose only accounted for a small fraction of the oxidation peak currents of the composite material-modified electrodes in the absence of glucose, implying that the ratio of Ni-MOF compositing with RuBpy and MWCNTs could be optimized to reduce the base current. The detection limit could thus be further improved compared to that of other enzymeless glucose sensors reported (Table 1). The actual measurements of glucose in a honey product from the local market with MWCNT-RuBpy@Ni-MOF/GCE was studied by the recovery experiment. The calibration plot based on the i-t curve (Figure 7) with the successive addition of glucose was used to determine the glucose concentration in honey samples. Using the same procedure for the calibration plot displayed in Figure 7, the diluted honey sample was first added in the testing solution, and subsequently, three glucose solutions were successively added under stirring, and the current responses were continuously recorded. Each current value was calculated based on the calibration plot. The experimental results are summarized in Table 2. The sample recoveries were in the range of 100.26–102.14%, which demonstrated the potential application prospects of MWCNT-RuBpy@Ni-MOF/GCE in the analysis of practical samples.

Table 1. Comparison of the performances of the electrochemical glucose sensors based on MWCNT-RuBpy@Ni-MOF and other Ni-based materials in the electrolyte 0.1 M NaOH.

Electrode	Linear Range (μM)	Detection Limit (μM)	Sensitivity (μA mM^{-1} cm^{-2})	Working Potential (V)	Ref.
CPO-27-Ni^{II}	40–500	1.46	585	+0.55	[1]
Ni(TPA)-SWCNT-CS	20–4400	4.6	-	+0.55	[3]
Au@Ni-BTC	5–7400	1.5	1447.1	+0.55	[14]
Ni$_3$(HHTP)$_2$	1–8000	0.66	21,744	+0.55	[27]
Ni/NiO@C	10–2000 2000–10,000	0.116	1291	+0.55	[51]
Ni/NCNs-500	0.1–533.6 533.6–3030	0.07	337.32 210.56	+0.55	[57]
Ni$_3$N@C	1–3000	0.3	1511.59	+0.6	[58]
Ni$_3$S$_2$@NCNT	0.46–3190	0.14	1447.64	+0.55	[59]
Ni/Co LDH/GNRs	5–800	0.82	344	+0.6	[60]
MWCNT-RuBpy@Ni-MOF	5–3500	1.7	1471.43	+0.50	This work

Table 2. Determination of glucose in the testing solution for honey samples (n = 3).

Sample	Original (μM)	Added (μM)	Found (μM)	Recovery (%)	RSD (%)
Honey	84.23	49.50	133.86	100.26	3.07
	26.20	39.64	66.69	102.14	4.50

4. Conclusions

In summary, the composite material of MWCNT-RuBpy@Ni-MOF was prepared and successfully applied to glucose oxidation and enzymeless electrochemical glucose sensing. In the alkaline condition, the catalytic oxidation peak of the redox pair of Ni^{2+}/Ni^{3+} from Ni-MOFs occurred around the working potential of +0.50 V (vs. Ag/AgCl). The catalytic current was enhanced by decorating the Ni-MOF material with the electron transfer mediator RuBpy, and the catalytic current was further enhanced after compositing with MWCNTs. Besides providing greater conductivity and a larger electrode area, MWCNTs could allow the aggregated RuBpy@Ni-MOF particles to be more dispersive on the electrode and thus expose more catalytic active sites for the oxidation of glucose and accordingly enhance the catalytic currents to some extent. The resulting enzymeless sensor based on MWCNT-RuBpy@Ni-MOF had an excellent performance in detecting glucose, with a wide range of linear responses, a low detection limit, high sensitivity and good selectivity. The sensor also had good stability and reproducibility and was successfully applied for the glucose measurement of the honey product. The results demonstrate that Ni-MOFs served as the catalytic and active center, while RuBpy and MWCNTs also promoted the glucose oxidation at the electrode and thus the sensor performance.

Author Contributions: Conceptualization, C.L. (Chenghong Lei) and C.L. (Chang Liu); methodology, Y.Z. and C.L. (Chang Liu); validation, Y.Z. and R.Y.; formal analysis, Y.Z., R.Y. and C.L. (Chang Liu); investi-gation, Y.Z., R.Y. and C.L. (Chang Liu); resources, C.L. (Chenghong Lei); data curation, Y.Z., R.Y. and C.L. (Chang Liu); writing—original draft preparation, C.L. (Chenghong Lei), Y.Z. and C.L. (Chang Liu); writing—review and editing, C.L. (Chenghong Lei), Y.Z. and R.Y.; supervision, C.L.

(Chenghong Lei); project administration, C.L. (Chenghong Lei); funding acquisition, C.L. (Chenghong Lei). All authors have read and agreed to the published version of the manuscript.

Funding: This work was funded by the National Natural Science Foundation of China (Grant number 32060521) and the Natural Science Foundation of Guangxi, China (Grant numbers AD22035016 and 2020GXNSFDA297023).

Institutional Review Board Statement: Not applicable.

Informed Consent Statement: Not applicable.

Data Availability Statement: The raw data supporting the conclusions of this article will be made available by the authors on request.

Conflicts of Interest: The authors report no conflicts of interest.

References

1. Lopa, N.S.; Rahman, M.; Ahmed, F.; Sutradhar, S.C.; Ryu, T.; Kim, W. A Ni-based redox-active metal-organic framework for sensitive and non-enzymatic detection of glucose. *J. Electroanal. Chem.* **2018**, *822*, 43–49. [CrossRef]
2. Chen, C.; Zhong, Y.; Cheng, S.; Huanga, Y.; Li, T.; Shi, T.; Liao, G.; Tang, Z. In Situ Fabrication of Porous Nanostructures Derived from Bimetal-Organic Frameworks for Highly Sensitive Non-Enzymatic Glucose Sensors. *J. Electrochem. Soc.* **2020**, *167*, 027531. [CrossRef]
3. Wang, F.; Chen, X.; Chen, L.; Yang, J.; Wang, Q. High-performance non-enzymatic glucose sensor by hierarchical flower-like nickel(II)-based MOF/carbon nanotubes composite. *Mater. Sci. Eng. C* **2019**, *96*, 41–50. [CrossRef]
4. Zheng, W.; Liu, Y.; Yang, P.; Chen, Y.; Tao, J.; Hu, J.; Zhao, P. Carbon nanohorns enhanced electrochemical properties of Cu-based metal organic framework for ultrasensitive serum glucose sensing. *J. Electroanal. Chem.* **2020**, *862*, 114018. [CrossRef]
5. Wang, H.-B.; Zhang, H.-D.; Chen, Y.; Li, Y.; Gan, T. H_2O_2-mediated fluorescence quenching of double-stranded DNA templated copper nanoparticles for label-free and sensitive detection of glucose. *RSC Adv.* **2015**, *5*, 77906–77912. [CrossRef]
6. Wang, H.-B.; Chen, Y.; Li, N.; Liu, Y.-M. A fluorescent glucose bioassay based on the hydrogen peroxide-induced decomposition of a quencher system composed of MnO_2 nanosheets and copper nanoclusters. *Microchim. Acta* **2017**, *184*, 515–523. [CrossRef]
7. Kreno, L.E.; Leong, K.; Farha, O.K.; Allendorf, M.; Van Duyne, R.P.; Hupp, J.T. Metal–Organic Framework Materials as Chemical Sensors. *Chem. Rev.* **2012**, *112*, 1105–1125. [CrossRef]
8. Lustig, W.P.; Mukherjee, S.; Rudd, N.D.; Desai, A.V.; Li, J.; Ghosh, S.K. Metal–organic frameworks: Functional luminescent and photonic materials for sensing applications. *Chem. Soc. Rev.* **2017**, *46*, 3242–3285. [CrossRef]
9. Yang, Q.; Xu, Q.; Jiang, H.-L. Metal–organic frameworks meet metal nanoparticles: Synergistic effect for enhanced catalysis. *Chem. Soc. Rev.* **2017**, *46*, 4774–4808. [CrossRef]
10. Jiao, L.; Wang, Y.; Jiang, H.-L.; Xu, Q. Metal–Organic Frameworks as Platforms for Catalytic Applications. *Adv. Mater.* **2018**, *30*, 1703663. [CrossRef]
11. Dhakshinamoorthy, A.; Li, Z.; Garcia, H. Catalysis and photocatalysis by metal organic frameworks. *Chem. Soc. Rev.* **2018**, *47*, 8134–8172. [CrossRef]
12. McKeithan, C.R.; Mayers, J.M.; Wojtas, L.; Larsen, R.W. Photophysical studies of Ru(II)tris(2,2'-bipyridine) encapsulated within the ZnHKUST-1 metal organic framework. *Inorg. Chim. Acta* **2018**, *483*, 1–5. [CrossRef]
13. Jin, X.; Li, G.; Xu, T.; Su, L.; Yan, D.; Zhang, X. Ruthenium-based Conjugated Polymer and Metal-organic Framework Nanocomposites for Glucose Sensing. *Electroanalysis* **2021**, *33*, 1902–1910. [CrossRef]
14. Chen, J.; Yin, H.; Zhou, J.; Wang, L.; Gong, J.; Ji, Z.; Nie, Q. Efficient Nonenzymatic Sensors Based on Ni-MOF Microspheres Decorated with Au Nanoparticles for Glucose Detection. *J. Electron. Mater.* **2020**, *49*, 4754–4763. [CrossRef]
15. Abbasi, A.R.; Yousefshahi, M.; Daasbjerg, K. Non-enzymatic Electroanalytical Sensing of Glucose Based on Nano Nickel-Coordination Polymers-Modified Glassy Carbon Electrode. *J. Inorg. Organomet. Polym. Mater.* **2020**, *30*, 2027–2038. [CrossRef]
16. Li, L.-X.; He, S.; Zeng, S.; Chen, W.-T.; Ye, J.-W.; Zhou, H.-L.; Huang, X.-C. Equipping carbon dots in a defect-containing MOF via self-carbonization for explosive sensing. *J. Mater. Chem. C* **2022**, *11*, 321–328. [CrossRef]
17. Elizbit Liaqat, U.; Hussain, Z.; Baig, M.M.; Khan, M.A.; Arif, D. Preparation of porous ZIF-67 network interconnected by MWCNTs and decorated with Ag nanoparticles for improved non-enzymatic electrochemical glucose sensing. *J. Korean Ceram. Soc.* **2021**, *58*, 598–605. [CrossRef]
18. Wen, Y.; Meng, W.; Li, C.; Dai, L.; He, Z.; Wang, L.; Li, M.; Zhu, J. Enhanced glucose sensing based on a novel composite Co^{II}-MOF/Acb modified electrode. *Dalton Trans.* **2018**, *47*, 3872–3879. [CrossRef]
19. Li, W.-H.; Lv, J.; Li, Q.; Xie, J.; Ogiwara, N.; Huang, Y.; Jiang, H.; Kitagawa, H.; Xu, G.; Wang, Y. Conductive metal–organic framework nanowire arrays for electrocatalytic oxygen evolution. *J. Mater. Chem. A* **2019**, *7*, 10431–10438. [CrossRef]
20. Lennon, S.J.; Robinson, F.P.A. The experimental determination of potential-pH diagrams for the $Ni-H_2O$ and low alloy steel-H_2O systems. *Corros. Sci.* **1986**, *26*, 995–1007. [CrossRef]
21. Camasso, N.M.; Sanford, M.S. Design, synthesis, and carbon-heteroatom coupling reactions of organometallic nickel(IV) complexes. *Science* **2015**, *347*, 1218–1220. [CrossRef]

22. Leem, Y.J.; Cho, K.; Oh, K.H.; Han, S.-H.; Nam, K.M.; Chang, J. A self-assembled Ni(cyclam)-BTC network on ITO for an oxygen evolution catalyst in alkaline solution. *Chem. Commun.* **2017**, *53*, 3454–3457. [CrossRef]
23. Rahmanifar, M.S.; Hesari, H.; Noori, A.; Masoomi, M.Y.; Morsali, A.; Mousavi, M.F. A dual Ni/Co-MOF-reduced graphene oxide nanocomposite as a high performance supercapacitor electrode material. *Electrochim. Acta* **2018**, *275*, 76–86. [CrossRef]
24. Gao, M.; Sheng, W.; Zhuang, Z.; Fang, Q.; Gu, S.; Jiang, J.; Yan, Y. Efficient Water Oxidation Using Nanostructured α-Nickel-Hydroxide as an Electrocatalyst. *J. Am. Chem. Soc.* **2014**, *136*, 7077–7084. [CrossRef] [PubMed]
25. Bediako, D.K.; Lassalle-Kaiser, B.; Surendranath, Y.; Yano, J.; Yachandra, V.K.; Nocera, D.G. Structure–Activity Correlations in a Nickel–Borate Oxygen Evolution Catalyst. *J. Am. Chem. Soc.* **2012**, *134*, 6801–6809. [CrossRef] [PubMed]
26. Xiao, X.; Zheng, S.; Li, X.; Zhang, G.; Guo, X.; Xue, H.; Pang, H. Facile synthesis of ultrathin Ni-MOF nanobelts for high-efficiency determination of glucose in human serum. *J. Mater. Chem. B* **2017**, *5*, 5234–5239. [CrossRef] [PubMed]
27. Qiao, Y.; Liu, Q.; Lu, S.; Chen, G.; Gao, S.; Lu, W.; Sun, X. High-performance non-enzymatic glucose detection: Using a conductive Ni-MOF as an electrocatalyst. *J. Mater. Chem. B* **2020**, *8*, 5411–5415. [CrossRef] [PubMed]
28. Zhang, X.; Bard, A.J. Electrogenerated chemiluminescent emission from an organized (L-B) monolayer of a tris(2,2′-bipyridine)ruthenium(2+)-based surfactant on semiconductor and metal electrodes. *J. Phys. Chem.* **1988**, *92*, 5566–5569. [CrossRef]
29. Miller, C.J.; McCord, P.; Bard, A.J. Study of Langmuir monolayers of ruthenium complexes and their aggregation by electrogenerated chemiluminescence. *Langmuir* **1991**, *7*, 2781–2787. [CrossRef]
30. Zholudov, Y.; Snizhko, D.; Kukoba, A.; Bilash, H.; Rozhitskii, M. Aqueous electrochemiluminescence of polycyclic aromatic hydrocarbons immobilized into Langmuir–Blodgett film at the electrode. *Electrochim. Acta* **2008**, *54*, 360–363. [CrossRef]
31. Sato, Y.; Uosaki, K. Electrochemical and electrogenerated chemiluminescence properties of tris(2,2′-bipyridine)ruthenium(II)-tridecanethiol derivative on ITO and gold electrodes. *J. Electroanal. Chem.* **1995**, *384*, 57–66. [CrossRef]
32. Obeng, Y.S.; Bard, A.J. Electrogenerated chemiluminescence. 53. Electrochemistry and emission from adsorbed monolayers of a tris(bipyridyl)ruthenium(II)-based surfactant on gold and tin oxide electrodes. *Langmuir* **1991**, *7*, 195–201. [CrossRef]
33. Collinson, M.M.; Novak, B.; Martin, S.A.; Taussig, J.S. Electrochemiluminescence of Ruthenium(II) Tris(bipyridine) Encapsulated in Sol−Gel Glasses. *Anal. Chem.* **2000**, *72*, 2914–2918. [CrossRef]
34. Sykora, M.; Meyer, T.J. Electrogenerated Chemiluminescence in SiO₂ Sol−Gel Polymer Composites. *Chem. Mater.* **1999**, *11*, 1186–1189. [CrossRef]
35. Matsui, K.; Momose, F. Luminescence Properties of Tris(2,2′-bipyridine)ruthenium(II) in Sol−Gel Systems of SiO₂. *Chem. Mater.* **1997**, *9*, 2588–2591. [CrossRef]
36. Choi, H.N.; Cho, S.-H.; Lee, W.-Y. Electrogenerated Chemiluminescence from Tris(2,2′-bipyridyl)ruthenium(II) Immobilized in Titania−Perfluorosulfonated Ionomer Composite Films. *Anal. Chem.* **2003**, *75*, 4250–4256. [CrossRef]
37. Lin, Z.; Chen, G. Determination of carbamates in nature water based on the enhancement of electrochemiluminescent of Ru(bpy)$_3^{2+}$ at the multi-wall carbon nanotube-modified electrode. *Talanta* **2006**, *70*, 111–115. [CrossRef]
38. Hun, X.; Zhang, Z. Electrogenerated chemiluminescence sensor for metoclopramide determination based on Ru(bpy)$_3^{2+}$-doped silica nanoparticles dispersed in Nafion on glassy carbon electrode. *J. Pharm. Biomed. Anal.* **2008**, *47*, 670–676. [CrossRef]
39. Martin, A.F.; Nieman, T.A. Chemiluminescence biosensors using tris(2,2′-bipyridyl)ruthenium(II) and dehydrogenases immobilized in cation exchange polymers. *Biosens. Bioelectron.* **1997**, *12*, 479–489. [CrossRef]
40. Zhao, C.-Z.; Egashira, N.; Kurauchi, Y.; Ohga, K. Electrochemiluminescence Sensor Having a Pt Electrode Coated with a Ru(bpy)$_3^{2+}$-Modified Chitosan/Silica Gel Membrane. *Anal. Sci.* **1998**, *14*, 439–441. [CrossRef]
41. Wang, H.; Xu, G.; Dong, S. Electrochemiluminescence of tris(2,2′-bipyridine)ruthenium(II) immobilized in poly(*p*-styrenesulfonate)–silica–Triton X-100 composite thin-films. *Analyst* **2001**, *126*, 1095–1099. [CrossRef]
42. Zhang, L.; Wang, F.; Dong, S. Layer-by-layer assembly of functional silica and Au nanoparticles for fabricating electrogenerated chemiluminescence sensor. *Electrochim. Acta* **2008**, *53*, 6423–6427. [CrossRef]
43. Lee, J.-K.; Lee, S.-H.; Kim, M.; Kim, H.; Kim, D.-H.; Lee, W.-Y. Organosilicate thin film containing Ru(bpy)$_3^{2+}$ for an electrogenerated chemiluminescence (ECL) sensor. *Chem. Commun.* **2003**, 1602–1603. [CrossRef]
44. Li, J.; Xu, Y.; Wei, H.; Huo, T.; Wang, E. Electrochemiluminescence Sensor Based on Partial Sulfonation of Polystyrene with Carbon Nanotubes. *Anal. Chem.* **2007**, *79*, 5439–5443. [CrossRef]
45. Guo, Z.; Dong, S. Electrogenerated Chemiluminescence from Ru(Bpy)$_3^{2+}$ Ion-Exchanged in Carbon Nanotube/Perfluorosulfonated Ionomer Composite Films. *Anal. Chem.* **2004**, *76*, 2683–2688. [CrossRef]
46. Kang, C.H.; Choi, Y.-B.; Kim, H.-H.; Choi, H.N.; Lee, W.-Y. Electrogenerated Chemiluminescence Sensor Based on a Self-Assembled Monolayer of Ruthenium(II)-bis(2,2′-bipyridyl)(aminopropyl imidazole) on Gold Deposited Screen Printed Electrode. *Electroanalysis* **2011**, *23*, 2131–2138. [CrossRef]
47. Xu, X.; Li, H.; Xu, Z. Multifunctional luminescent switch based on a porous PL-MOF for sensitivity recognition of HCl, trace water and lead ion. *Chem. Eng. J.* **2022**, *436*, 135028. [CrossRef]
48. Ando, Y.; Zhao, X.; Shimoyama, H.; Sakai, G.; Kaneto, K. Physical properties of multiwalled carbon nanotubes. *Int. J. Inorg. Mater.* **1999**, *1*, 77–82. [CrossRef]
49. Wei, B.Q.; Vajtai, R.; Ajayan, P.M. Reliability and current carrying capacity of carbon nanotubes. *Appl. Phys. Lett.* **2001**, *79*, 1172–1174. [CrossRef]

50. Zeraati, M.; Alizadeh, V.; Kazemzadeh, P.; Safinejad, M.; Kazemian, H.; Sargazi, G. A new nickel metal organic framework (Ni-MOF) porous nanostructure as a potential novel electrochemical sensor for detecting glucose. *J. Porous Mater.* **2022**, *29*, 257–267. [CrossRef]
51. Ma, X.; Tang, K.-l.; Yang, M.; Shi, W.; Zhao, W. Metal–organic framework-derived yolk–shell hollow Ni/NiO@C microspheres for bifunctional non-enzymatic glucose and hydrogen peroxide biosensors. *J. Mater. Sci.* **2021**, *56*, 442–456. [CrossRef]
52. Morgan, D.J. Resolving ruthenium: XPS studies of common ruthenium materials. *Surf. Interface Anal.* **2015**, *47*, 1072–1079. [CrossRef]
53. Al-Hinaai, M.; Khudaish, E.A. Electrochemical Construction of a Polymer-Metal Complex Surface Network for Selective Determination of Dopamine in Blood Serum. *Anal. Lett.* **2022**, *55*, 1249–1268. [CrossRef]
54. Ren, Z.; Mao, H.; Luo, H.; Liu, Y. Glucose sensor based on porous Ni by using a graphene bottom layer combined with a Ni middle layer. *Carbon* **2019**, *149*, 609–617. [CrossRef]
55. Cao, M.; Wang, H.; Ji, S.; Zhao, Q.; Pollet, B.G.; Wang, R. Hollow core-shell structured $Cu_2O@Cu_{1.8}S$ spheres as novel electrode for enzyme free glucose sensing. *Mater. Sci. Eng. C* **2019**, *95*, 174–182. [CrossRef]
56. Li, G.; Chen, D.; Chen, Y.; Dong, L. MOF Ni-BTC Derived Ni/C/Graphene Composite for Highly Sensitive Non-Enzymatic Electrochemical Glucose Detection. *ECS J. Solid State Sci. Technol.* **2020**, *9*, 121014. [CrossRef]
57. Jia, H.; Shang, N.; Feng, Y.; Ye, H.; Zhao, J.; Wang, H.; Wang, C.; Zhang, Y. Facile preparation of Ni nanoparticle embedded on mesoporous carbon nanorods for non-enzymatic glucose detection. *J. Colloid Interface Sci.* **2021**, *583*, 310–320. [CrossRef]
58. Chen, J.; Yin, H.; Zhou, J.; Wang, L.; Ji, Z.; Zheng, Y.; Nie, Q. Hybrid Ni_3N-nitrogen-doped carbon microspheres (Ni_3N@C) in situ derived from Ni-MOFs as sensitive non-enzymatic glucose sensors. *Mater. Technol.* **2021**, *36*, 286–295. [CrossRef]
59. Li, G.; Xie, G.; Chen, D.; Gong, C.; Chen, X.; Zhang, Q.; Pang, B.; Zhang, Y.; Li, C.; Hu, J.; et al. Facile synthesis of bamboo-like Ni_3S_2@NCNT as efficient and stable electrocatalysts for non-enzymatic glucose detection. *Appl. Surf. Sci.* **2022**, *585*, 152683. [CrossRef]
60. Asadian, E.; Shahrokhian, S.; Iraji Zad, A. Highly sensitive nonenzymetic glucose sensing platform based on MOF-derived NiCo LDH nanosheets/graphene nanoribbons composite. *J. Electroanal. Chem.* **2018**, *808*, 114–123. [CrossRef]

Disclaimer/Publisher's Note: The statements, opinions and data contained in all publications are solely those of the individual author(s) and contributor(s) and not of MDPI and/or the editor(s). MDPI and/or the editor(s) disclaim responsibility for any injury to people or property resulting from any ideas, methods, instructions or products referred to in the content.

Article

A Host–Guest Platform for Highly Efficient, Quantitative, and Rapid Detection of Nitroreductase

Wen Si, Yang Jiao *, Xianchao Jia, Meng Gao, Yihao Zhang, Ye Gao, Lei Zhang and Chunying Duan

State Key Laboratory of Fine Chemicals, School of Chemistry, Dalian University of Technology, Dalian 116024, China; siwen93@mail.dlut.edu.cn (W.S.); xcjia@mail.dlut.edu.cn (X.J.); gm172407@163.com (M.G.); 1329731948@mail.dlut.edu.cn (Y.Z.); gaoye@mail.dlut.edu.cn (Y.G.); aizome@mail.dlut.edu.cn (L.Z.); cyduan@dlut.edu.cn (C.D.)
* Correspondence: jiaoyang@dlut.edu.cn

Abstract: Nitroreductase (NTR) is an enzyme expressed at an abnormally high level in solid tumors, which is associated with the hypoxia level in tumors. The establishment of a high-performance and convenient fluorescent platform for the fast monitoring of NTR is of pivotal importance. Herein, a novel host–guest complex was created by encapsulating a fluorescent substrate **GP-NTR** within a metal–organic capsule **Zn-MPB** that included a NADH mimic for the detection of hypoxia via responding to nitroreductase (NTR) with fast responsiveness and good fluorescence imaging. Notably, the double-substrate process was streamlined to a single–substrate process by the host–guest supramolecular method in the catalytic process of NTR, which enabled the reaction to be independent of the cofactor NADH supply and shortened the distance between the substrate and the active site of NTR. The increasing fluorescence intensity of **Zn-MPB⊃GP-NTR** exhibits a linear relationship with NTR concentration and shows a fast response toward NTR in solution in tens of seconds. **Zn-MPB⊃GP-NTR** also displays high sensitivity to NTR with a low detection limit of 6.4 ng/mL. Cells and in vivo studies have confirmed that **Zn-MPB⊃GP-NTR** could be successfully applied for the fast imaging of NTR in NTR-overexpressed tumor cells and tumor-bearing animals. The host–guest platform not only provides a new avenue for the design and optimization of a fluorescence detection platform for the rapid and quantitative detection of NTR activity, but also offers an imaging tool for the early diagnosis of hypoxia-related tumors.

Keywords: host–guest interaction; hypoxia; bioimaging; rapid detection; nitroreductase

Citation: Si, W.; Jiao, Y.; Jia, X.; Gao, M.; Zhang, Y.; Gao, Y.; Zhang, L.; Duan, C. A Host–Guest Platform for Highly Efficient, Quantitative, and Rapid Detection of Nitroreductase. *Chemosensors* **2024**, *12*, 145. https://doi.org/10.3390/chemosensors12080145

Received: 21 May 2024
Revised: 17 June 2024
Accepted: 4 July 2024
Published: 30 July 2024

Copyright: © 2024 by the authors. Licensee MDPI, Basel, Switzerland. This article is an open access article distributed under the terms and conditions of the Creative Commons Attribution (CC BY) license (https://creativecommons.org/licenses/by/4.0/).

1. Introduction

Hypoxia is a common physiological manifestation in the growth of malignant tumors [1–5], and it could result in the overexpression of several intracellular reductase enzymes that typically catalyze redox reactions in the tumor microenvironment with impressive selectivity and specificity [6,7]. As one type of reductive enzyme in hypoxic tumors, nitroreductase (NTR) could effectively reduce nitroaromatic compounds to the corresponding arylamine with reduced nicotinamide adenine dinucleotide (NADH) as the coenzyme [8–10]. It is of great significance to monitor the expression of NTR in hypoxic tumors owing to its crucial role in malignant tumor progression, invasion, angiogenesis, and prodrug activation for directed anticancer therapies [11–13]. Several analytical methods have been developed for NTR detection, including positron emission tomography (PET) [14], magnetic resonance imaging (MRI) [15], ultrasound imaging [16], colorimetric methods [17], and fluorescence spectroscopy [18]. Among them, numerous fluorescent probes have recently been reported with the ability to visualize NTR owing to their high sensitivity, easy operation, and unrivaled spatiotemporal resolution [19,20]. They mainly depend on the changes in the fluorescence intensity of NTR activity to the fluorescent substrate, and the intensity of its emission may be closely related to the activity of the enzyme and the local concentrations of the probe and the coenzyme. Based on the 'ping-pong' mechanism, a stable relationship rarely exists between the fluorescence intensity and enzyme content when the NADH concentration is not fixed. The uncertain levels of enzymes and coenzymes lead to the variability in probe emission over time. In the conventional detection system of NTR, a large excess of NADH is often added to the reaction solution, which could reduce the interference of the concentration of NADH, ensure the reaction reaches fluorescence emission equilibrium quickly, and improve the accuracy of the detection [21,22]. Considering that the expression levels of hypoxia enzymes and coenzymes in living system are indefinite, the in vivo detection of hypoxic enzymes is highly dependent on the corresponding coenzyme; therefore, it is worth paying attention to improving the detection efficiency of hypoxic enzymes [23]. It is challenging to maintain the coenzyme supply via a covalent molecule bond with the coenzyme because of the intricate molecular design and tedious synthesis. Thus, developing a platform with sufficient coenzymes for highly efficient, quantitative, and rapid detection of NTR in biological systems is highly desired.

Supramolecular systems based on host–guest interactions have been evolving immensely in recent years, and they are widely used in various biomedical applications, such as biological imaging, clinical medicine, and drug delivery due to the functional modification features and dynamic properties of host–guest self-assembly [24]. Well-known macrocyclic molecules such as cyclodextrins, crown ethers, calixarenes, and pillararenes have been established as 'hosts' to accommodate 'guests' inside their cavities through noncovalent interactions to construct host–guest complexes [25]. Metal–organic capsules have attracted extensive attention due to their adjustable cavity and characteristics of easy assembly and could solve the problem of difficult covalent modification [26,27]. A metal–organic capsule incorporated by the active site of NADH and its mimics has been postulated to be an effective host molecule for constructing an artificial catalytic platform, and it could emulate the environment of an enzyme pocket that changes the reactivity of the substrate and achieves a significant increase in the reaction rate [28–30]. The host–guest platform consists of a cofactor-modified metal–organic capsule and a substrate through intermolecular interactions, featuring the advantages of strong contact, stimulus response and dynamic characteristics. This platform provides a simple and effective approach for the detection of NTR, which could enable the redox reaction between the substrate, cofactor, and enzyme to occur effectively. When the host–guest platform interacts with NTR, it enables the efficient proton/electron transport of the biological microenvironment, eliminates the diffusion of cofactors during the process, and accelerates the detection efficiency of NTR.

Herein, a host–guest complex was reported by encapsulating a substrate **GP-NTR** within a NADH mimic-containing metal–organic capsule **Zn-MPB** for the detection of NTR (Scheme 1). The substrate, **GP-NTR**, was well established with naphthalimide as the fluorophore and a nitro group as the recognition moiety. **Zn-MPB** serves not only as a substitute for NADH, but also as a carrier to efficiently transport the substrate to NTR, which accelerates the process of the reaction. Its water solubility and biocompatibility have been improved, which was attributed to the host–guest system. The positively charged **Zn-MPB** would contribute to the accumulation of the host–guest complex in tumor cells. It is particularly worth noting that the host–guest strategy could simplify the original double–substrate enzymatic process of a substrate-based probe into a simpler pseudo-single-substrate kinetic, with the Michaelis–Menten equation. In accordance with this feature, the enzymatic reaction process was independent of the cofactor NADH supply, which accelerated the reaction rate and shortened the equilibrium time. The fluorescence intensity of **Zn-MPB⊃GP-NTR** was linearly related to the concentration of NTR. **Zn-MPB⊃GP-NTR** exhibited superior performance in vitro, including fast NTR response time, low detection limit, and good NTR selectivity. The specific recognition of NTR by **Zn-MPB⊃GP-NTR** was verified by HPLC. **Zn-MPB⊃GP-NTR** shows low cytotoxicity and could be used to evaluate the hypoxia levels in both MCF-7 and HepG2 cells. In vitro and in vivo experiments have proven the good NTR detection ability of **Zn-MPB⊃GP-NTR**. It could serve as a trustworthy platform for the effective bio-tracing of NTR in an hypoxic microenvironment and has great potential for cancer diagnostic applications.

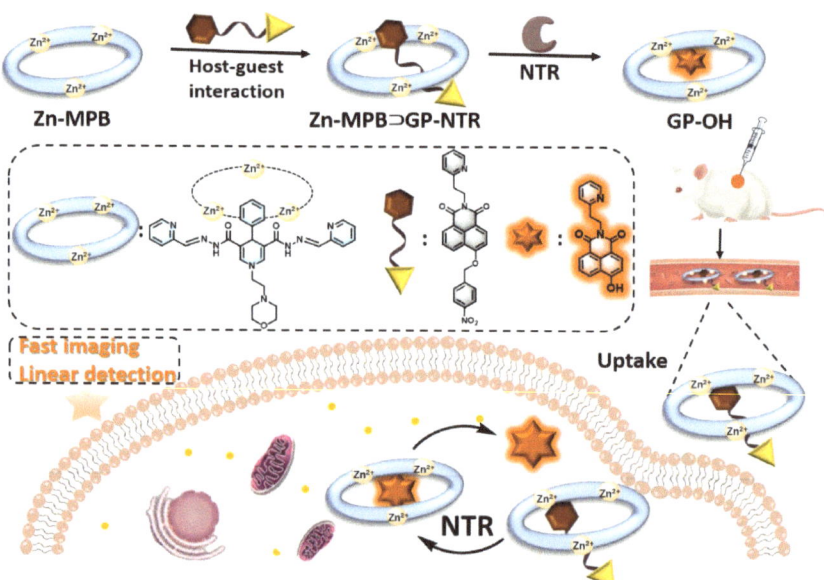

Scheme 1. Schematic diagram of preparation and application of **Zn-MPB⊃GP-NTR** for NTR detection.

2. Materials and Methods

2.1. Reagents and Instrumentation

All reagents and solvents were purchased commercially and used as received. Nitroreductase from *Escherichia coli* and NADH were purchased from Sigma-Aldrich (St. Louis, MO, USA). NMR spectra were acquired using a Bruker AVANCE III 400 MHz spectrometer and a Bruker AVANCE III 500 MHz spectrometer (Bruker, Ferranden, Switzerland). Mass spectrometric data were collected using an LTQ Orbitrap XL mass spectrometer (Thermo Fisher Scientific, Waltham, MA, USA). Titration calorimetry (ITC) was performed on a Nano ITC (TA Instruments Inc. Waters LLC, New Castle County, DE, USA). Fluorescent spectra were obtained with a Steady State, Fluorescence lifetime Spectrometer (Edinburgh Instruments (FLS1000), Livingston, Scotland). The UV–vis spectra were measured on a TU 1900 UV–vis spectrometer (Persee, Shanghai, China). HPLC analyses were performed using Agilent 1100 high-performance liquid chromatography (Agilent, Santa Clara, CA, USA). The BD Accuri C6 Plus flow cytometer was used to sort out specific cells (BD, Franklin Lake, NJ, USA). Confocal fluorescence imaging was performed using an OLYMPUS FV1000 confocal microscopy (OLYMPUS, Tokyo, Japan). The mice were imaged using Berthold Night Owl LB 983 NC100 systems (Berthold Technologies, Baden-wurttemberg, Germany).

2.2. Synthesis of Compound 2

4-Bromo-1,8-naphthalic anhydride (1100 mg, 4 mmol) and 2-(2-aminoethyl)pyridine (526 μL, 4.4 mmol) were dissolved in 15 mL of ethanol [31]. The reaction mixture was heated at 75 °C for 8 h. After cooling to room temperature, the solvent was removed under vacuum, and the obtained solid was washed with ethanol to give 4-bromo-N-2-aminoethylpyridine-1,8-naphthalimide (compound **2**) as a white solid product. Yield: 82%.
^{1}H NMR (400 MHz, DMSO) δ 8.51 (d, *J* = 4.7 Hz, 2H), 8.44 (d, *J* = 4.3 Hz, 1H), 8.27 (d,

J = 6.4 Hz, 1H), 8.21–8.15 (m, 1H), 7.96 (t, J = 7.3 Hz, 1H), 7.70 (t, J = 7.5 Hz, 1H), 7.31 (d, J = 7.7 Hz, 1H), 7.25–7.17 (m, 1H), 4.37 (t, J = 7.4 Hz, 2H), 3.08 (t, J = 7.5 Hz, 2H).

2.3. Synthesis of Compound GP-OH

Compound **2** (343 mg, 0.9 mmol) and N-Hydroxyphthalimide (163 mg, 1 mmol) were dissolved in 5 mL dimethyl sulfoxide, then potassium carbonate (414 mg, 3 mmol) was added, and the mixture was refluxed at 80 °C for 6 h. After cooling to room temperature, the reaction solution was added dropwise to 100 mL of ultrapure water and then stirred. The pH was adjusted to about 3 by concentrated hydrochloric acid. The pure product 4-hydro-N-2-aminoethylpyridine-1,8-naphthalimide (compound **GP-OH**) was obtained by filtering reaction mixture without further purification. Yield: 76%. ^{1}H NMR (400 MHz, DMSO) δ 11.85 (s, 1H), 8.52 (dd, J = 8.3, 0.8 Hz, 1H), 8.49–8.42 (m, 2H), 8.33 (d, J = 8.2 Hz, 1H), 7.79–7.65 (m, 2H), 7.29 (d, J = 7.8 Hz, 1H), 7.21 (dd, J = 7.0, 5.2 Hz, 1H), 7.15 (d, J = 8.2 Hz, 1H), 4.42–4.32 (m, 2H), 3.10–3.03 (m, 2H). HR–MS m/z: [M+H]$^+$ calculated for [C$_{19}$H$_{15}$N$_2$O$_3$]$^+$: 319.1086, found: 319.1093.

2.4. Synthesis of Compound GP-NTR

The synthetic route of **GP-NTR** was described in Scheme 2. Compound **GP-OH** (318 mg, 1 mmol) and K$_2$CO$_3$ (420 mg, 3 mmol) were dissolved in 5 mL acetonitrile, and the mixture was stirred for 30 min, and then, 1-(bromomethyl)-4-nitrobenzene (216 mg, 1 mmol) was added and refluxed at 75 °C for 8 h. After cooling to room temperature, the solvent was removed under vacuum; then, the crude product was purified by column chromatography on silica gel using eluent (dichloromethane/methanol = 100/1) to give **GP-NTR**. Yield: 84%. ^{1}H NMR (400 MHz, CDCl$_3$/CD$_3$OD) δ 8.71 (d, J = 8.3 Hz, 1H), 8.66 (d, J = 7.2 Hz, 1H), 8.58 (d, J = 8.2 Hz, 1H), 8.54 (d, J = 4.1 Hz, 1H), 8.38 (d, J = 8.6 Hz, 2H), 7.82 (dd, J = 12.9, 8.4 Hz, 3H), 7.72 (t, J = 7.6 Hz, 1H), 7.42–7.36 (m, 2H), 7.19 (d, J = 8.3 Hz, 1H), 5.57 (s, 2H), 4.72–4.53 (m, 2H), 3.36–3.21 (m, 2H). ^{13}C NMR (126 MHz, CDCl$_3$/CD$_3$OD) δ 168.30, 167.71, 163.20, 162.55, 152.65, 152.56, 151.83, 146.85, 141.06, 137.31, 137.22, 135.75, 135.68, 133.34, 132.53, 131.85, 130.27, 127.89, 127.57, 127.44, 126.14, 125.78, 119.41, 110.53, 73.40, 43.87, 39.99. HR–MS m/z: [M+H]$^+$ calculated for [C$_{26}$H$_{20}$N$_3$O$_5$]$^+$: 454.1402, found: 454.1391.

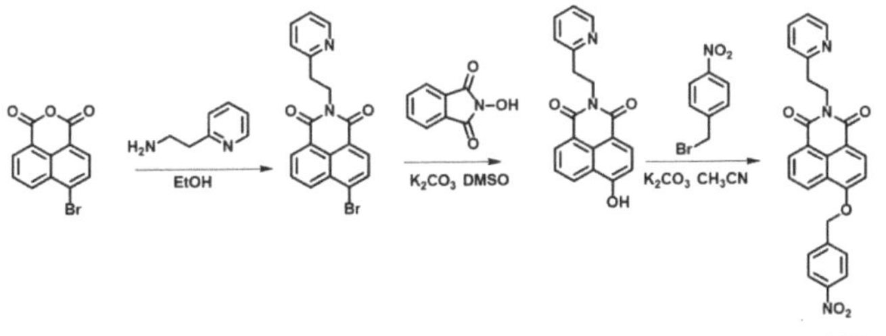

Scheme 2. The synthetic route of **GP-NTR**.

2.5. The General Procedure for In Vitro Spectra Measurement

The spectroscopic measurements were performed in Tris–HCl at 25 °C. A stock solution of **GP-NTR** (2 mM) or **Zn-MPB** (2 mM) was prepared in DMSO and diluted to the required concentration when it was used in vitro measurement. Other analytes were prepared by dissolving the relevant analytes with the same amount of distilled water to obtain the final concentrations, including NaCl, KCl, L-Tyr, L-Cys, β-Ala, L-Pro, L-Phe, L-Pro, H$_2$O$_2$, AA (Ascorbic Acid), NQO1 (NAD(P)H quinone oxidoreductase 1), GSR (Glutathione Reductase), and Escherichia coli NTR (Nitroreductase) (5 μg/mL in distilled water) (λ_{ex} = 440 nm, λ_{em} = 520 nm). The solution was swiftly combined, and then, the absorbance or fluorescence spectrum was measured by pouring it into a quartz colorimetric dish.

*2.6. Kinetic Study of **GP-NTR** and **Zn-MPB⊃GP-NTR** Reacting with NTR*

The catalytic activity of NTR toward the reduction of **Zn-MPB⊃GP-NTR** was investigated by fluorescence spectra at 25 °C. By using the Michaelis–Menten equation, the kinetic parameters $V_{max}^{Zn\text{-}MPB \supset GP\text{-}NTR}/V_{max}^{GP\text{-}NTR}$ and $k_{cat}^{Zn\text{-}MPB \supset GP\text{-}NTR}/k_{cat}^{GP\text{-}NTR}$ were estimated [32,33]. The catalytic number ratio between **GP-NTR** and **Zn-MPB⊃GP-NTR** reacting with NTR is $k_{cat}^{Zn\text{-}MPB \supset GP\text{-}NTR}/k_{cat}^{GP\text{-}NTR}$. The maximum reaction rate ratio between **GP-NTR** and **Zn-MPB⊃GP-NTR** reacting with NTR is $V_{max}^{Zn\text{-}MPB \supset GP\text{-}NTR}/V_{max}^{GP\text{-}NTR}$

2.7. ITC Experiments

The isothermal titration microcalorimeter was applied in the ITC experiments (at atmospheric pressure and 25 °C) [34], and it provided data for the association constant (K) and thermodynamic parameters. A solution of **GP-NTR** (1 mM) in a 0.25 mL syringe was sequentially injected with stirring at 250 rpm into a solution of **Zn-MPB** (0.1 mM) in the sample cell (1.30 mL volume). The experiments were conducted in DMF–H_2O solution (DMF/H_2O = 98/2, v/v). Using the 'independent' model, all of the thermodynamic parameters given in this study were determined.

2.8. Molecular Docking Preparation

The structure of NTR was obtained from the Protein Crystal Structure Database (https://www.rcsb.org/3d-view/1ICR/1 (accessed on July 7, 2024)) (PDB code: 1ICR for NTR). The basic structure of **Zn-MPB** and **GP-NTR** were appropriately optimized and adjusted. For the docking calculations, the models of NTR were refined by eliminating water molecules, adding hydrogen atoms, and adding Gasteiger charges, fragmental volumes, and atomic solvation parameters to adhesive through AutoDock Tools (AutoDock Tools-1.5.6) [35]. For the ligand, the molecule was refined by removing and subsequently adding hydrogen atoms in a similar manner. PyMOL 2.5.4 was used to render the AutoDock docking calculation results following optimization.

2.9. HPLC Analysis

The substrate **GP-NTR** was dissolved in PBS, and NTR was added into the solution of NADH or **Zn-MPB** for 30 min, and the resultant solution (NTR: 10 µg/mL, **GP-NTR** or **GP-OH**: 10 µM) was filtered before sample injection. Mobile phase A was water containing 0.2% acetic acid and 0.2% triethylamine, and phase B was pure methanol. The flow rate was 0.8 mL/min.

2.10. Cell Culture and Cell Cytotoxicity Studies

The Michigan Cancer Foundation-7 (MCF-7) cells and HepG2 cells were cultured in DMEM supplemented with 10% (v/v) fetal bovine serum (FBS) and 1% (v/v) penicillin–streptomycin in a 5% CO_2 humidified incubator at 37 °C.

The cytotoxicity of **GP-NTR** and **Zn-MPB** to MCF-7 cells and HepG2 cells was measured by the MTT method. In brief, cells were seeded in 96-well microplates and incubated in complete DMEM culture medium for 24 h. Then, the cells were washed with PBS and replaced with a fresh culture medium and then cultured in medium with 0, 1, 2, 5, 10, and 20 µM of **GP-NTR** and **Zn-MPB** for 24 h. After that, the medium was removed, and the cells were washed 3 times with fresh PBS. The MTT reagent with the fresh medium was added into each well and then incubated for another 4 h. The culture medium was taken out, and 100 µL DMSO was added to dissolve the precipitate. Finally, the Thermo MK3 ELISA Microplate Reader was used to measure the absorbance at 570 nm, and the cell viability of MCF-7 cells and HepG2 cells was estimated.

2.11. Confocal Fluorescence Imaging for Cells

The MCF-7 cells and HepG2 cells were treated under normoxic (20% O_2) or different hypoxic conditions (O_2 concentration of 8% or 0.1%) for 6 h at 37 °C. After culture, the cells were further incubated with **GP-NTR** (2 µM) and **Zn-MPB⊃GP-NTR** (2 µM) biomimetic catalytic platform for 1 min. Kinetic fluorescence imaging was carried out by incubating MCF-7 cells and HepG2 cells with **GP-NTR** (2 µM) and **Zn-MPB⊃GP-NTR** (2 µM) for 0, 1, 5, 10, or 20 min after treatment under hypoxic conditions (0.1% O_2) for 6 h at 37 °C. The cells were rinsed three times with PBS. An Olympus FV1000 fluorescent microscope was used to take fluorescence photographs of the cells. The wavelength of the laser source was 488 nm, and the cell fluorescence signals were collected at 510–600 nm.

2.12. Flow Cytometry Analyses

MCF-7 cells were cultured with different degrees of hypoxia (20% O_2 and 0.1% O_2) for 6 h and treated with 5 µM **Zn-MPB⊃GP-NTR** for 15 min. The cells were washed 3 times with PBS buffer, and then trypsinized, centrifugated, and resuspended in 1 mL PBS medium, and subjected to flow cytometric analysis. In another experiment, the MCF-7 cells and HepG2 cells were incubated under 0.1% O_2 condition for 6 h. Then, they were treated with 5 µM **GP-NTR** or 5 µM **Zn-MPB⊃GP-NTR** at 37 °C for 15 min and washed 3 times with PBS buffer, and then trypsinized, centrifugated, and resuspended in 1 mL PBS medium, and subjected to flow cytometric analysis.

2.13. Establishment of Mice Model

A Berthold Night Owl LB 983 NC100 system was employed for in vivo imaging. Mice were anesthetized before fluorescence imaging. Then, the mice were placed in the in vivo imaging system for fluorescence imaging after giving injection of **GP-NTR** (100 µL, 100 µM) or **Zn-MPB⊃GP-NTR** (100 µL, 100 µM) in saline (λ_{ex} = 450 nm, λ_{em} = 520 nm).

3. Results and Discussion

*3.1. Preparation and Characterization of **Zn-MPB⊃GP-NTR***

The well-known NTR-sensitive moiety, p-nitrobenzyl, was attached to the hydroxyl part of naphthalimide to construct the substrate, **GP-NTR**. The metal–organic capsule **Zn-MPB** consists of three Zn^{2+} ion and three alternating connected ligands, and each Zn^{2+} ion was combined with two tridentate chelators from two different ligands. The NADH mimics was located on the surface of **Zn-MPB**, and it could make the capsule function as the coenzyme beta-nicotinamide adenine dinucleotide (NADH) that receives and transfers protons and electrons directly from enzymatic processes when the capsule enters the catalytic domain of the enzyme and then provides adequate electrons and protons to reduce the nitro group of **GP-NTR** [36,37]. The cyclic voltammogram experiment of **Zn-MPB** has been performed in CH_3CN (Figure S33), confirming that the redox activity of **Zn-MPB** was enhanced compared to the ligand H_2MPB [29]. It is noteworthy that the morpholine moieties in **Zn-MPB** increase the water solubility and biocompatibility of the supramolecular system, and the positively charged **Zn-MPB** contributes to the accumulation of host–guest complex in the tumor. **GP-NTR** and H_2MPB were easily synthesized and purified by column chromatography. The detailed synthetic rote of **GP-NTR** was laid out in Scheme 2. The synthesis of **Zn-MPB** was reported in the supporting information. The chemical structures of **GP-NTR** and **Zn-MPB** were characterized by ^1H NMR and mass spectrometry.

The assembly of **Zn-MPB⊃GP-NTR** was monitored via ESI–MS, ^1H NMR spectroscopy, ^1H diffusion ordered spectroscopy (^1H DOSY), and ^1H–^1H NOESY spectroscopy. Results from the ESI–MS analysis of **Zn-MPB** in acetonitrile revealed an intense peak at m/z = 942.2663 that was attributed to $[H_2Zn_3(MPB)_3]^{2+}$ species (Figure 1a), showing the integrity of **Zn-MPB** [38]. After adding an equimolar amount of **GP-NTR** to **Zn-MPB**, a $H_2Zn_3(MPB)_3(GP-NTR)]^{2+}$ species appeared at m/z = 1169.8331 (Figure 1b). The establishment of a 1:1 host–guest species **Zn-MPB⊃GP-NTR** in solution was proposed by a comparison of the experimental peaks and those obtained via simulation based on natural isotopic abundances. With the ^1H NMR spectra of the free **GP-NTR**, the free **Zn-MPB**, and the mixture of **GP-NTR** and **Zn-MPB**, the encapsulation of **GP-NTR** in **Zn-MPB** could be signaled by the proton chemical shift changes in peaks for the host and guest molecules with respect to the corresponding signal of the starting material (Figures 1c and S10) [39,40]. The peak related to the protons (H12 and H15) of **Zn-MPB** moved ~0.05 ppm upfield, which indicated the encapsulation of **GP-NTR** into **Zn-MPB** [28,41]. The diffusion-ordered NMR spectroscopy (DOSY) spectra of the 1:1 mixture of **GP-NTR** with **Zn-MPB** measured in DMSO-d_6 indicated that **GP-NTR** and **Zn-MPB** could belong to a single assembly species with a weight average diffusion coefficient of 2.104×10^{-10} m^2 s^{-1} (Figure 1c) [42]. The formation of **Zn-MPB⊃GP-NTR** species was demonstrated by NOESY spectroscopy of the phenyl ring H$_c$ of **GP-NTR** with the H$_{12}$ and H$_{14}$ of **Zn-MPB** (Figure S12). These results suggest that the host–guest system forms steadily. The formation of host–guest complexation species was examined by UV–vis spectra [43,44]. Job's plot between **Zn-MPB** and **GP-NTR** with different molar ratios clearly showed a 1:1 stoichiometry between **Zn-MPB** and **GP-NTR** (Figures 2a and S15). Isothermal titration calorimetry (ITC) measurements were conducted in DMF-H$_2$O solution (DMF/H$_2$O = 98/2, v/v) to further characterize the host–guest combination (Figure 2c). The titration results showed that, upon the addition of **GP-NTR**, **Zn-MPB** exhibited the absolute values of enthalpic and entropic changes of ΔH = 11.16 kJ mol^{-1} and TΔS = 43.26 kJ mol^{-1}, respectively [45,46]. The free energy of binding between **Zn-MPB** and **GP-NTR** was calculated as -7.66 kcal mol^{-1}, and the binding constant (K_a) was calculated as 4.22×10^5 M^{-1}, indicating the favored formation and steady existence of the 1:1 host–guest system [28,41]. As shown in Figure S32, the solubility of the host–guest system was also improved due to the self-assembly of the host and guest, which made **Zn-MPB⊃GP-NTR** more suitable for biological detection.

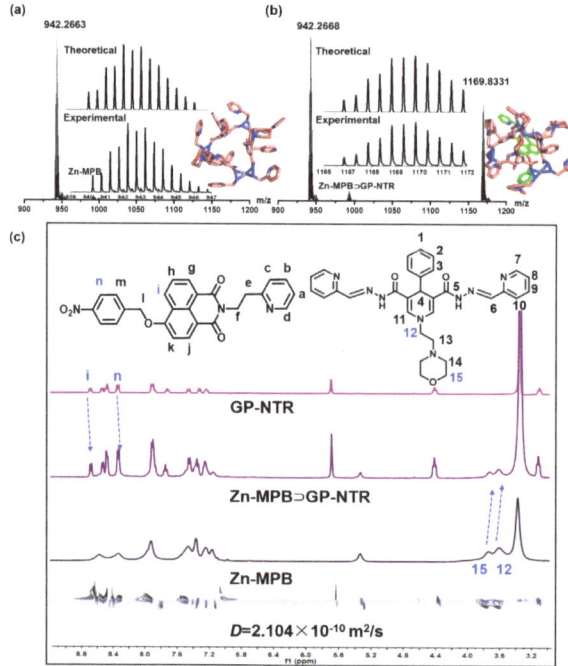

Figure 1. Characterization of the interactions between **Zn–MPB** and **GP-NTR**. ESI–MS spectra of **Zn-MPB** (**a**) and **Zn-MPB** following the addition of 1.0 equiv of **GP-NTR** (**b**) in CH$_3$CN solution. (**c**) ^{1}H DOSY spectra of **Zn-MPB⊃GP-NTR** (DMSO-d_6, 298 K).

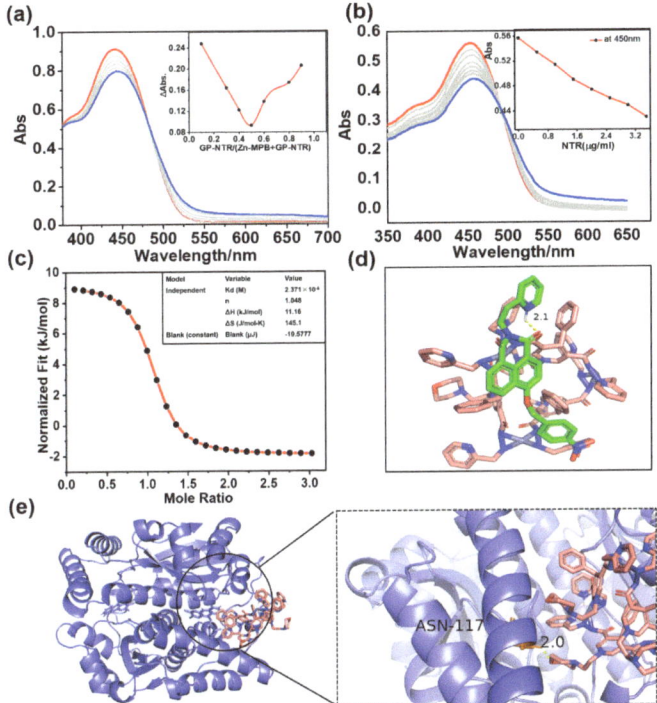

Figure 2. (a) UV titration spectra of **Zn-MPB** (10 μM) with **GP-NTR** (0–20 μM) in DMSO/Tris–HCl solution (red line: the initial titration curve; blue line: the final titration curve). The inset shows a Job's plot curve of **GP-NTR**/**Zn-MPB** mixtures with different molar ratios. (b) UV titration spectra of **Zn-MPB** (10 μM) with NTR (0–3.5 μg/mL) in Tris–HCl solution (red line: the initial titration curve; blue line: the final titration curve). The inset shows the changes at 450 nm. (c) ITC experiments of **Zn-MPB** upon the addition of **GP-NTR** in the DMF/H$_2$O (98:2) solution. (d) Molecular docking results of **GP-NTR** with **Zn-MPB**. (e) Molecular docking results of **Zn-MPB** with NTR.

3.2. Molecular Docking

In molecular design and the simulation of chemical and protein binding processes, molecular docking is a helpful technique for predicting the binding mode of receptors and ligands [47–49]. Small molecule ligands and receptor molecules were docked with a semi-flexible docking method. The docking results demonstrated that a possible hydrogen bond was formed between the N atom of the pyridine ring of **GP-NTR** and the O atom of **Zn-MPB** (labeled in yellow), and the distance of hydrogen bonding was 2.1 Å. The free energy of binding between **Zn-MPB** and **GP-NTR** was about −8.3 kcal mol^{-1}, which indicated the stable binding behavior of **Zn-MPB** with **GP-NTR** (Figure 2d). The free energy of binding between **GP-NTR** and NTR was about −6.47 kcal mol^{-1} (Figure S16). As shown in Figure 2e, the docking results revealed that strong hydrogen bonds were formed between the O atom of the pyridine ring of **Zn-MPB** and the nearby arginine residues (ASN-117) of NTR, which further enabled **Zn-MPB** to enter the hydrophobic protein pocket of NTR and bind to it closely. The binding free energy between **Zn-MPB** and NTR was about −3.34 kcal mol^{-1}. The above results demonstrated that **Zn-MPB** may have good binding ability to NTR.

*3.3. Fluorescence Responses of **Zn−MPB⊃GP-NTR** toward NTR*

The spectroscopic responses of **Zn-MPB⊃GP-NTR** or **GP-NTR** were investigated in vitro to verify the response ability of **Zn-MPB⊃GP-NTR** toward NTR under model conditions (Tris–HCl buffer, pH 6.5, and 25 °C) [50]. Upon the addition of NTR to **Zn-MPB⊃GP-NTR**, there was a noticeable increase in fluorescence intensity at 520 nm within 40 s. The fluorescence intensity and NTR content (0–5 µg/mL) exhibited a good linear relationship, which ensures the quantitative analysis of NTR (Figure 3a,b,e). According to the slope of the curve, the limit of detection (LOD) value of **Zn-MPB⊃GP-NTR** could be obtained by using the formula 3σ/k. And the LOD was calculated to be 6.4 ng/mL (R^2 = 0.9936) (Figure S17). In comparison, traditional probe **GP-NTR** was investigated by varying the amount of NTR added to the mixture of **GP-NTR** and NADH, which displayed similar emission peaks to those of the above-mentioned solution (Figure 3c). However, **GP-NTR** needed over 10 min to reach equilibrium, which was an over 20-fold longer reaction time than that of **Zn-MPB⊃GP-NTR** (Figure 3d). Due to the plots and slopes of the intensity versus NTR content change over time during the slower reaction of **GP-NTR** with NTR, it is difficult for a linear relationship to form between the enhancement in fluorescence intensity and the concentration of enzyme (Figure S18b). For better understanding the reduction progress of **Zn-MPB⊃GP-NTR** to NTR, HPLC analysis was performed (Figure S27). A new chromatographic peak (retention time, TR = 5.13 min/5.15 min) was observed after **GP-NTR** was incubated with NTR for 20 min in the presence of NADH or **Zn-MPB**, which was exactly matched with the resultant **GP-OH** fluorophore. In addition, principal component analysis (PCA) was applied to further distinguish the reaction modes of **GP-NTR** and **Zn-MPB⊃GP-NTR** with NTR. As shown in Figure 3f, there was little overlap between the two sets of data. The projection data of **Zn-MPB⊃GP-NTR** on the third component axis were narrowly clustered between about −0.4 and −0.7, while the data of **GP-NTR** on the third component axis were scattered between −0.3 and 1.2. It was further confirmed that the intensity of the host–guest complex changed little after the reaction equilibrium was obtained rapidly from 30 s to 100 s. The host–guest complex significantly improved NTR detection with a substantially shorter response time than that of the conventional probe, which indicated that the host–guest complex can be applied for the fast detection of NTR [51]. Further investigation exemplified that **GP-NTR** reduction by NTR was significantly influenced by the increased concentration of NADH (Figure S19a), implying that the supply of NADH was necessary for NTR detection. Little difference was seen in the reaction between **Zn-MPB⊃GP-NTR** and NTR with or without the addition of NADH before or after the reaction (Figure S20). It could be seen from the fluorescence comparison experiments that **Zn-MPB⊃GP-NTR** could improve the fluorescence response rate of NTR and shorten the equilibrium time, and the fluorescence intensity showed a linear response to the concentration of NTR. In addition, we tested the photostability of **Zn-MPB⊃GP-NTR** and **Zn-MPB⊃GP-NTR** coexisting with NTR (5 µg/mL) and the influence of temperature change on the efficiency of the host–guest system. The results proved that the photostability of the host–guest system was better, and the influence of temperature on the system was relatively small in 0–40 °C before and after the addition of NTR (Figures S21 and S22). When the host–guest platform was compared with some published NTR probes, the fluorescence response time of **Zn-MPB⊃GP-NTR** was relatively short, and the detection limit was also relatively low (Table S1). These results showed that the host–guest platform **Zn-MPB⊃GP-NTR** could be a suitable turn-on biosensor for the quantitative detection of NTR with a significantly shorter response time.

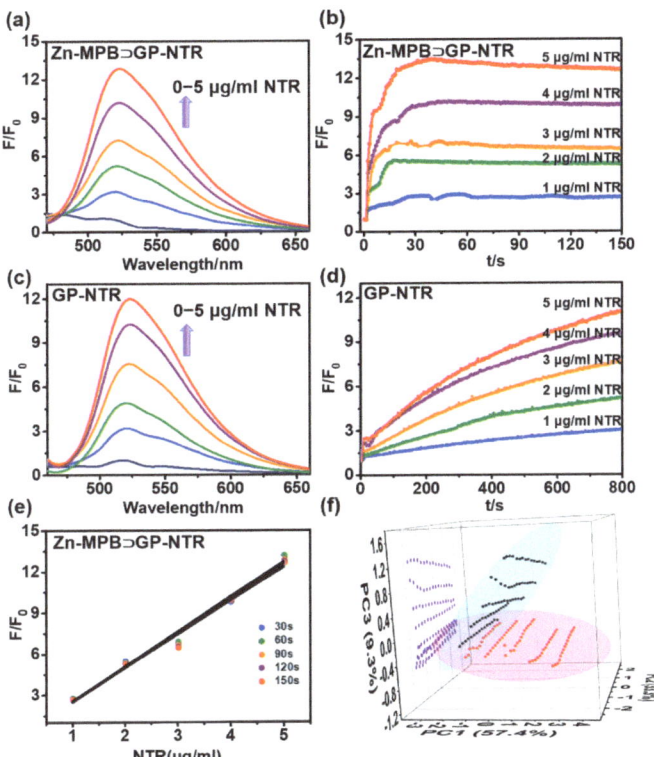

Figure 3. Fluorescence spectra of **GP-NTR** (5 μM) in response to different concentrations of NTR with **Zn-MPB** (5 μM) (**a**) or NADH (15 μM) (**c**). (**b**,**d**) Time-dependent intensity variation in **GP-NTR** (5 μM) with **Zn-MPB** (5 μM) or NADH (15 μM) with different concentrations of NTR (1–5 μg/mL). (**e**) Intensity vs. NTR levels in 5 μM **Zn-MPB⊃GP-NTR** at different times. (**f**) 3D PCA plot of 5 μM **Zn-MPB⊃GP-NTR** (red) and 5 μM **GP-NTR** (black) reaction with 1–5 μg/mL NTR in 30–100 s and their YZ plane projection (purple). 95% confidence ellipses surround each sample cluster. λ_{ex} = 440 nm, λ_{em} = 520 nm.

Encouraged by the excellent fluorescence capability of **Zn-MPB⊃GP-NTR**, the kinetic parameters of the NTR-catalyzed cleavage reaction were also determined (Figures S23 and S24). It is well known that the reduction of nitro group by NADH and NTR is a typical enzyme process involving two substrates [52]. However, the response of **Zn-MPB⊃GP-NTR** to NTR concentration was not affected by NADH, and the pseudo-intramolecular signal communication reduced the complexity of the double-substrate mechanism into a single-substrate free collision process. By using the Michaelis–Menten equation, the kinetic parameter value of $V_{max}^{Zn-MPB⊃GP-NTR}/V_{max}^{GP-NTR}$ and $k_{cat}^{Zn-MPB⊃GP-NTR}/k_{cat}^{GP-NTR}$ is about 28 [32,33]. In comparison to a natural catalytic system, **Zn-MPB⊃GP-NTR** enhanced catalytic efficiency by changing the catalytic kinetics. These above-mentioned results demonstrated that **Zn-MPB⊃GP-NTR** was capable of the rapid detection of NTR with great potential in biological applications.

On account of the complexity of cellular milieux, selectivity experiments were performed via recording the fluorescence spectra changes after treatment with various potential interfering species (NaCl, KCl, L-Tyr, L-Cys, β-Ala, L-Pro, L-Phe, L-Pro, H_2O_2, AA, NQO1, GSR) to verify the specificity of **Zn-MPB⊃GP-NTR** for detecting NTR (Figure S25) [53]. Upon the addition of analytes to the Tris–HCl solution with **GP-NTR** or **Zn-MPB⊃GP-NTR**, only NTR caused an obvious change in fluorescence spectra, whereas the variations with the other reactive species were negligible. It is worth noting that the addition of NTR to the solution containing interfering analytes and **Zn-MPB⊃GP-NTR** quickly produced a fluorescence response identical to that of the **Zn-MPB⊃GP-NTR** solution, suggesting that NTR could be detected effectively even in complex situations. This feature is extremely beneficial for monitoring hypoxic regions in vivo.

*3.4. Imaging NTR in Cancer Cells by **Zn-MPB⊃GP-NTR***

The cytotoxicity of **GP-NTR** and **Zn-MPB** to living cells was evaluated with MCF-7 and HepG2 cells by the standard MTT assay (Figure S28) [54]. It is demonstrated that, after 24 h of incubation at various concentrations, the viability levels of MCF-7 cells and HepG2 cells maintained above 85%, indicating that both **GP-NTR** and **Zn-MPB** were suitable for NTR detection in living cells. No detectable toxicity was observed in the case of MCF-7 cell line and HepG2 cell line. These results suggest that **Zn-MPB⊃GP-NTR** has acceptable toxicological properties and excellent biocompatibility for NTR detection in cancer cells and mice models.

The ability of **Zn-MPB⊃GP-NTR** was evaluated to detect hypoxia by monitoring NTR in MCF-7 and HepG2 cells after incubating under normoxic (20% O_2) and different hypoxic (8% and 0.1% O_2) conditions for 6 h, respectively, followed by treatment with **GP-NTR** or **Zn-MPB⊃GP-NTR** for 1 min under the respective conditions (Figures 4a and S29a) [55]. For the **Zn-MPB⊃GP-NTR** group, negligible fluorescence was observed in MCF-7 cells and HepG2 cells under the normoxic condition. Significantly enhanced fluorescence was observed under hypoxia conditions, and the signal became stronger when the O_2 concentration decreased. In the comparison group, it could be seen that cells incubated with **GP-NTR** had much lower fluorescence intensity than that of the group incubated with **Zn-MPB⊃GP-NTR** at the same oxygen content. These findings revealed that **Zn-MPB⊃GP-NTR** was suitable for the detection of NTR under different hypoxia levels in a short incubation time. It is confirmed that the host–guest system was highly responsive to intracellular NTR changes and could be used to monitor the hypoxia in living cells.

Inspired by the superiority of **Zn-MPB⊃GP-NTR** as a new molecular tool for NTR imaging in living cells, the tests were conducted for MCF-7 and HepG2 cells regarding the capacity of intracellular NTR imaging with various incubation times (Figures 4b and S29b). The cells were incubated in a hypoxic environment (0.1% O_2) for 6 h and then treated with **Zn-MPB⊃GP-NTR** or **GP-NTR** for different times (0–20 min). The MCF-7 and HepG2 cells exhibited no fluorescence when not incubated with **Zn-MPB⊃GP-NTR** or **GP-NTR**. The fluorescence intensity of MCF-7 and HepG2 cells in the red channel became stronger with a longer incubation time. In particular, the **GP-NTR** group took longer to reach equilibrium (approximately 20 min), while the fluorescence intensity of cells in the **Zn-MPB⊃GP-NTR** groups reached equilibrium quickly in 5 min after incubation and exhibited no significant change even after a longer period (until 20 min). In comparison, the fluorescence intensity of MCF-7 and HepG2 cells treated with **GP-NTR** was weaker than the **Zn-MPB⊃GP-NTR** group after reaching fluorescence equilibrium. These results illustrated that **Zn-MPB⊃GP-NTR** could rapidly monitor NTR in cells with the improvement of the double-substrate mechanism to a single-substrate one. Hence, **Zn-MPB⊃GP-NTR** was applicable for the rapid and effective detection of NTR in hypoxic cancer cells.

Flow cytometry analysis has been regarded as a high-throughput assay technique that is frequently employed for the analysis of various samples [56,57]. Flow cytometry analysis was preferred to verify the fluorescence imaging results of the host−guest complex. The fluorescence intensity of MCF-7 cells treated with **Zn-MPB⊃GP-NTR** under hypoxic conditions (0.1% O_2) for 6 h was significantly higher than that of the control group (Figure 5c). Then, a comparative experiment between the **Zn-MPB⊃GP-NTR** group and the **GP-NTR** group was performed under hypoxic conditions (0.1% O_2). As shown in Figures 5d and S31, the fluorescence intensity of the **Zn-MPB⊃GP-NTR** group and the **GP-NTR** group treated with 0.1% O_2 for 6 h in MCF-7 and HepG2 cells was significantly increased than that of the control group, while the fluorescence intensity of **Zn-MPB⊃GP-NTR** group was slightly higher. These results indicate that the host–guest complex **Zn-MPB⊃GP-NTR** could achieve the good detection of NTR with fast imaging. The flow cytometry results were consistent with cell confocal imaging results, demonstrating that **Zn-MPB⊃GP-NTR** could be applied to monitoring NTR in hypoxia cancer cells.

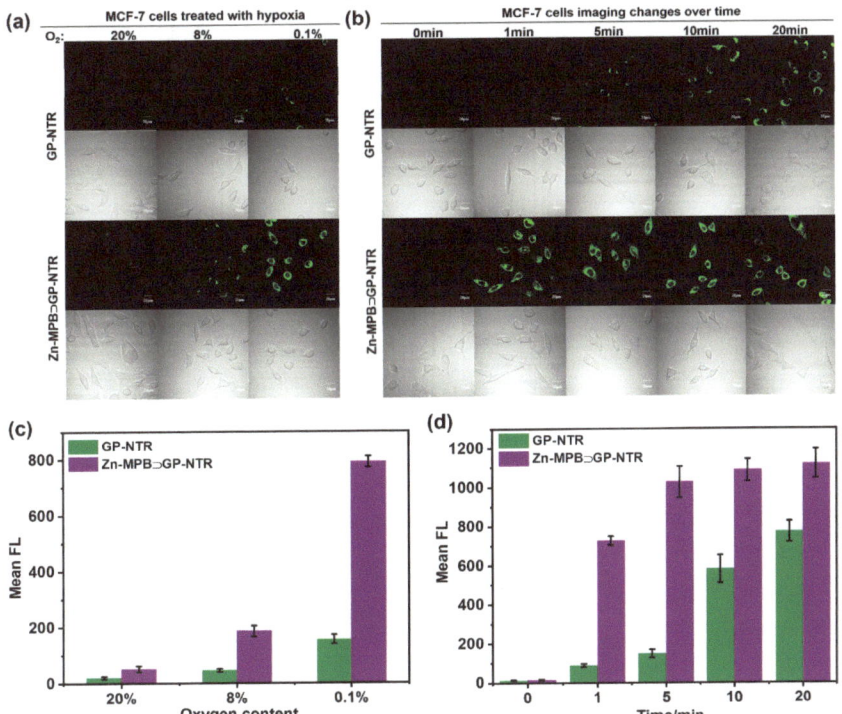

Figure 4. (**a**) Imaging of MCF-7 cells under different conditions. Cells were treated with 2 μM **GP-NTR** or 2 μM **Zn-MPB⊃GP-NTR** for 1 min after 6 h incubation under normoxic (20% O_2) and different hypoxic (8% and 0.1% O_2) conditions. (**b**) Imaging of MCF-7 cells under 0.1% O_2 condition incubated with 2 μM **GP-NTR** or 2 μM **Zn-MPB⊃GP-NTR** for different times (0, 1, 5, 10, 20 min). Scale bar: 20 μm. (**c**) Relative fluorescence intensity of the corresponding images of MCF-7 cells after being incubated with 2 μM **GP-NTR** or 2 μM **Zn-MPB⊃GP-NTR** under different oxygen contents for 1 min. (**d**) Relative fluorescence intensity of the corresponding images of MCF-7 cells at varied time points after being incubated with 2 μM **GP-NTR** or 2 μM **Zn-MPB⊃GP-NTR** under hypoxia conditions. The results were presented as mean ± SE with replicates, n = 3.

3.5. In Vivo Fluorescence Imaging of NTR

Based on the good selectivity and rapid response to NTR, the ability of **Zn-MPB⊃GP-NTR** for the in vivo real-time tracking imaging of NTR was studied. In consideration of the adaptability of **Zn-MPB⊃GP-NTR**, MCF-7 tumor-bearing model was used to trace the NTR catalytic kinetics in vivo due to the NTR overexpression in hypoxic tumors. The mice injected with **GP-NTR** were selected as the control group. As shown in Figure 5a,b, the mice injected with **GP-NTR** or **Zn-MPB⊃GP-NTR** showed no obvious background fluorescence. Mice treated with **GP-NTR** showed a small fluorescence signal initially, and this signal increased with the time. By contrast, after the intratumoral injection of **Zn-MPB⊃GP-NTR** was administered to mice, a certain fluorescence response appeared in a relatively short time and showed no obvious alterations over a longer incubation time. These findings revealed that the fluorescence signal of **Zn-MPB⊃GP-NTR** in the tumor area was strong, and the rapid reaction of **Zn-MPB⊃GP-NTR** with NTR occurred in the tumor site. Taken together, the interaction of **Zn-MPB⊃GP-NTR** with NTR could lead to the rapid detection of NTR, making it appropriate for the in vivo imaging of NTR in hypoxic tumors.

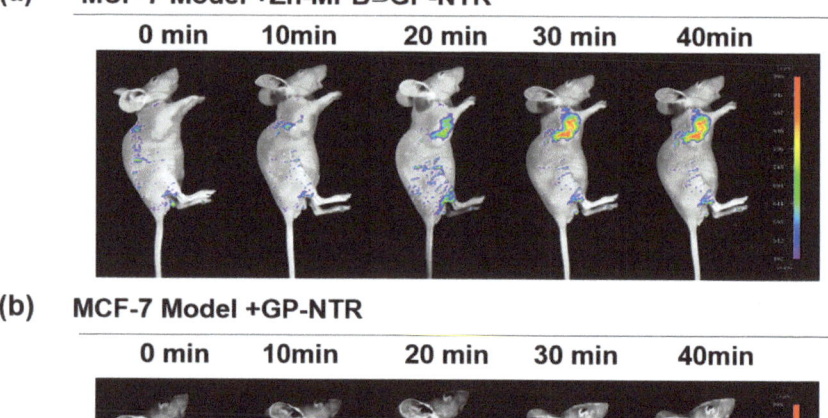

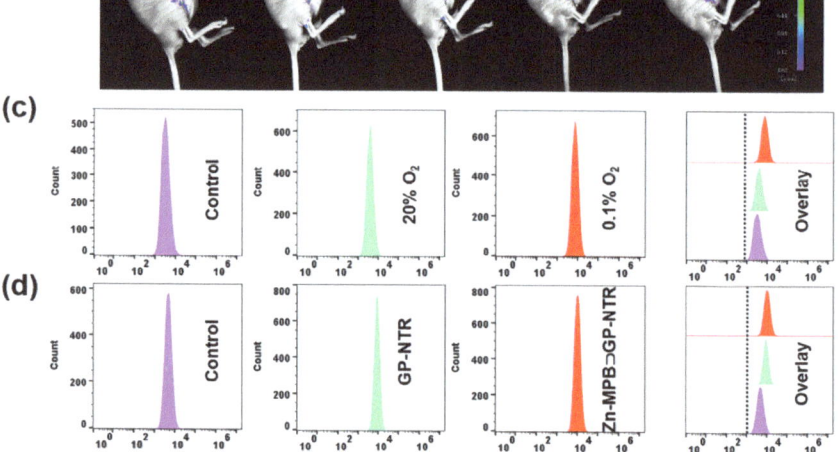

Figure 5. Real-time in vivo fluorescence imaging of hypoxia in MCF-7 tumor-bearing mice after an injection of **Zn-MPB⊃GP-NTR** (**a**) or **GP-NTR** (**b**). (**c**) Flow cytometry analyses of MCF-7 cells treated with 5 μM **Zn-MPB⊃GP-NTR** under hypoxic or normoxic conditions. (**d**) Flow cytometry analysis of MCF-7 cells treated with 5 μM **GP-NTR** or 5 μM **Zn-MPB⊃GP-NTR** under hypoxic conditions.

4. Conclusions

In summary, a host–guest complex, **Zn-MPB⊃GP-NTR**, was developed for quantitative, sensitive, and rapid bio-tracking of NTR. **Zn-MPB⊃GP-NTR** was encapsulated by the inclusion of a fluorescent substrate **GP-NTR** in a NADH mimic-containing metal–organic capsule **Zn-MPB** to improve its responsiveness to NTR. The resulting host–guest platform made the detection of NTR independent of NADH, while the response process was switched from the original complex double-substrate process to a single-substrate one.

It not only demonstrated a fast response toward NTR with a low detection limit, but also could been employed for the quantitative detection of NTR. Confocal microscopy, flow cytometry analysis, and tumor imaging in the tumor-bearing mouse model demonstrated that **Zn-MPB⊃GP-NTR** showed an excellent detection performance for NTR. The host–guest platform described here might present significant opportunities for the development of molecular tools for in vivo hypoxia monitoring and could be used for the early detection of cancers associated with hypoxia.

Supplementary Materials: The following supporting information can be downloaded at: https://www.mdpi.com/article/10.3390/chemosensors12080145/s1, Figure S1: ^{1}H NMR spectrum of compound **2** (400 MHz, 298 K, DMSO-d_6); Figure S2: ^{1}H NMR spectrum of **GP-OH** (400 MHz, 298 K, DMSO-d_6); Figure S3: ^{1}H NMR spectrum of **GP-NTR** (400 MHz, 298 K, CDCl$_3$/CD$_3$OD (v/v = 6/1)); Figure S4: ^{13}C NMR spectrum of **GP-NTR** (400 MHz, 298 K, CDCl$_3$/CD$_3$OD (v/v = 6/1)); Figure S5: ^{1}H NMR spectrum of **H$_2$MPB** (400 MHz, 298 K, DMSO-d_6); Figure S6: ^{1}H NMR spectrum of **Zn-MPB** (400 MHz, 298 K, DMSO-d_6); Figure S7: HR–MS spectrum of **GP-OH**; Figure S8: HR–MS spectrum of **GP-NTR**; Figure S9: HR–MS spectrum of **H$_2$MPB**; Figure S10: (a) ^{1}H NMR spectra of the free **GP-NTR**, **Zn-MPB**, and an equimolar mixture of **Zn-MPB** and **GP-NTR** in DMSO-d_6 solution. (b) Uphill of the protons of subtract **GP-NTR** (H$_i$, H$_d$, H$_g$, H$_j$, and H$_n$). (c) Uphill of the protons of the molecular square (H$_{12}$ and H$_{15}$); Figure S11: ^{1}H COSY spectroscopy of intermolecular H-H interaction of **GP-NTR** with equimolar **Zn-MPB** (v/v, 1/1) in a DMSO-d_6 solution; Figure S12: ^{1}H NOESY spectroscopy of intermolecular H-H interaction of **GP-NTR** with equimolar **Zn-MPB** (v/v, 1/1) in a DMSO-d_6 solution; Figure S13: (a) UV titration experiments of **Zn-MPB** with **GP-NTR** in DMSO/Tris–HCl solution. (b) UV titration experiments of **Zn-MPB** with NTR in Tris–HCl solution; Figure S14: (a) Time-dependent UV tests for the stability of **GP-NTR** (5 µM) in Tris–HCl. (b) Time-dependent UV tests for the stability of **Zn-MPB⊃GP-NTR** (5 µM) in Tris–HCl; Figure S15: (a) UV–vis difference spectrometry of guest inclusion in DMSO/Tris–HCl solution. (b) Job-plot of changes in the absorbance of host–guest complex at 370 nm in DMSO/Tris–HCl solution; Figure S16: Theoretical docking study optimized model of **GP-NTR** with NTR; Figure S17: The Linear fitting curve of the fluorescence intensity of **Zn-MPB⊃GP-NTR** at 520 nm versus the concentration of NTR from 0 to 5 µg/mL; Figure S18: Intensity vs. NTR concentration after reacting with (a) **Zn-MPB⊃GP-NTR** (5 µM) or (b) **GP-NTR** (5 µM) and NADH (15 µM) for different times; Figure S19: Solvent kinetics tests of **Zn-MPB⊃GP-NTR**. (a) NTR (1 µg/mL) reacted with **GP-NTR** (5 µM) at various concentrations of NADH (1 and 15 µM). (b) NTR (1 µg/mL) reacted with **Zn-MPB⊃GP-NTR** (5 µM) at various concentrations of NADH (1, 15, or 30 µM); Figure S20: The reaction kinetics between **Zn-MPB⊃GP-NTR** (5 µM) and NTR (1 µg/mL) with/without the addition of NADH (15 µM) before or after the reaction; Figure S21: The photostability of **Zn-MPB⊃GP-NTR** and **Zn-MPB⊃GP-NTR** coexisting with NTR (5 µg/mL), λ_{ex} = 440 nm, λ_{em} = 520 nm; Figure S22: The fluorescence intensity of **Zn-MPB⊃GP-NTR** and **Zn-MPB⊃GP-NTR** coexisting with NTR (5 µg/mL) toward different temperatures, λ_{ex} = 440 nm, λ_{em} = 520 nm; Figure S23: (a) Kinetics of NTR (1 µg/mL) with various concentrations of **Zn-MPB⊃GP-NTR**. (b) Kinetics of NTR (1 µg/mL) with NADH (15 µM) and various concentrations of **GP-NTR**; Figure S24: Plot of 1/v against 1/S according to the Michaelis–Menten equation, where the black line was derived from **Zn-MPB⊃GP-NTR** (0-5 µM) reacting with NTR (1 µg/mL), and the blue and red lines were obtained from the kinetic curves of NTR reacting with various concentrations of **GP-NTR** in 1 or 15 µM NADH; Figure S25: Fluorescence response of 5 µM **GP-NTR** (with 15 µM NADH, black) or **Zn-MPB⊃GP-NTR** (blue) in the presence of different analytes for 10 min, until **GP-NTR** groups reached equilibrium. Then, adding NTR (5 µg/mL) to the **Zn-MPB⊃GP-NTR** (5 µM) in the presence of various interfering analyte mixture solutions, data recorded equilibrium in seconds (red); Figure S26: Fitting results of lifetime for **Zn-MPB⊃GP-NTR** in DMSO; Figure S27: HPLC analysis to confirm the reaction mechanism of **GP-NTR** towards NTR detection; Figure S28: Cell viability of MCF-7 cells treated with various concentrations of **GP-NTR** (a) and **Zn-MPB** (b) (from 0 to 20 µM) for 24 h. Cell viability of HepG2 cells treated with various concentrations of **GP-NTR** (c) and **Zn-MPB** (d) (from 0 to 20 µM) for 24 h; Figure S29: (a) Imaging of HepG2 cells under different conditions. (b) Imaging of HepG2 cells under 0.1% O$_2$ condition incubated with 2 µM **GP-NTR** or 2 µM **Zn-MPB⊃GP-NTR** for different times (0, 1, 5, 10, 20 min); Figure S30: (a) Relative fluorescence intensity of the corresponding images of HepG2 cells after being incubated with 2 µM **GP-NTR** or 2 µM **Zn-MPB⊃GP-NTR** under different oxygen contents for 1 min. (b) Relative fluorescence intensity of the corresponding images of HepG2 cells at varied time points after being incubated with 2 μM **GP-NTR** or 2 μM **Zn-MPB⊃GP-NTR** under hypoxia conditions; Figure S31: Flow cytometry analysis of HepG2 cells treated with 5 μM **GP-NTR** or 5 μM **Zn-MPB⊃GP-NTR** under hypoxic conditions; Figure S32: The corresponding image of **GR-NTR**, **Zn-MPB⊃GP-NTR**, and **Zn-MPB** in water; Figure S33: Cyclic voltammogram of **Zn-MPB** (0.1 mM) and H$_2$MPB (0.2 mM) in CH$_3$CN containing TBAPF$_6$ (0.1 M). Table S1: Comparison of fluorescent probes for NTR detection. References [33,48,58,59] are cited in the Supplementary Materials.

Author Contributions: Conceptualization, W.S., Y.J. and C.D.; validation, W.S., Y.J., M.G., Y.G. and L.Z.; investigation, W.S., X.J., M.G. and Y.Z.; resources, X.J., Y.Z. and Y.G.; data curation, W.S. and X.J.; formal analysis, W.S., X.J. and L.Z.; writing—original draft preparation, W.S.; writing—review and editing, Y.J.; visualization, Y.G., Y.Z. and L.Z.; supervision, Y.J. and C.D.; project administration, C.D.; funding acquisition, Y.J. All authors have read and agreed to the published version of the manuscript.

Funding: This research was supported by the National Natural Science Foundation of China (No. 21977015) and the Fundamental Research Funds for the Central Universities (No. DUT23YG208).

Institutional Review Board Statement: Not applicable.

Informed Consent Statement: Not applicable.

Data Availability Statement: The original contributions presented in the study are included in the article/Supplementary Material, and further inquiries can be directed to the corresponding author.

Conflicts of Interest: The authors declare no conflicts of interest.

References

1. Janczy-Cempa, E.; Mazuryk, O.; Kania, A.; Brindell, M. Significance of Specific Oxidoreductases in the Design of Hypoxia-Activated Prodrugs and Fluorescent Turn off–on Probes for Hypoxia Imaging. *Cancers* **2022**, *14*, 2686–2711. [CrossRef] [PubMed]
2. Multhoff, G.; Radons, J.; Vaupel, P. Critical Role of Aberrant Angiogenesis in the Development of Tumor Hypoxia and Associated Radioresistance. *Cancers* **2014**, *6*, 813–828. [CrossRef] [PubMed]
3. Kheshtchin, N.; Hadjati, J. Targeting hypoxia and hypoxia-inducible factor-1 in the tumor microenvironment for optimal cancer immunotherapy. *J. Cell. Physiol.* **2022**, *237*, 1285–1298. [CrossRef] [PubMed]
4. Wilson, W.R.; Hay, M.P. Targeting hypoxia in cancer therapy. *Nat. Rev. Cancer* **2011**, *11*, 393–410. [CrossRef] [PubMed]
5. Lee, P.; Chandel, N.S.; Simon, M.C. Cellular adaptation to hypoxia through hypoxia inducible factors and beyond. *Nat. Rev. Mol. Cell Biol.* **2020**, *21*, 268–283. [CrossRef] [PubMed]
6. Sidhu, J.S.; Kaur, N.; Singh, N. Trends in small organic fluorescent scaffolds for detection of oxidoreductase. *Biosens. Bioelectron.* **2021**, *191*, 113441. [CrossRef] [PubMed]
7. Dias, G.G.; King, A.; de Moliner, F.; Vendrell, M.; da Silva Júnior, E.N. Quinone-based fluorophores for imaging biological processes. *Chem. Soc. Rev.* **2018**, *47*, 12–27. [CrossRef] [PubMed]
8. Pitsawong, W.; Hoben, J.P.; Miller, A.F. Understanding the Broad Substrate Repertoire of Nitroreductase Based on Its Kinetic Mechanism. *J. Biol. Chem.* **2014**, *289*, 15203–15214. [CrossRef]
9. Liu, Y.F.; Li, J.Y.; Huang, H.J.; Shu, Y. A fluorescent probe for imaging nitroreductase with signal amplification in high-viscosity environments. *J. Mater. Chem. B* **2023**, *11*, 9509–9515. [CrossRef]
10. Li, H.D.; Kim, D.Y.; Yao, Q.C.; Ge, H.Y.; Chung, J.W.; Fan, J.L.; Wang, J.Y.; Peng, X.J.; Yoon, J. Activity-Based NIR Enzyme Fluorescent Probes for the Diagnosis of Tumors and Image-Guided Surgery. *Angew. Chem. Int. Ed.* **2021**, *60*, 17268–17289. [CrossRef]
11. Qi, Y.L.; Guo, L.; Chen, L.L.; Li, H.; Yang, Y.S.; Jiang, A.Q.; Zhu, H.L. Recent progress in the design principles, sensing mechanisms, and applications of small-molecule probes for nitroreductases. *Coord. Chem. Rev.* **2020**, *421*, 213460. [CrossRef]
12. Yang, Q.; Wang, S.; Li, D.; Yuan, J.; Xu, J.; Shao, S. A mitochondria-targeting nitroreductase fluorescent probe with large Stokes shift and long-wavelength emission for imaging hypoxic status in tumor cells. *Anal. Chim. Acta* **2020**, *1103*, 202–211. [CrossRef] [PubMed]
13. Jin, Y.B.; Hu, D.H.; Yin, J.M.; Sun, K.S.; Chen, L.J.; Liu, S.J.; Li, F.Y.; Zhao, Q. An iridium complex-based probe for phosphorescent lifetime-elongated imaging of nitroreductase in living cells. *Sens. Actuators B Chem.* **2024**, *401*, 134960. [CrossRef]
14. Xu, Y.M.; Hu, B.; Cui, Y.J.; Li, L.; Nian, F.; Zhang, Z.X.; Wang, W.T. A highly selective ratio-metric fluorescent sensor for visualizing nitroreductase in hypoxic cells. *Chem. Commun.* **2024**, *60*, 83–86. [CrossRef] [PubMed]
15. Brennecke, B.; Wang, Q.; Zhang, Q.; Hu, H.Y.; Nazare, M. An Activatable Lanthanide Luminescent Probe for Time-Gated Detection of Nitroreductase in Live Bacteria. *Angew. Chem. Int. Ed.* **2020**, *59*, 8512–8516. [CrossRef] [PubMed]
16. Sarkar, S.; Lee, H.; Ryu, H.G.; Singha, S.; Lee, Y.M.; Reo, Y.J.; Jun, Y.W.; Kim, K.H.; Kim, W.J.; Ahn, K.H. A Study on Hypoxia Susceptibility of Organ Tissues by Fluorescence Imaging with a Ratiometric Nitroreductase Probe. *ACS Sens.* **2021**, *6*, 148–155. [CrossRef]

17. Li, T.; Gu, Q.S.; Chao, J.J.; Liu, T.; Mao, G.J.; Li, Y.F.; Li, C.Y. An intestinal-targeting near-infrared probe for imaging nitroreductase in inflammatory bowel disease. *Sens. Actuators B Chem.* **2024**, *403*, 135181. [CrossRef]
18. Li, M.R.; Zhang, Y.; Ren, X.J.; Niu, W.C.; Yuan, Q.; Cao, K.; Zhang, J.C.; Gao, X.Y.; Su, D.D. Activatable fluorogenic probe for accurate imaging of ulcerative colitis hypoxia in vivo. *Chem. Commun.* **2022**, *58*, 819–822. [CrossRef] [PubMed]
19. Zhang, J.J.; Chai, X.H.; He, X.P.; Kim, H.J.; Yoon, J.; Tian, H. Fluorogenic probes for disease-relevant enzymes. *Chem. Soc. Rev.* **2019**, *48*, 683–722. [CrossRef]
20. Meng, T.J.; Ma, W.B.; Fan, M.Y.; Tang, W.; Duan, X.R. Enhancing the Contrast of Tumor Imaging for Image-Guided Surgery Using a Tumor-Targeting Probiotic with the Continuous Expression of a Biomarker. *Anal. Chem.* **2022**, *94*, 10109–10117. [CrossRef]
21. Fu, Y.X.; Guo, W.Y.; Wang, N.; Dai, Y.J.; Zhang, Z.Y.; Sun, X.L.; Yang, W.C.; Yang, G.F. Diagnosis of Bacterial Plant Diseases via a Nitroreductase-Activated Fluorescent Sensor. *Anal. Chem.* **2022**, *94*, 17692–17699. [CrossRef] [PubMed]
22. Kaur, A.; New, E.J. Bioinspired Small-Molecule Tools for the Imaging of Redox Biology. *Acc. Chem. Res.* **2019**, *52*, 623–632. [CrossRef] [PubMed]
23. Chen, W.W.; Freinkman, E.; Wang, T.; Birsoy, K.; Sabatini, D.M. Absolute quantification of matrix metabolites reveals the dynamics of mitochondrial metabolism. *Cell* **2016**, *166*, 1324–1337. [CrossRef] [PubMed]
24. Dai, D.H.; Yang, J.; Yang, Y.W. Supramolecular Assemblies with Aggregation-Induced Emission Properties for Sensing and Detection. *Chem. Eur. J.* **2022**, *28*, e202103185. [CrossRef] [PubMed]
25. Wu, Q.; Lei, Q.; Zhong, H.C.; Ren, T.B.; Sun, Y.; Zhang, X.B.; Yuan, L. Fluorophore-based host–guest assembly complexes for imaging and therapy. *Chem. Commun.* **2023**, *59*, 3024–3039. [CrossRef]
26. Tarzia, A.; Jelfs, K.E. Unlocking the computational design of metal–organic cages. *Chem. Commun.* **2022**, *58*, 3717–3730. [CrossRef] [PubMed]
27. Ghosh, A.; Slappendel, L.; Nguyen, B.T.; von Krbek, L.K.S.; Ronson, T.K.; Castilla, A.M.; Nitschke, J.R. Light-Powered Reversible Guest Release and Uptake from Zn_4L_4 Capsules. *J. Am. Chem. Soc.* **2023**, *145*, 3828–3832. [CrossRef]
28. Zhang, L.; Jiao, Y.; Yang, H.; Jia, X.C.; Li, H.Y.; He, C.; Si, W.; Duan, C.Y. Supramolecular Host–Guest Strategy for the Accelerating Detection of Nitroreductase. *ACS Appl. Mater. Interfaces* **2023**, *15*, 21198–21209. [CrossRef]
29. Zhao, L.; Cai, J.K.; Li, Y.N.; Wei, J.W.; Duan, C.Y. A host–guest approach to combining enzymatic and artificial catalysis for catalyzing biomimetic monooxygenation. *Nat. Commun.* **2020**, *11*, 2903. [CrossRef]
30. Wei, J.W.; Zhao, L.; Zhang, Y.; Han, G.; He, C.; Wang, C.; Duan, C.Y. Enzyme Grafting with a Cofactor-Decorated Metal–Organic Capsule for Solar-to-Chemical Conversion. *J. Am. Chem. Soc.* **2023**, *145*, 6719–6729. [CrossRef]
31. Zang, S.P.; Shu, W.; Shen, T.J.; Gao, C.C.; Tian, Y.; Jing, J.; Zhang, X.L. Palladium-triggered ratiometric probe reveals CO's cytoprotective effects in mitochondria. *Dye. Pigment.* **2020**, *173*, 107861. [CrossRef]
32. Crofts, T.S.; Sontha, P.; King, A.O.; Wang, B.; Biddy, B.A.; Zanolli, N.; Gaumnitz, J.; Dantas, G. Discovery and characterization of a nitroreductase capable of conferring bacterial resistance to chloramphenicol. *Cell Chem. Biol.* **2019**, *26*, 559–570. [CrossRef] [PubMed]
33. Bhakta, S.; Nayek, A.; Roy, B.; Dey, A. Induction of enzyme-like peroxidase activity in an iron porphyrin complex using second sphere interactions. *Inorg. Chem.* **2019**, *58*, 2954–2964. [CrossRef] [PubMed]
34. Cai, J.K.; Zhao, L.; Li, Y.N.; He, C.; Wang, C.; Duan, C.Y. Binding of Dual-Function Hybridized Metal–Organic Capsules to Enzymes for Cascade Catalysis. *JACS Au* **2022**, *2*, 1736–1746. [CrossRef] [PubMed]
35. Liu, S.Y.; Wang, H.L.; Nie, G. Ultrasensitive Fibroblast Activation Protein-α-Activated Fluorogenic Probe Enables Selective Imaging and Killing of Melanoma In Vivo. *ACS Sens.* **2022**, *7*, 1837–1846. [CrossRef]
36. Fang, Y.; Powell, J.A.; Li, E.; Wang, Q.; Perry, Z.; Kirchon, A.; Yang, X.; Xiao, Z.; Zhu, C.; Zhang, L.; et al. Catalytic reactions within the cavity of coordination cages. *Chem. Soc. Rev.* **2019**, *48*, 4707–4730. [CrossRef] [PubMed]
37. Yang, Y.; Jing, X.; Shi, Y.P.; Wu, Y.C.; Duan, C.Y. Modifying Enzymatic Substrate Binding within a Metal–Organic Capsule for Supramolecular Catalysis. *J. Am. Chem. Soc.* **2023**, *145*, 10136–10148. [CrossRef]
38. Zhang, M.M.; Saha, M.L.; Wang, M.; Zhou, Z.X.; Song, B.; Lu, C.J.; Yan, X.Z.; Li, X.P.; Huang, F.H.; Yin, S.C.; et al. Multicomponent Platinum(II) Cages with Tunable Emission and Amino Acid Sensing. *J. Am. Chem. Soc.* **2017**, *139*, 5067–5074. [CrossRef] [PubMed]
39. Wang, L.J.; Bai, S.; Han, Y.F. Water-Soluble Self-Assembled Cage with Triangular Metal–Metal-Bonded Units Enabling the Sequential Selective Separation of Alkanes and Isomeric Molecules. *J. Am. Chem. Soc.* **2022**, *144*, 16191–16198. [CrossRef]
40. Mei, Y.X.; Zhang, Q.W.; Gu, Q.Y.; Liu, Z.C.; He, X.; Tian, Y. Pillar[5]arene-Based Fluorescent Sensor Array for Biosensing of Intracellular Multi-neurotransmitters through Host–Guest Recognitions. *J. Am. Chem. Soc.* **2022**, *144*, 2351–2359. [CrossRef]
41. Li, Q.; Zhou, J.; Sun, J.F.; Yang, J. Host–guest interactions of a twisted cucurbit[15]uril with paraquat derivatives and bispyridinium salts. *Tetrahedron Lett.* **2019**, *60*, 151022. [CrossRef]
42. Sobiech, T.A.; Zhong, Y.L.; Miller, D.P.; McGrath, J.K.; Scalzo, C.T.; Redington, M.C.; Zurek, E.; Gong, B. Ultra-Tight Host-Guest Binding with Exceptionally Strong Positive Cooperativity. *Angew. Chem. Int. Ed.* **2022**, *61*, e202213467. [CrossRef]
43. Bobylev, E.O.; Poole, D.A.; Bruin, B.; Reek, J.N.H. M_6L_{12} Nanospheres with Multiple C70 Binding Sites for $^1O_2$ Formation in Organic and Aqueous Media. *J. Am. Chem. Soc.* **2022**, *144*, 15633–15642. [CrossRef] [PubMed]
44. Yao, S.Y.; Yue, Y.X.; Ying, A.K.; Hu, X.Y.; Li, H.B.; Cai, K.; Guo, D.S. An Antitumor Dual-Responsive Host-Guest Supramolecular Polymer Based on Hypoxia-Cleavable Azocalix[4]arene. *Angew. Chem. Int. Ed.* **2023**, *62*, e202213578. [CrossRef] [PubMed]
45. Demers, J.; Mittermaier, A. Binding Mechanism of an SH_3 Domain Studied by NMR and ITC. *J. Am. Chem. Soc.* **2009**, *131*, 4355–4367. [CrossRef] [PubMed]

46. Altmann, P.J.; Pöthig, A. Pillarplexes: A Metal–Organic Class of Supramolecular Hosts. *J. Am. Chem. Soc.* **2016**, *138*, 13171–13174. [CrossRef] [PubMed]
47. Zhu, M.; Liu, R.R.; Zhai, H.L.; Meng, Y.J.; Han, L.; Ren, C.L. The binding mechanism of nitroreductase fluorescent probe: Active pocket deformation and intramolecular hydrogen bonds. *Int. J. Biol. Macromol.* **2020**, *150*, 509–518. [CrossRef] [PubMed]
48. Qi, Y.L.; Wang, H.R.; Kang, Q.J.; Chen, L.L.; Qi, P.F.; He, Z.X.; Yang, Y.S.; Zhu, H.L. A versatile fluorescent probe for simultaneously detecting viscosity, polarity and nitroreductases and its application in bioimaging. *Sens. Actuators B Chem.* **2022**, *352*, 130989. [CrossRef]
49. Chen, S.J.; Ma, X.D.; Wang, L.; Wu, Y.Y.; Wang, Y.P.; Hou, S.C.; Fan, W.K. Construction of an intelligent fluorescent probe that can accurately track β-galactosidase activity in fruits and living organisms. *Sens. Actuators B Chem.* **2023**, *387*, 133787. [CrossRef]
50. Guo, H.W.; Yang, K.P.; Fan, X.P.; Chen, M.; Ke, G.L.; Ren, T.B.; Yuan, L.; Zhang, X.B. Designing a brightness-restored rhodamine derivative by the ortho-compensation effect for assessing drug-induced acute kidney injury. *Anal. Chem.* **2023**, *95*, 6863–6870. [CrossRef]
51. Tang, Z.X.; Yan, Z.; Gong, L.L.; Zhang, L.; Yin, X.M.; Sun, J.; Wu, K.; Yang, W.J.; Fan, G.W.; Li, Y.L.; et al. Precise Monitoring and Assessing Treatment Response of Sepsis-Induced Acute Lung Hypoxia with a Nitroreductase-Activated Golgi-Targetable Fluorescent Probe. *Anal. Chem.* **2022**, *94*, 14778–14784. [CrossRef] [PubMed]
52. Race, P.R.; Lovering, A.L.; Green, R.M.; Ossor, A.; White, S.A.; Searle, P.F.; Wrighton, C.J.; Hyde, E.I. Structural and Mechanistic Studies of Escherichia coli Nitroreductase with the Antibiotic Nitrofurazone. *J. Biol. Chem.* **2005**, *280*, 13256–13264. [CrossRef] [PubMed]
53. Yadav, A.K.; Zhao, Z.X.; Weng, Y.R.; Gardner, S.H.; Brady, C.J.; Peguero, O.D.P.; Chan, J. Hydrolysis-Resistant Ester-Based Linkers for Development of Activity-Based NIR Bioluminescence Probes. *J. Am. Chem. Soc.* **2023**, *145*, 1460–1469. [CrossRef]
54. Du, W.; Wang, J.Q.; Fang, H.X.; Ji, W.H.; Liu, Y.; Qu, Y.W.; Zhang, D.T.; Shao, T.; Hou, X.Y.; Wu, Q.; et al. Mitochondria-specific two-photon fluorogenic probe for simultaneously visualizing nitroreductase and viscosity in cancer cells. *Sens. Actuators B Chem.* **2022**, *370*, 132456. [CrossRef]
55. Chen, S.Z.; Xiao, L.; Li, Y.; Qiu, M.S.; Yuan, Y.P.; Zhou, R.; Li, C.G.; Zhang, L.; Jiang, Z.X.; Liu, M.L.; et al. In Vivo Nitroreductase Imaging via Fluorescence and Chemical Shift Dependent ^{19}F NMR. *Angew. Chem. Int. Ed.* **2022**, *61*, e202213495. [CrossRef]
56. Zwicker, V.E.; Oliveira, B.L.; Yeo, J.H.; Fraser, S.T.; Bernardes, G.J.L.; New, E.J.; Jolliffe, K.A. A Fluorogenic Probe for Cell Surface Phosphatidylserine Using an Intramolecular Indicator Displacement Sensing Mechanism. *Angew. Chem. Int. Ed.* **2019**, *58*, 3087–3091. [CrossRef]
57. Wang, Q.Y.; Li, Z.; Hao, Y.T.; Zhang, Y.; Zhang, C.X. Near-Infrared Fluorescence Probe with a New Recognition Moiety for Specific Detection and Imaging of Aldehyde Dehydrogenase Expecting the Identification and Isolation of Cancer Stem Cells. *Anal. Chem.* **2022**, *94*, 17328–17333. [CrossRef]
58. Jiao, Y.; Zhang, L.; Gao, X.; Si, W.; Duan, C.Y. Cofactor-substrate-based Reporter for Enhancing Signaling Communications towards Hypoxia Enzyme Expression. *Angew. Chem. Int. Ed.* **2020**, *59*, 6021–6027. [CrossRef]
59. Zhou, J.; Fang, S.J.; Li, J.; Du, W.; Wu, Q. A novel pyrimidine-based two-photon fluorogenic probe for rapidly visualizing nitroreductase activity in hypoxic cancer cells and in vivo. *Sens. Actuators B Chem.* **2023**, *390*, 134015. [CrossRef]

Disclaimer/Publisher's Note: The statements, opinions and data contained in all publications are solely those of the individual author(s) and contributor(s) and not of MDPI and/or the editor(s). MDPI and/or the editor(s) disclaim responsibility for any injury to people or property resulting from any ideas, methods, instructions or products referred to in the content.

Article

The Efficient and Sensitive Detection of Serum Dopamine Based on a MOF-199/Ag@Au Composite SERS Sensing Structure

Yuyu Peng, Chunyan Wang *, Gen Li, Jianguo Cui, Yina Jiang, Xiwang Li, Zhengjie Wang and Xiaofeng Zhou

College of Pharmacy and Bioengineering, Chongqing University of Technology, Chongqing 400054, China; pengyuyu@stu.cqut.edu.cn (Y.P.); ligen1990@cqut.edu.cn (G.L.); cui_jianguo@cqut.edu.cn (J.C.); ynjiang@stu.cqut.edu.cn (Y.J.); 52231013127@stu.cqut.edu.cn (X.L.); zjwang@stu.cqut.edu.cn (Z.W.); xiaofengzhou@stu.cqut.edu.cn (X.Z.)
* Correspondence: cywang@cqut.edu.cn

Abstract: In this study, a MOF-199/Ag@Au SERS sensing structure was successfully synthesized by combining metal–organic frameworks (MOFs) with surface-enhanced Raman scattering (SERS) technology for the efficient detection of dopamine (DA), a biomarker for neurological diseases, in serum. Using electrochemical methods, a copper-based MOF (MOF-199) was synthesized in situ on copper substrates and further deposited with silver nanoparticles (AgNPs). Subsequently, gold nanoshells were encapsulated around these silver cores by in situ chemical deposition. This preparation process is simple, controllable, and inexpensive. Furthermore, a novel Azo reaction-based DA SERS method was proposed to detect 1 pM DA, which represents an improvement in sensitivity by two orders of magnitude compared to previous unlabeled SERS detection methods and by four orders of magnitude compared to another SERS approach proposed in this work. There was an excellent linear relationship (R^2 = 0.976) between the SERS signal at 1140 cm^{-1} and the DA concentration (0.001 M~1 pM). The results indicate that the MOF-199/Ag@Au sensor structure can successfully achieve both the qualitative and quantitative detection of DA in serum, thus providing a robust technical basis for the application of SERS technology in the field of clinical neurological disease screening.

Keywords: dopamine; surface-enhanced Raman spectroscopy (SERS); Azo reaction; MOF-199; neurological disease diagnosis

1. Introduction

Neurotransmitters are a series of messenger molecules in the body that transmit information and maintain the balance of physiological functions [1]. They play important roles in emotion regulation and behavior, learning and cognitive activities, homeostasis in the body, and individual motor control and coordination [2–4]. Dopamine (DA) is one of the most common of these neurotransmitters, regulating physiological processes in the nervous, hormonal, and cardiovascular systems [5]. The deficiency of DA leads to clinical symptoms such as Parkinson's disease, depression, and schizophrenia [6–10].

Clinically, the main methods of DA detection include fluorescence, colorimetric, and electrochemical methods [11–14]. Fluorescence and colorimetric methods require complex sample pretreatment and are not highly sensitive [15,16]; electrochemical methods, although sensitive, are difficult to accurately quantify DA in complex biological samples such as serum and urine due to the similarity of the electrode potentials of ascorbic acid, uric acid and DA [17,18]. Therefore, a rapid and simple analytical method with good specificity and high sensitivity for DA is urgently needed [19].

Surface-enhanced Raman scattering (SERS), as a powerful surface-sensitive analytical tool, shows great potential for highly specific and sensitive analyses in the fields of food safety, environmental protection, and biological research [20–24]. Gold and silver nanoparticles (Au and Ag NPs) are widely used as SERS substrates due to their remarkable SERS enhancement [25,26]. Efficient SERS substrates need to satisfy two basic requirements: excellent adsorption properties to adsorb molecules onto the surface of metal nanoparticles and the formation of abundant Raman hotspots between neighboring nanoparticles [27–29]. By preparing hybrid materials such as silicon nanowire-coated silver (SiNWs@Ag), cerium oxide-coated titanium dioxide (CeO_2@TiO_2) [30], and silver nanoparticles modified graphene oxide (AgNPs@GO) [31], these SERS substrates exhibit enhanced performance due to their increased surface area and superior adsorption properties, facilitating efficient target molecule capture. Additionally, the synergistic interactions at the interfaces of different materials enhance the local electromagnetic fields, resulting in increased Raman signal enhancement and the formation of abundant hotspots, thereby improving the sensitivity and efficacy of SERS applications.

Wang et al. [32] reported a study on the selective label-free SERS detection of DA using graphene–gold nanopyramidal heterostructures, further expanding the potential of SERS technology in the analytical field. Although the method performed well in terms of a detection limit of about 10^{-10} M, it was not suitable for the specific detection of DA. Next, Jin's group [33] proposed a surface acoustic wave-based SERS active sensing technique for silver nanoparticle clustering to achieve the rapid, label-free detection of DA; however, its sensitivity to the operating power may lead to nanoparticle sintering and nanogap loss, thus limiting the reproducibility and reliability of the method [34]. Jiang et al. [35] employed silver nanoparticles incorporated into a metal–organic framework Material of Institute Lavoisier-101 (MIL-101) for the SERS detection of DA. This composite structure demonstrated excellent sensitivity and selectivity toward DA. However, the complex synthesis process and the instability of the nanosilver colloids used in the preparation present challenges to the method's simplicity and stability.

The selective detection of low concentrations of DA in complex biological samples is challenging, and utilizing probe molecules that can selectively recognize DA [36] offers a feasible solution [36]. Zhang et al. [37] captured DA by adsorption of 3,3′-dithiodipropionic acid di(N-hydroxysuccinimide ester) (DSP) on a gold thin film and introduced 3-mercapto phenylboronic acid (3-MPBA) and functionalized AgNPs to form a plasma 'hotspots' to enhance the SERS signals, but the sparse hotspots limited the detection of low concentrations of DA. To solve this problem, Lu et al. [38] used mercaptopropionic acid (MPA) functionalized nanoporous silver film (AgNF) and modified silver nanocubes (AgNCs) with 4-mercapto phenylboronic acid (4-MPBA) as a Raman reporter, which enhances the SERS signal through specific response and improves the detection sensitivity. However, this method is susceptible to complex interferences such as glucose in real serum samples, and it still has specific detection difficulties.

In conclusion, this paper addresses the challenges encountered in the SERS detection of serum DA by emphasizing the innovative design and optimization of SERS sensing structures. The design and efficient fabrication of a copper-based MOF (MOF-199)/silver–gold (Ag@Au) composite SERS sensing structure, enables the integration of the high porosity, excellent adsorption properties, and large specific surface area of MOF-199 with the potent SERS enhancement effect of Ag@Au NPs. This integration not only enhances the capture efficiency of DA molecules but also intensifies the SERS signal strength, thereby achieving a dual improvement in detection sensitivity and stability. On this foundation, this paper further introduces a method for the detection of DA by SERS, which is based on the Azo reaction. The Azo reaction entails the coupling of aromatic amines with nitrite ions under alkaline conditions, resulting in the formation of Azo compounds. These compounds exhibit strong and distinctive Raman signals due to their large Raman scattering cross-sections. This reaction specifically labels DA molecules, enhancing their detection by providing robust and characteristic Raman-active sites, thereby significantly increasing both the sensitivity and specificity of SERS. The improved sensing performance is attributed to the efficient amplification of the SERS signal and the reduction in interference from non-target molecules, which is crucial for detecting DA in complex biological samples. Consequently, this research not only possesses significant scientific value but also offers novel technical support for the diagnosis and treatment of DA-related diseases.

2. Materials and Methods

2.1. Materials and Apparatus

Anhydrous ethanol (C_2H_6O, 99.7%) was purchased from Shanghai Titan Technology company (Shanghai, China). Homotrimellitic acid ($C_9H_6O_6$), ammonium perchlorate (NH_4ClO_4), and silver nitrate ($AgNO_3$, 99.7%) were obtained from Aladdin (Shanghai, China).Sodium hydroxide (NaOH) and ammonia ($NH_3 \cdot H_2O$, 25%) were purchased from Chongqing Chemical Reagent company (Chongqing, China). Sodium nitrite ($NaNO_2$), sodium carbonate ($NaCO_3$), 4-aminothiophenol (PATP), p-aminothiophenol (C_6H_7NS), cystine ($C_6HN_2O_4S_2$), cysteine ($C_3H_7NO_2S$), alanine ($C_3H_7NO_2$), ascorbic acid ($C_6H_8O_6$), lysine ($C_6H_{14}N_2O_2$), uric acid ($C_5H_4N_4O_3$), aspartic acid ($C_4H_4NO_4$), glucose ($C_6H_{12}O_6$), chlortetracycline ($AuCl_3 \cdot HCl \cdot 4H_2O$), were purchased from Shanghai Titan Technology company (Shanghai, China). Copper flakes (Cu) was purchased from Hefei Shushan District, special nuclear metal materials merchant company (Hefei, China). All aqueous solutions were prepared using ultrapure water (18.2 MΩ-cm) extracted from a Milli-Q water purification system.

The deposition of AgNPs on the surface of a copper sheet was conducted via the chronoamperometric method, employing a CHI600E electrochemical workstation (CHI600E CH Shanghai, China). Raman detection was conducted using an ATR8300 micro-Raman spectrometer (ATR8300 Optosky Xiamen, China). The excitation laser was 785 nm. Morphological and compositional analyses were conducted using a field-emission scanning electron microscope with an energy-dispersive spectroscopy (EDS) system (FE-SEM/EDS, JEOL JSM-7800F, Akishima, Japan).

2.2. Preparation of MOF-199

The electrochemical preparation of MOF-199 was successfully carried out in accordance with the existing literature [39]. The solvent was composed of 40 mL of deionized water and 120 mL of anhydrous ethanol, and the electrolyte was formulated by dissolving 840.6 mg of homobenzoic acid and 2735.2 mg of ammonium perchlorate. The cleaned copper sheet was employed as the anode of the electrochemical workstation, the saturated calomel electrode (SCE) was utilized as the reference electrode, and the platinum sheet was utilized as the auxiliary electrode to construct a three-electrode system and introduce the electrolyte. The electrochemical deposition was conducted via cyclic voltammetry, with the scanning voltage set between +1 V and −1 V, a scanning speed of 0.08 V/s, a resting time of 100 s, and a sensitivity of 0.01 A/V. This process yielded the desired MOF-199 product.

2.3. Preparation of MOF-199/Ag

Subsequently, MOF-199/Ag was successfully prepared via a two-step timed-current method. A solution of 170 mg of silver nitrate in 10 mL of deionized water was prepared, yielding a concentration of 0.010 M. Subsequently, 0.001% ammonia was added until the resulting brown precipitate was no longer visible, thereby producing the silver–ammonia electrolyte. The chronoamperometric method was initiated, the nucleation time was set to 9 s, the nucleation voltage was set to −1.6 volts, the growth time was set to 360 s, the growth voltage was set to −0.9 volts, and the sensitivity was set to 0.01 amperes per volt to ultimately yield MOF-199/Ag.

2.4. Preparation of MOF-199/Ag@Au

The MOF-199/Ag@Au was prepared by chemical deposition, and the MOF-199/Ag substrate was immersed in a 0.2 mM chloroauric acid solution for a period of 6 min in order to prepare the MOF-199/Ag composite substrate. Once the reaction was complete, the resulting MOF-199/Ag@Au was rinsed with deionized water, blown dry, and set aside for the final preparation of MOF-199/Ag@Au. The aforementioned reaction involves immersing the MOF-199/Ag substrate in a 0.2 mM chloroauric acid solution (HAuCl$_4$) for 6 min. This process is a chemical deposition reaction, whereby gold ions (Au^{3+}) in the chloroauric acid solution are reduced and deposited onto the surface of the MOF-199/Ag composite. This process was performed to create a core–shell structure, whereby a thin layer of Au was deposited on the surface of the Ag NPs. The deposition of Au onto Ag is crucial because it enhances the stability of the silver nanoparticles by protecting them from oxidation and corrosion. Furthermore, the gold coating enhances the overall performance of the SERS sensing structure. This is due to the excellent plasmonic properties of gold, which contribute to stronger electromagnetic fields at the surface, thereby improving the Raman signals. The combination of silver's high enhancement factor and gold's stability resulted in an optimized composite material, MOF-199/Ag@Au, which exhibited both high sensitivity and improved durability for detecting molecules such as DA in complex biological samples.

2.5. Testing of MOF-199/Ag@Au NPs

The unspiked and DA-spiked serum samples were combined with a specific volume (1:4) of methanol and then subjected to centrifugation at 6000 rpm for 15 min. Subsequently, 1 mL of the supernatant was removed and placed in an ice water bath. To the supernatant, 100 μL of NaNO$_2$ solution (50 g/L), 100 μL of PATP solution (1.00 mmol/L), and 100 μL of Na$_2$CO$_3$ solution (100 g/L) were added, and the derivatization was carried out for 5 min to obtain the derivatized product, Azo. The MOF-199/Ag@Au substrate, prepared as previously described, was then placed on a confocal Raman spectrometer under a 20× objective lens at 785 nm. The SERS patterns of the obtained Azo compound and the reactant DA were acquired under the following conditions: 785 nm excitation light, 300 mW excitation light power, 5 s integration time, and 1 accumulation number under the objective lens.

3. Results

3.1. Design of Composite SERS Sensing Structure

This paper presents an innovative proposal for the combination of MOF-199 with core–shell structured Ag@Au NPs to construct a MOF-199/Ag@Au composite SERS sensing structure. The objective is to construct an efficient, sensitive, and specific detection structure for DA in serum. The proposal is based on the integration of materials science, nanotechnology, and chemical analysis methods. This design employs the three-dimensional porous structure of MOF-199, which exhibits high porosity and a large specific surface area, to provide abundant adsorption sites for DA molecules and facilitate their effective enrichment at the sensing interface [40]. Moreover, it functions as a carrier to support a greater number of Ag@Au NPs, thereby forming denser SERS-active hotspots. Concurrently, as the principal functional unit, the core–shell configuration of Ag@Au NPs not only preserves the elevated SERS enhancement capacity of silver but also markedly enhances the structural durability and antioxidant attributes through the gold coating. The electromagnetic coupling effect between Ag and Au serves to amplify the SERS signal of DA, as evidenced by the comparative analysis of SERS spectra measured before and after gold deposition. As illustrated in Figure 1, the experimental outcomes substantiate a considerable augmentation in the SERS signal subsequent to gold deposition, thereby corroborating the pivotal role of Au in enhancing the detection sensitivity. The combination of electromagnetic and chemical enhancement effects ensures high sensitivity in detection. Furthermore, this paper introduces the Azo reaction as a specific recognition method, which precisely and efficiently identifies DA molecules through its specific labeling and the large Raman scattering cross-section of the reaction product. This effectively addresses the issues of non-specific adsorption and signal interference that may exist in traditional SERS detection.

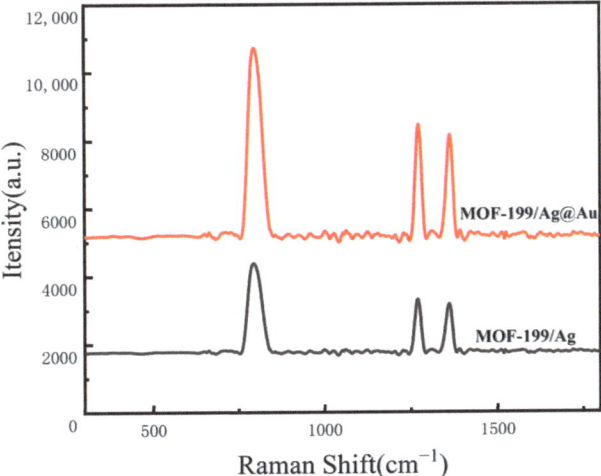

Figure 1. SERS spectra of 10^{-3} M DA measured on the surfaces of MOF-199/Ag@AuNPs, MOF-199/Ag.

3.2. Characterization Results and Analysis of Composite SERS Sensing Structure

In order to investigate the influence of the surface morphology, size, and material type of the composite sensing structure on the sensing effect, the surface morphology and size of the electrochemically prepared MOF-199 were first characterized by SEM, and the results are shown in Figure 2a. It can be observed that the prepared MOF-199 has an ellipsoidal distribution with an average size of 1~2 μm. To ensure the accuracy of the particle size statistics, only nanoparticles facing the lens were selected for measurement for our analyses in Figure 2b,c, while those particles with obvious tilts or incomplete displays were excluded from the statistics. This selection minimizes the effect of viewing angle bias on the particle size measurement results, thus ensuring that the obtained particle size data are more reliable and representative. Thus, the particle size analysis of the MOF-199/Ag composites in Figure 2b shows that the average particle size of Ag nanoparticles is 208 nm, whereas in Figure 2c, the average particle size of MOF-199/Ag@Au nanoparticles increases to 233 nm after Au deposition. This increase in size confirms the successful deposition of gold shells on Ag nanoparticles, which not only improves the structural stability of the composites and the oxidation resistance but also enhances its SERS activity. The formation of dendritic structures is a result of the combination of electrodeposition conditions and crystal growth kinetics [41–44].

The dendritic structure resembles sequoia branches with a length of about 1.5 μml the diameters of the main trunk and the branches are about 500 nm; and the growth angle of the side branches relative to the main trunk is 60 degrees, which suggests that the silver dendritic crystals have a tendency to grow preferentially in specific directions [45]. The decrease in surface free energy drives the directional attachment, which promotes the transition from polycrystalline to single-crystalline structures [46,47]. In addition, the growth time significantly affects the crystallinity of the crystals, with a shorter growth time leading to an imperfect polycrystalline structure at the tip, while a longer growth time contributes to further growth and improved crystallinity [48–50]. Taken together, specific electrodeposition conditions and thermodynamic driving forces contributed to the formation of dendritic silver nanostructures. The dendritic structure has a significant role in enhancing the surface plasmon resonance [51–53]. Figure 2d shows the EDS elemental mapping of the composite nanostructure, revealing the coexistence of elements such as Cu, O, Ag, and Au, which indicates the successful preparation of the MOF-199/Ag@Au composite nanostructure.

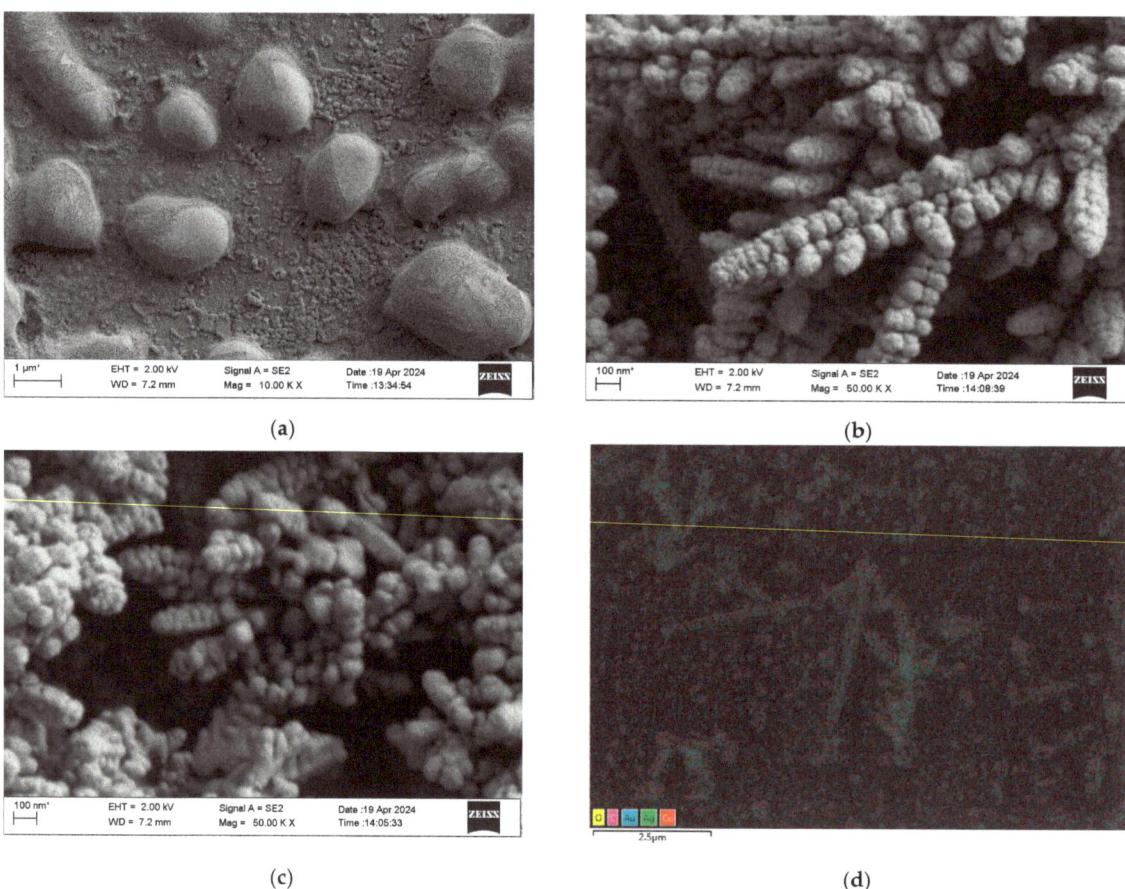

Figure 2. (**a**) SEM characterization of MOF-199; (**b**) SEM characterization of MOF-199/Ag; (**c**) SEM characterization of MOF-199/Ag@Au; (**d**) EDS elemental mapping of MOF-199/Ag@Au. The asterisk (*) on the scale bar in the SEM image is a standard notation of the SEM system and does not typically affect the accuracy of the measurements.

3.3. Improvement in and Analysis of DA SERS Detection Efficiency by MOF-199

Conventional noble metal SERS sensing structures have been observed to exhibit deficiencies in sensitivity, signal attenuation, and selectivity for DA SERS detection [54]. MOF-199 has been demonstrated to effectively address these shortcomings due to its large specific surface area, excellent pore structure, and strong optical enhancement properties. To substantiate the beneficial impact of MOF-199 in DA SERS detection, Ag@Au and MOF-199/Ag@Au were selected as sensing structures in this study. The SERS test was conducted using a consistent DA concentration as the test samples under identical test conditions, and the outcomes are illustrated in Figure 3a. Significant enhancement of the SERS signals was observed, with the appearance of Raman peaks at 780 cm^{-1}, 1265 cm^{-1}, 1330 cm^{-1}, and so forth. These peaks corresponded to the benzene ring skeleton vibration (780 cm^{-1}), the C-O bending vibration (1265 cm^{-1}) [55], and the C-OH vibration (1330 cm^{-1}) of DA, respectively [56]. The composite nanostructures integrated with MOF-199 are more effective in enhancing the SERS signals of DA, which can be attributed to the high specific surface area of MOF-199, which provides more active sites and allows for the loading of more noble metal nanoparticles [57]. The combined effects of electromagnetic and chemical enhancement enable the detection of DA by SERS with superior efficacy.

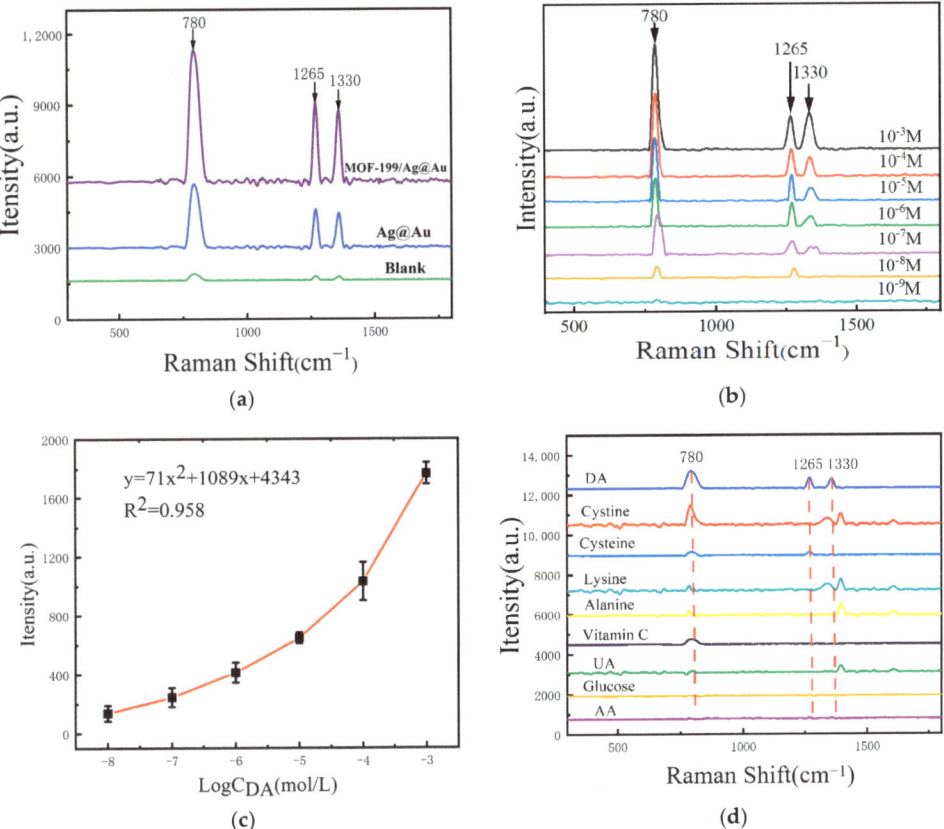

Figure 3. (a) SERS spectra of 10^{-3} M DA measured on the surfaces of MOF-199/Ag@AuNPs, Ag@AuNPs, and glass slides without any enhancement material; (b) SERS spectra of DA solutions at different concentrations measured on the surface of MOF-199/Ag@AuNPs; (c) the relationship curve between the intensity of the Raman peak at 1330 cm^{-1} and the concentration of DA; (d) the SERS spectra measured by adsorbing nine sample solutions of the same concentration on the surface of MOF-199/Ag@Au nanocomposites.

3.4. Evaluation of Composite SERS Sensing Structures for the Label-Free SERS Detection of DA

Label-free detection has the advantages of not needing to introduce additional markers, easy operation, simplified sample processing, etc. Detection sensitivity, specificity, and assay reproducibility are three commonly used SERS assay evaluation indexes. In order to further evaluate the sensitivity of the composite SERS sensing structures designed and prepared in this study for the label-free SERS detection of DA, several composite SERS sensing structures were prepared under the same optimization conditions, and the same volume of DA solutions with concentrations from 10^{-3} mol/L to 10^{-9} mol/L were added onto the surfaces. The SERS signals obtained under the same test conditions are shown in Figure 3b. It can be seen that the SERS signal intensity gradually decreased with the decrease in the DA concentration, and the detection limit was 10^{-8} mol/L. Figure 3c demonstrates the fitting curve between the SERS signal intensity and the logarithm of the DA concentration at 1330 cm^{-1}, with a correlation coefficient of R^2 of 0.958, which indicates that the composite SERS sensing structure designed and prepared in this study can perform the label-free SERS detection, but the detection sensitivity is low and the linearity is poor.

To further demonstrate the specificity of the composite sensing structure for DA SERS detection, several substances commonly found in serum, such as cystine, cysteine, lysine, alanine, ascorbic acid, uric acid, glucose, and aspartic acid, were selected and subjected to SERS testing. The results are shown in Figure 3d. It can be observed that some Raman characteristic peaks of these substances exhibit Raman shifts that are extremely close to those of DA. When actual serum samples were tested, it was difficult to identify the specific signals of DA from the composite SERS signals. This indicates that the unlabeled SERS detection method for DA suffers from poor detection specificity.

3.5. Principles and Effect Evaluation of DA SERS Detection Based on the Azo Reaction

The previous results found that the DA-unlabeled SERS assay suffers from low detection sensitivity, as well as poor linear correlation and specificity. In order to solve this problem, this study proposes an Azo reaction-based SERS method for the detection of DA in serum, which is schematically shown in Figure 4.

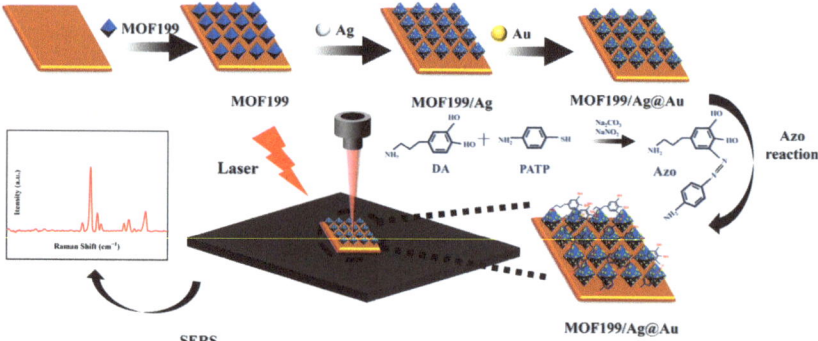

Figure 4. The schematic diagram of DA SERS detection based on the Azo reaction.

The principle of the Azo reaction is through the process of aromatic amines reacting with nitrite under alkaline conditions to produce Azo compounds. Specifically, 100 μL of $NaNO_2$ solution (50 g/L) and 100 μL of PATP solution (1.00 mmol/L) were added to 100 μL of DA solution in the presence of 100 μL of a Na_2CO_3 solution (100 g/L). In this study, $NaNO_2$ reacts with the amino group (-NH_2) in PATP to produce a diazonium salt intermediate (-N_2^+). Subsequently, this diazonium salt intermediate reacts with the phenolic hydroxyl group (-OH) on the benzene ring of the DA molecule via a nucleophilic coupling reaction of its diazonium group (-N_2^+) to form a C-N bond and produce the Azo compound Azo (-N=N-), a product that contains an Azo group directly linked to the DA benzene ring in the molecular structure. Through the analysis of SERS signals of Azo reaction products, DA was indirectly identified qualitatively and quantitatively. Additionally, due to the large Raman scattering cross-section of Azo reaction products, DA was identified with higher sensitivity. To further evaluate the detection effect of the Azo reaction-based SERS assay for DA, the feasibility of the method for DA SERS detection was first verified. Under the same testing conditions, using the same SERS sensing structure, SERS tests were performed on the same concentration of DA, $NaNO_2$, Na_2CO_3, and PATP involved in the Azo reaction, as well as the reaction's product Azo, and the results are shown in Figure 5a. It was found that the Azo reaction product had the same Raman characteristic peak as DA at the Raman shift of 1265 cm^{-1}. In addition, new Raman peaks appeared at the Raman shifts of 1140 cm^{-1}, 1385 cm^{-1}, and 1434 cm^{-1}, respectively. This indicates that SERS detection based on the Azo reaction can be used to identify the presence of DA.

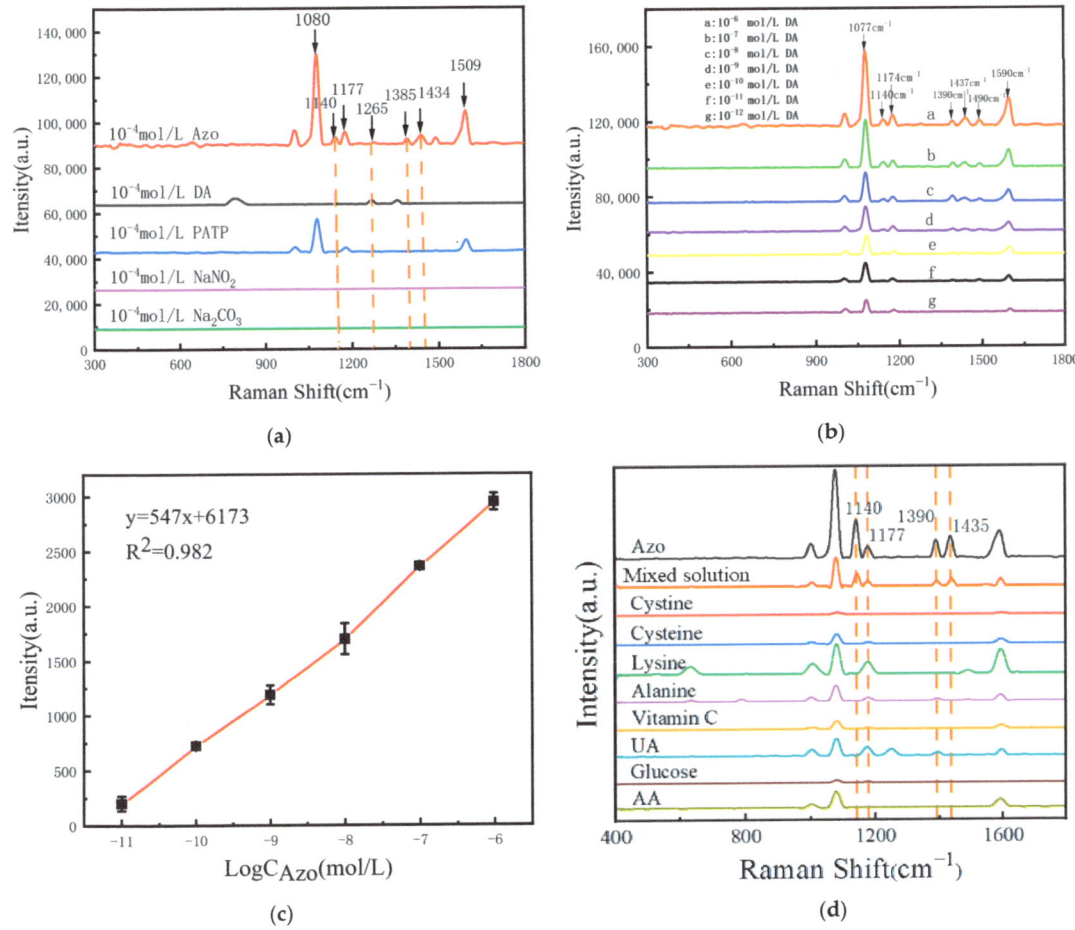

Figure 5. (a) SERS spectra of Azo compounds, DA, PATP based on MOF-199/Ag@Au; (b) the SERS spectra of the individual products resulting from the Azo reactions between different concentrations of DA and the same reactant; (c) the relationship curve between the intensity of the Raman peak at 1140 cm^{-1} and the concentration of DA; (d) The SERS spectra measured by adsorbing nine sample solutions of the same concentration and their mixtures on the surface of MOF-199/Ag@Au nanocomposites.

In this study, equal volumes of DA at varying concentrations were reacted with PATP, NaNO$_2$, and Na$_2$CO$_3$ for the same duration. The reaction products were then tested for SERS, and the results are presented in Figure 5b. It was observed that the intensity of the SERS signal of the Azo reaction products decreased with the decrease in the DA concentration. Furthermore, the detection limit of DA could be as low as 10^{-13} mol/L, which was five orders of magnitude higher than that of the unlabeled detection method. Furthermore, the relationship curve between the intensity of the Raman peak at 1140 cm^{-1} Raman shift and the logarithm of the DA concentration was plotted (Figure 5c), with a linear correlation coefficient of $R^2 = 0.982$, indicating a strong linear correlation between the two variables. Furthermore, the specificity of the Azo reaction-based SERS assay for DA must be evaluated.

The control substances cystine, cysteine, lysine, alanine, ascorbic acid, uric acid, glucose, and aspartic acid were selected and reacted with PATP, NaNO$_2$, and Na$_2$CO$_3$ under identical conditions. The resulting reaction products were then subjected to SERS testing, and the results are presented in Figure 5d. In contrast to the results observed with DA, the products generated from the reaction of these control substances with PATP, $NaNO_2$, and Na_2CO_3 did not demonstrate notable changes or an enhancement of the characteristic peaks in the SERS spectra. In contrast, the Azo compounds generated by the reaction of DA with PATP exhibited distinctive Raman peaks at 1140 cm^{-1}, 1177 cm^{-1}, 1390 cm^{-1}, and 1435 cm^{-1}. In comparison to the reaction products of the other eight interfering substances, the characteristic peaks of the DA reaction products exhibited a notable distinction, indicating that this SERS detection method based on the Azo reaction has high specificity for DA and can effectively distinguish DA from other interfering substances.

3.6. Performance Evaluation of Azo Reaction-Based DA SERS Detection in Serum

To verify the practicality of the composite sensing structure and the Azo reaction-based DA SERS detection method in serum, we chose fetal bovine serum as the solvent, configured a 10^{-3} M~10^{-13} M DA serum solution, carried out the Azo reaction under the same conditions, and performed the SERS test on their reaction products. The results are shown in Figure 6a, which shows that the detection limit for DA can be as low as 10^{-12} mol/L. The SERS signal intensity of the Azo reaction products decreased with the decrease in the DA concentration in serum.

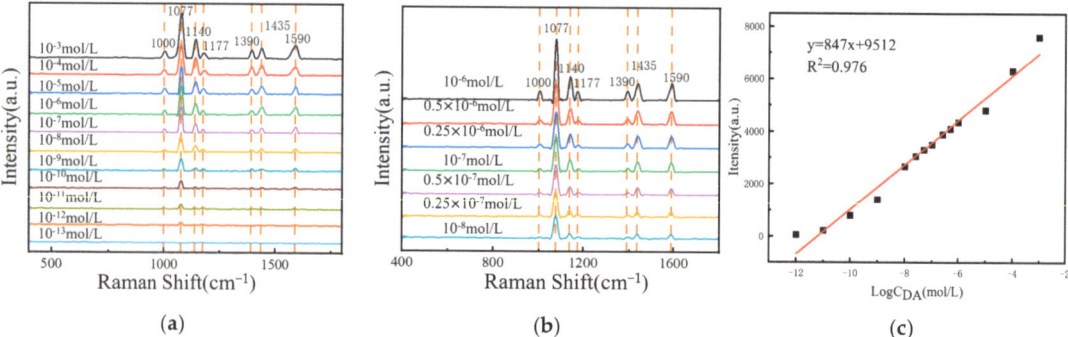

Figure 6. (a) The SERS spectra of Azo reaction products of DA serum solutions at concentrations ranging from 10^{-3} M to 10^{-13} M were measured on the surface of MOF-199/Ag@Au composite structure; (b) the SERS spectra of Azo reaction products of DA serum solutions at concentrations ranging from 10^{-6} M to 10^{-8} M were measured on the surface of MOF-199/Ag@Au composite structure; (c) the relationship curve between Raman signal intensity at 1140 cm^{-1} and the logarithm of the DA concentration.

In actual serum, the concentration range of DA in normal human serum is typically 10^{-6} M~10^{-7} M, while the DA content in patients with neurological disorders is usually 10^{-7} M~10^{-8} M [58]. To further evaluate the practical quantitative detection performance, fetal bovine serum was also selected as the solvent, and DA solutions with concentrations of 10^{-6} M, 0.5×10^{-6} M, 0.25×10^{-6} M, 10^{-7} M, 0.5×10^{-7} M, 0.25×10^{-7} M, and 10^{-8} M were prepared. The Azo reactions were carried out under the same conditions, and the SERS signals of their Azo reaction products were tested. The results are shown in Figure 6b. It was found that the SERS signal intensity of the Azo reaction products decreased with the decrease in the DA concentration in serum.

As illustrated in Figure 6c, the SERS signal intensities at 1140 cm^{-1} of DA serum solutions in the concentration ranges of 10^{-3} M~10^{-12} M and 10^{-6} M~10^{-8} M were plotted against the corresponding logarithms of the concentrations on the same standard curve. The results demonstrate a strong positive correlation between the SERS signal intensity and the DA concentration in both the high concentration range of 10^{-3} M~10^{-12} M and the practical application concentration range of 10^{-6} M~10^{-8} M, with $R^2 = 0.976$. This indicates that the composite SERS sensing membrane and the Azo reaction-based SERS DA detection method proposed in this paper have excellent linear response over a wide range of concentrations, thereby demonstrating the potential of this method for quantitative detection in practical applications and highlighting the reliability of serum as a screening tool for neurological diseases.

A comprehensive comparison of the MOF-199/Ag@Au-based SERS assay for DA with the methodologies previously documented in the literature is presented in Table 1. It is evident that the MOF-199/Ag@Au-based SERS method exhibits a superior linear range and a low LOD for DA. This is attributed to the porous MOF-199, which facilitates physical adsorption, thereby trapping a greater quantity of DA and MOFs and increasing the number of available SPR hotspots.

Table 1. Comparison of MOF-199/Ag@Au-based SERS assay for DA with other methods.

Method	Linear Range (mol/L)	LOD (mol/L)	Sample	Reference
SERS	10^{-3}–10^{-12}	10^{-12}	serum	This work
SERS	2.5×10^{-6}–5×10^{-9}	6.7×10^{-10}	-	[59]
Bioprobe	6×10^{-6}–10^{-7}	6×10^{-8}	serum	[60]
Electrochemistry	-	9.3×10^{-6}	urine	[61]
Green-emitting carbon dots	4×10^{-5}–2×10^{-6}	5.2×10^{-7}	serum	[62]
SERS	10^{-6}–10^{-10}	10^{-10}	urine	[63]

4. Conclusions

In this paper, a novel MOF-199/Ag@Au composite SERS sensing structure was proposed, and its controllable preparation was conveniently realized using a combination of electrochemical and chemical reduction methods. The results demonstrate that the material exhibits excellent SERS enhancement capabilities. In comparison to previous unlabeled SERS methods, the detection limit is enhanced by two orders of magnitude [32]. Moreover, the novel approach exhibits an enhancement of 4–5 orders of magnitude in comparison to another SERS approach developed in this study. Moreover, the SERS detection of DA was conducted in conjunction with the Azo reaction analysis method. In comparison to the unlabeled method, the detection limit is enhanced by five orders of magnitude, and both the linear correlation (R^2 = 0.982) and specificity exhibit notable improvement. When performing the SERS detection of DA in serum, the detection limit can be as low as 10^{-12} M, and there is also a good linear correlation (R^2 = 0.976) between the possible concentrations of DA in actual serum and the SERS signals. The combination of the MOF-199/Ag@Au composite sensing structure with the Azo reaction analysis method resulted in the efficient, highly sensitive, and specific quantitative detection of DA, thereby demonstrating promising prospective applications in the early screening of clinical DA-related diseases.

Author Contributions: Conceptualization, Y.P. and C.W.; methodology, Y.P.; software, Y.P.; validation, Y.P., C.W., G.L. and J.C.; formal analysis, Y.P. and G.L.; investigation, Y.P. and J.C.; resources, C.W.; data curation, Y.P. and J.C.; writing—original draft preparation, Y.P.; writing—review and editing, Y.P., Z.W. and G.L.; visualization, Y.P. and X.Z.; supervision, C.W.; project administration, C.W.; funding acquisition, C.W. and Y.J.; technical support, X.L.; discussion and interpretation of results, G.L., Z.W. and Y.J.; chart preparation, X.Z. All authors have read and agreed to the published version of the manuscript.

Funding: This work was supported by the Chongqing University of Technology Research and Innovation Team Cultivation Program (No. 2023TDZ012), the Horizontal Project of Chongqing University of Technology (No. 2022Q535, 2023Q645), and the Natural Science Foundation of Chongqing (No. 2024NSCQ-MSX1731), and the Special funding project of Army Medical University under Grant (No. 2022XJS07X).

Institutional Review Board Statement: Not applicable.

Informed Consent Statement: Not applicable.

Data Availability Statement: The data that support the findings of this study are available from the corresponding author upon reasonable request. The data are not publicly available due to privacy.

Conflicts of Interest: The authors declare no conflicts of interest.

References

1. Teleanu, R.I.; Niculescu, A.G.; Roza, E.; Vlădâcenco, O.; Grumezescu, A.M.; Teleanu, D.M. Neurotransmitters—key factors in neurological and neurodegenerative disorders of the central nervous system. *Int. J. Mol. Sci.* **2022**, *23*, 5954. [CrossRef]
2. Reis, H.J.; Guatimosim, C.; Paquet, M.; Santos, M.; Ribeiro, F.M.; Kummer, A.; Schenatto, G.; Salgado, J.V.; Vieira, L.B.; Teixeira, A.L.; et al. Neuro-transmitters in the central nervous system & their implication in learning and memory processes. *Curr. Med. Chem.* **2009**, *16*, 796–840.
3. Nimgampalle, M.; Chakravarthy, H.; Sharma, S.; Shree, S.; Bhat, A.R.; Pradeepkiran, J.A.; Devanathan, V. Neurotransmitter systems in the etiology of major neurological disorders: Emerging insights and therapeutic implications. *Ageing Res. Rev.* **2023**, *89*, 101994. [CrossRef]
4. Hodo, T.W.; De Aquino, M.T.P.; Shimamoto, A.; Shanker, A. Critical neurotransmitters in the neuroimmune network. *Front. Immunol.* **2020**, *11*, 1869. [CrossRef]
5. Klein, M.O.; Battagello, D.S.; Cardoso, A.R.; Hauser, D.N.; Bittencourt, J.C.; Correa, R.G. Dopamine: Functions, signaling, and association with neurological diseases. *Cell. Mol. Neurobiol.* **2019**, *39*, 31–59. [CrossRef]
6. Mehler-Wex, C.; Riederer, P.; Gerlach, M. Dopaminergic dysbalance in distinct basal ganglia neurocircuits: Implications for the pathophysiology of Parkinson's disease, schizophrenia and attention deficit hyperactivity disorder. *Neurotox. Res.* **2006**, *10*, 167–179. [CrossRef]
7. Swerdlow, N.R.; Koob, G.F. Dopamine, schizophrenia, mania, and depression: Toward a unified hypothesis of cortico-striatopallido-thalamic function. *Behav. Brain. Sci.* **1987**, *10*, 197–208. [CrossRef]
8. Connolly, B.; Fox, S.H. Treatment of cognitive, psychiatric, and affective disorders associated with Parkinson's disease. *Neurotherapeutics* **2014**, *11*, 78–91. [CrossRef]
9. Van Praag, H.M.; Korf, J.; Lakke, J.; Lakke, J.P.W.F.; Schut, T. Dopamine metabolism in depressions, psychoses, and Parkinson's disease: The problem of the specificity of biological variables in behaviour disorders. *Psychol. Med.* **1975**, *5*, 138–146. [CrossRef] [PubMed]
10. Dallé, E.; Mabandla, M.V. Early life stress, depression and Parkinson's disease: A new approach. *Mol. Brain.* **2018**, *11*, 1–13. [CrossRef] [PubMed]
11. Wei, X.; Zhang, Z.; Wang, Z.A. simple dopamine detection method based on fluorescence analysis and dopamine polymerization. *Microchem. J.* **2019**, *145*, 55–58. [CrossRef]
12. Liu, C.; Gomez, F.A.; Miao, Y.; Cui, P.; Lee, W. A colorimetric assay system for dopamine using microfluidic paper-based analytical devices. *Talanta* **2019**, *194*, 171–176. [CrossRef] [PubMed]
13. Kamal Eddin, F.B.; Wing Fen, Y. Recent advances in electrochemical and optical sensing of dopamine. *Sensors.* **2020**, *20*, 1039. [CrossRef]
14. Wang, H.B.; Zhang, H.D.; Chen, Y.; Huang, K.J.; Liu, Y.M. A label-free and ultrasensitive fluorescent sensor for dopamine detection based on double-stranded DNA templated copper nanoparticles. *Sens. Actuators, B* **2015**, *220*, 146–153. [CrossRef]
15. Wang, J.; Hu, Y.; Zhou, Q.; Hu, L.; Fu, W.; Wang, Y. Peroxidase-like activity of metal–organic framework [Cu (PDA)(DMF)] and its application for colorimetric detection of dopamine. *ACS. Appl. Mater. Interfaces* **2019**, *11*, 43799–44932. [CrossRef]
16. Zhang, X.; Yin, J.; Yoon, J. Recent advances in development of chiral fluorescent and colorimetric sensors. *Chem. Rev.* **2014**, *114*, 4918–4959. [CrossRef]
17. Lakard, S.; Pavel, I.A.; Lakard, B. Electrochemical Biosensing of Dopamine Neurotransmitter: A Review. *Biosensors* **2021**, *11*, 179. [CrossRef]
18. Tukimin, N.; Abdullah, J.; Sulaiman, Y. Electrochemical detection of uric acid, dopamine and ascorbic acid. *J. Electrochem. Soc.* **2018**, *165*, B258. [CrossRef]
19. Abrantes Dias, A.S.; Amaral Pinto, J.C.; Magalhães, M.; Mendes, V.M.; Manadas, B. Analytical methods to monitor dopamine metabolism in plasma: Moving forward with improved diagnosis and treatment of neurological disorders. *J. Pharmaceut. Biomed.* **2020**, *187*, 113323. [CrossRef] [PubMed]
20. Wu, L.; Tang, X.; Wu, T.; Zeng, W.; Zhu, X.; Hu, B.; Zhang, S. A review on current progress of Raman-based techniques in food safety: From normal Raman spectroscopy to SESORS. *Food Res. Int.* **2023**, *169*, 112944. [CrossRef]
21. Hassan, M.M.; Zareef, M.; Xu, Y.; Li, H.; Chen, Q. SERS based sensor for mycotoxins detection: Challenges and improvements. *Food Chem.* **2021**, *344*, 128652. [CrossRef] [PubMed]
22. Logan, N.; Cao, C.; Freitag, S.; Haughey, S.A.; Krska, R.; Elliott, C.T. Advancing Mycotoxin Detection in Food and Feed: Novel Insights from Surface-Enhanced Raman Spectroscopy (SERS). *Adv. Mater.* **2024**, *36*, 2309625. [CrossRef]
23. Huang, Z.; Peng, J.; Xu, L.; Liu, P. Development and Application of Surface-Enhanced Raman Scattering (SERS). *Nanomaterials* **2024**, *14*, 1417. [CrossRef]
24. Bernat, A.; Samiwala, M.; Albo, J.; Jiang, X.; Rao, Q. Challenges in SERS-based pesticide detection and plausible solutions. *J. Agric. Food. Chem.* **2019**, *67*, 12341–12347. [CrossRef]

25. Herrera, G.M.; Padilla, A.C.; Hernandez-Rivera, S.P. Surface enhanced Raman scattering (SERS) studies of gold and silver nanoparticles prepared by laser ablation. *Nanomaterials* **2013**, *3*, 158–172. [CrossRef]
26. Mosier-Boss, P.A. Review of SERS substrates for chemical sensing. *Nanomaterials* **2017**, *7*, 142. [CrossRef]
27. Cao, Y.; Zhang, J.; Yang, Y.; Huang, Z.; Long, N.V.; Fu, C. Engineering of SERS substrates based on noble metal nanomaterials for chemical and biomedical applications. *Appl. Spectrosc. Rev.* **2015**, *50*, 499–525. [CrossRef]
28. Lai, H.; Xu, F.; Zhang, Y.; Wang, L. Recent progress on graphene-based substrates for surface-enhanced Raman scattering applications. *J. Mater. Chem.* **2018**, *6*, 4008–4028. [CrossRef]
29. Xu, K.; Zhou, R.; Takei, K.; Hong, M. Toward flexible surface-enhanced Raman scattering (SERS) sensors for point-of-care diagnostics. *Adv. Sci. Lett.* **2019**, *6*, 1900925. [CrossRef]
30. Gwon, Y.; Kim, J.-H.; Lee, S.-W. Quantification of Plasma Dopamine in Depressed Patients Using Silver-Enriched Silicon Nanowires as SERS-Active Substrates. *ACS Sens.* **2024**, *9*, 870–882. [CrossRef]
31. Kayalık, A.; Saçmacı, Ş. A novel dopamine platform based on CeO2@ TiO2 nanocomposite modified AuNPs/AgNPs nanoparticle and SERS application. *J. Mol. Struct.* **2024**, *1304*, 137644. [CrossRef]
32. Wang, P.; Xia, M.; Liang, O.; Sun, K.; Cipriano, A.F.; Schroeder, T.; Liu, H.; Xie, Y.H. Label-free SERS selective detection of dopamine and serotonin using graphene-Au nanopyramid heterostructure. *Anal. Chem.* **2015**, *87*, 10255–10261. [CrossRef]
33. Park, J.O.; Choi, Y.; Ahn, H.M.; Lee, C.K.; Chun, H.; Park, Y.M.; Kim, K.B. Aggregation of Ag nanoparticle based on surface acoustic wave for surface-enhanced Raman spectroscopy detection of dopamine. *Anal. Chim. Acta* **2024**, *1285*, 342036. [CrossRef]
34. Sibug-Torres, S.M.; Grys, D.B.; Kang, G.; Niihori, M.; Wyatt, E.; Spiesshofer, N.; Ruane, A.; de Nijs, B.; Baumberg, J.J. In situ electrochemical regeneration of nanogap hotspots for continuously reusable ultrathin SERS sensors. *Nat. Commun.* **2024**, *15*, 2022. [CrossRef]
35. Jiang, Z.; Gao, P.; Yang, L.; Huang, C.; Li, Y. Facile in situ synthesis of silver nanoparticles on the surface of metal–organic framework for ultrasensitive surface-enhanced Raman scattering detection of dopamine. *Anal. Chem.* **2015**, *87*, 12177–12182. [CrossRef]
36. Yang, Y.; Li, M.; Zhu, Z. A novel electrochemical sensor based on carbon nanotubes array for selective detection of dopamine or uric acid. *Talanta* **2019**, *201*, 295–300. [CrossRef]
37. Zhang, K.; Liu, Y.; Wang, Y.; Zhang, R.; Liu, J.; Wei, J.; Qian, H.; Qian, K.; Chen, R.; Liu, B. Quantitative SERS detection of dopamine in cerebrospinal fluid by dual-recognition-induced hot spot generation. *ACS Appl. Mater. Interfaces* **2018**, *10*, 15388–15394. [CrossRef]
38. Lu, D.; Fan, M.; Cai, R.; Huang, Z.; You, R.; Huang, L.; Feng, S.; Lu, Y. Silver nanocube coupling with a nanoporous silver film for dual-molecule recognition based ultrasensitive SERS detection of dopamine. *Analyst* **2020**, *145*, 3009–3016. [CrossRef]
39. Dokhan, S.; di Caprio, D.; Taleb, A.; Reis, F.D.A.A. Effects of Adsorbate Diffusion and Edges in a Transition from Particle to Dendritic Morphology during Silver Electrodeposition. *ACS Appl. Mater. Interfaces* **2022**, *14*, 49362–49374. [CrossRef] [PubMed]
40. Li, L.; Chen, Y.; Yang, L.; Wang, Z.; Liu, H. Recent advances in applications of metal–organic frameworks for sample preparation in pharmaceutical analysis. *Coord. Chem. Rev.* **2020**, *411*, 213235. [CrossRef]
41. Nazemi, M.; Soule, L.; Liu, M.; El-Sayed, M.A. Ambient ammonia electrosynthesis from nitrogen and water by incorporating palladium in bimetallic gold–silver nanocages. *J. Electrochem. Soc.* **2020**, *167*, 054511. [CrossRef]
42. Rezaei, B.; Damiri, S. Electrodeposited silver nanodendrites electrode with strongly enhanced electrocatalytic activity. *Talanta* **2010**, *83*, 197–204. [CrossRef] [PubMed]
43. Bahadori, S.R.; Mei, L.; Athavale, A.; Chiu, Y.J.; Pickering, C.S.; Hao, Y. New insight into single-crystal silver dendrite formation and growth mechanisms. *Cryst. Growth Des.* **2020**, *20*, 7291–7299. [CrossRef]
44. Li, S.; Wang, Z.; Shao, Y.; Zhang, K.; Mei, L.; Wang, J. In situ detection of fluid media based on a three-dimensional dendritic silver surface-enhanced Raman scattering substrate. *New J. Chem.* **2022**, *46*, 1785–1790. [CrossRef]
45. Zhang, G.; Sun, S.; Banis, M.N.; Li, R.; Cai, M.; Sun, X. Morphology-Controlled Green Synthesis of Single Crystalline Silver Dendrites, Dendritic Flowers, and Rods, and Their Growth Mechanism. *Cryst. Growth Des.* **2011**, *11*, 2493–2499. [CrossRef]
46. Ha, M.; Kim, J.H.; You, M.; Li, Q.; Fan, C.; Nam, J.M. Multicomponent plasmonic nanoparticles: From heterostructured nanoparticles to colloidal composite nanostructures. *New J. Chem.* **2019**, *119*, 12208–12278. [CrossRef]
47. Ko, J.; Berger, R.; Lee, H.; Yoon, H.; Cho, J.; Char, K. Electronic effects of nano-confinement in functional organic and inorganic materials for optoelectronics. *Chem. Soc. Rev.* **2021**, *50*, 3585–3628. [CrossRef]
48. Zhang, J.; Lin, L.; Jia, K.; Sun, L.; Peng, H.; Liu, Z. Controlled growth of single-crystal graphene films. *Adv. Mater.* **2020**, *32*, 1903266. [CrossRef]
49. Shtukenberg, A.G.; Punin, Y.O.; Gujral, A.; Kahr, B. Growth actuated bending and twisting of single crystals. *Angew. Chem. Int. Ed.* **2014**, *53*, 672–699. [CrossRef]
50. Gao, Y.; Peng, X. Crystal structure control of CdSe nanocrystals in growth and nucleation: Dominating effects of surface versus interior structure. *J. Am. Chem. Soc.* **2014**, *136*, 6724–6732. [CrossRef]
51. Ceballos, M.; Arizmendi-Morquecho, A.; Sánchez-Domínguez, M.; López, I. Electrochemical growth of silver nanodendrites on aluminum and their application as surface-enhanced Raman spectroscopy (SERS) substrates. *Mater. Chem. Phys.* **2020**, *240*, 122025. [CrossRef]
52. Cai, W.F.; Pu, K.B.; Ma, Q.; Wang, Y.H. Insight into the fabrication and perspective of dendritic Ag nanostructures. *J. Exp. Nanosci.* **2017**, *12*, 319–337. [CrossRef]

53. Xu, H.; Song, P.; Fernandez, C.; Wang, J.; Shiraishi, Y.; Wang, C.; Du, Y. Surface plasmon enhanced ethylene glycol electrooxidation based on hollow platinum-silver nanodendrites structures. *J. Taiwan Inst. Chem. Eng.* **2018**, *91*, 316–322. [CrossRef]
54. Niihori, M.; Földes, T.; Readman, C.A.; Arul, R.; Grys, D.B.; Nijs, B.; Rosta, E.; Baumberg, J.J. SERS sensing of dopamine with Fe (III)-sensitized nanogaps in recleanable AuNP monolayer films. *Small* **2023**, *19*, 2302531. [CrossRef]
55. Figueiredo, M.L.; Martin, C.S.; Furini, L.N.; Rubira, R.J.; Batagin-Neto, A.; Alessio, P.; Constantino, C.J. Surface-enhanced Raman scattering for dopamine in Ag colloid: Adsorption mechanism and detection in the presence of interfering species. *Appl. Surf. Sci.* **2020**, *522*, 146466. [CrossRef]
56. Nam, W.; Kim, W.; Zhou, W.; You, E.A. A digital SERS sensing platform using 3D nanolaminate plasmonic crystals coupled with Au nanoparticles for accurate quantitative detection of dopamine. *Nanoscale* **2021**, *13*, 17340–17349. [CrossRef]
57. Meng, X.; Dai, Z.; Jia, C.Q.; Yang, L.; Jiang, W.; Yao, L.; Zhou, Q.; Xu, B. Hierarchical Porous MOF-199 and Zeolite Composites with High Adsorption Performance for Both Toluene and Acetone. *Ind. Eng. Chem. Res.* **2023**, *62*, 19702–19714. [CrossRef]
58. Shi, J.; Li, J.; Liang, A.; Jiang, Z. Highly catalysis MOFCe supported Ag nanoclusters coupled with specific aptamer for SERS quantitative assay of trace dopamine. *Talanta* **2022**, *245*, 123468. [CrossRef]
59. Huang, H.N.; Wang, S.Y.; Chiang, W.H. Microplasma-engineered Ag/GONR-based nanocomposites for selective and label-free SERS-sensitive detection of dopamine. *ACS Appl. Nano Mater.* **2021**, *4*, 10360–10369. [CrossRef]
60. Tang, Z.; Jiang, K.; Sun, S.; Qian, S.; Wang, Y.; Lin, H. A conjugated carbon-dot–tyrosinase bioprobe for highly selective and sensitive detection of dopamine. *Analyst* **2019**, *144*, 468–473. [CrossRef]
61. Shukla, R.P.; Aroosh, M.; Matzafi, A.; Ben-Yoav, H. Partially functional electrode modifications for rapid detection of dopamine in urine. *Adv. Funct. Mater.* **2021**, *31*, 2004146. [CrossRef]
62. Wei, M.X.; Wei, N.; Pang, L.F.; Guo, X.F.; Wang, H. Determination of dopamine in human serum based on green-emitting fluorescence carbon dots. *Opt. Mater.* **2021**, *118*, 111257. [CrossRef]
63. Ansah, I.B.; Lee, W.; Mun, C.; Rha, J.; Jung, H.S.; Kang, M.; Park, S.; Kim, D. In situ electrochemical surface modification of Au electrodes for simultaneous label-free SERS detection of ascorbic acid, dopamine and uric acid. *Sens. Actuators B* **2022**, *353*, 131196. [CrossRef]

Disclaimer/Publisher's Note: The statements, opinions and data contained in all publications are solely those of the individual author(s) and contributor(s) and not of MDPI and/or the editor(s). MDPI and/or the editor(s) disclaim responsibility for any injury to people or property resulting from any ideas, methods, instructions or products referred to in the content.

Article

A Zinc Oxide Nanorod-Based Electrochemical Aptasensor for the Detection of Tumor Markers in Saliva

Junrong Li [1,†], Yihao Ding [1,†], Yuxuan Shi [1], Zhiying Liu [1], Jun Lin [1], Rui Cao [1], Miaomiao Wang [1], Yushuo Tan [1], Xiaolin Zong [2], Zhan Qu [1,*], Liping Du [1,*] and Chunsheng Wu [1,*]

[1] Institute of Medical Engineering, Department of Biophysics, School of Basic Medical Sciences, Health Science Center, Xi'an Jiaotong University, Xi'an 710061, China; leejr@stu.xjtu.edu.cn (J.L.); dingyihao@stu.xjtu.edu.cn (Y.D.); 2206124511xuan@stu.xjtu.edu.cn (Y.S.); zyliu-or@stu.xjtu.edu.cn (Z.L.); linjun625@stu.xjtu.edu.cn (J.L.); der1116@stu.xjtu.edu.cn (R.C.); wmm15029418463@stu.xjtu.edu.cn (M.W.); tanys_3509@stu.xjtu.edu.cn (Y.T.)

[2] Jiashan JunYuan New Material Sci&Tech Co., Ltd., Jiashan 314100, China; zongxl@zju.edu.cn

* Correspondence: zhan.qu@xjtu.edu.cn (Z.Q.); duliping@xjtu.edu.cn (L.D.); wuchunsheng@xjtu.edu.cn (C.W.)

† These authors contributed equally to this work.

Abstract: Biosensors have emerged as a promising tool for the early detection of oral squamous cell carcinoma (OSCC) due to their rapid, sensitive, and specific detection of cancer biomarkers. Saliva is a non-invasive and easy-to-obtain biofluid that contains various biomarkers of OSCC, including the carcinoembryonic antigen (CEA). In this study, an electrochemical aptasensor for the detection of CEA in saliva has been developed towards the diagnosis and early screening of OSCC. This aptasensor utilized a CEA-sensitive aptamer as sensitive elements. A fluorine-doped Tin Oxide (FTO) chip with a surface modification of a zinc oxide nanorod was employed as a transducer. Electrochemical measurements were carried out to detect the responsive signals originating from the specific binding between aptamers and CEAs. The measurement results indicated that this aptasensor was responsive to different concentrations of CEA ranging from 1 ng/mL to 80 ng/mL in a linear relationship. The limit of detection (LOD) was 0.75 ng/mL. This aptasensor also showed very good specificity and regenerative capability. Stability testing over a 12-day period showed excellent performance of this aptasensor. All the results demonstrated that this aptasensor has great potential to be used for the detection of CEA in the saliva of OSCC patients. This aptasensor provides a promising method for the rapid detection of CEA with convenience, which has great potential to be used as a new method for clinical diagnoses and early screening of OSCC.

Keywords: carcinoembryonic antigen; oral squamous cell carcinoma; electrochemical; biosensor; saliva

1. Introduction

Oral squamous cell carcinoma (OSCC) is one of the most common head and neck malignancies, accounting for more than 90% of all oral cancers [1,2]. Because the early symptoms of OSCC are not obvious or individuals are even asymptomatic, most OSCC is not detected and diagnosed until the late stage, and the mortality rate is high [3]. Survey data from the National Cancer Institute (NCI) show that the 5-year survival rate for patients with OCSCC is about 63% overall, with 83% in the early stage and 38% in the late stage [4,5]. Therefore, an early and accurate diagnosis of OSCC is of great significance to improve the survival rate of patients and reduce the mortality rate.

At present, OSCC is mainly diagnosed by oral examination and tissue biopsy [6]. As the gold standard in diagnoses, tissue biopsy has irreplaceable value, but it is not suitable for the early diagnosis of OSCC due to its invasive nature, high cost, and possible infection close to normal tissues [6–9]. In recent years, liquid biopsy has shown important clinical value in the early diagnosis of tumors through the detection and analysis of biomarkers [10–12]. The carcinoembryonic antigen (CEA) is one of the most widely used tumor biomarkers.

Studies have shown that the increase in CEA concentration in blood and saliva is related to the occurrence and development of OSCC, which can be used for the early identification, efficacy judgment, and prognosis monitoring of OSCC [13,14]. Compared with blood testing, saliva testing has the advantages of non-invasive and painless sampling, simplicity and convenience, and low risk of contamination, and is more suitable for point-of-care testing (POCT) [15,16]. In addition, because saliva is more stable than blood, there are fewer interfering substances, and it is in direct contact with cancerous tissue, so the accuracy and specificity of the test are also higher [6,17]. Traditional CEA detection methods include an enzyme-linked immunosorbent assay (ELISA), fluorescence spectrometry, colorimetry, a chemiluminescence immunoassay, etc., which have many shortcomings such as low sensitivity and selectivity, time consumption, complex operation, and high cost [18–20]. Therefore, it is particularly important and essential to develop a sensitive, accurate, simple, rapid, and low-cost CEA detection method towards early diagnoses and screening of OSCC.

Electrochemical biosensors have great prospects for CEA detection due to their high sensitivity, convenience, and significant specificity [21,22]. For example, antibody-based electrochemical immunosensors have been widely studied in the determination of CEA. Although they are specific, they have disadvantages such as high cost and poor stability [23–25]. In contrast, aptamers are specially modified oligonucleotides, which have the advantages of a small size, high affinity, strong stability, non-immunogenicity, easy preparation, low cost of synthesis, and modification. Therefore, electrochemical biosensors based on aptamers have attracted much attention [26–28].

In recent years, nanomaterials have been widely used in the field of biosensors because of their superior physical and chemical properties [29–31]. Among them, zinc oxide (ZnO) nanomaterials have attracted much attention in the field of electrochemical biosensors due to their good biocompatibility, electron transfer properties, large specific surface area, and non-toxicity, which effectively improve the sensitivity and responsiveness of biosensors [32,33]. In order to continuously optimize the performance of ZnO nanosensors, 3D ZnO nanostructures have attracted extensive attention in recent years [34–36]. The three-dimensional structure of ZnO nanocrystals has a higher specific surface area, which can effectively increase the contact area between the tested substance and the semiconductor material, and improve the electron transfer rate and the fixing efficiency of the aptamer [33,37,38]. To effectively monitor CEA levels in saliva, aptamers need to be fixed to substrates such as carbon nanotubes and metal–organic frameworks [39,40]. In recent years, gold (Au), silver (Ag), and other metal nanoparticles (NPs) have been widely used to fix aptamers on the surface of sensor electrodes because of their good electrical conductivity and biocompatibility, large surface area, and easy synthesis [41,42]. Studies have shown that 3D biosensors based on 3D ZnO nanostructures and gold nanoparticles can effectively improve the sensitivity of detection [43].

In this study, an electrochemical aptasensor based on an aptamer-modified fluorine-doped Tin Oxide (FTO)-ZnO-Au structure was developed to detect CEA in saliva for the early diagnosis and screening of OSCC. Firstly, the FTO-ZnO-Au structure was formed by modifying ZnO nanorods with gold nanoparticles, and the signal amplification system of the aptamer-based biosensor was constructed. Then, the 3′-SH CEA-sensitive aptamer was covalently immobilized onto the surface of the electrode gold nanoparticle by the Au-S bond to achieve the high-efficient coupling of sensitive elements with transducers. Electrochemical impedance spectroscopy (EIS) and cyclic voltammetry (CV) measurement were employed to characterize the aptasensor preparation process as well as the performance testing of this biosensor for CEA detection. This electrochemical biosensor based on the FTO-ZnO-Au structure utilizes the three-dimensional structure constructed by ZnO-NRs (zinc oxide nanorods) to bind more aptamer molecules, and the highly specific binding of the aptamer to CEA also facilitates rapid and inexpensive CEA detection. The obtained results demonstrated the good performance of this electrochemical biosensor for CEA detection. This biosensor provides a new method for clinical diagnoses and early screening of OSCC by realizing the rapid detection of CEA in a rapid and convenient manner. It is worthwhile to note that FTO chips and CEA-sensitive aptamers used in this study are only for the demonstration of the technical feasibility of the novel approach for CEA detection in saliva.

2. Materials and Methods

2.1. Materials and Reagents

The following reagents were used as received: 6-mercapto-1-hexanol (MCH), tris-(2-carboxyethyl) phosphine hydrochloride (TCEP), and bovine serum albumin (BSA) (purchased from Sigma-Aldrich, St. Louis, MO, USA). We also used the AREG Antigen (AREG-Ag), C-reactive protein (CRP), and the carcinoembryonic antigen (CEA) (purchased from Sangon Biotech, Shanghai, China). The wash buffer was phosphate buffer saline (PBS) (0.1 M, pH 7.4). We used sulfuric acid (H_2SO_4), hydrochloric acid (HCl), sodium hydroxide (NaOH), hydrogen peroxide (H_2O_2), potassium ferricyanide ($K_3[Fe(CN)_6]$), potassium ferrocyanide ($K_4[Fe(CN)_6]$), and potassium chloride (KCl) (purchased from Sinopharm Chemical Reagent Co, Ltd., Shanghai, China). All reagents were analytical grade and used directly without further purification.

The CEA-sensitive aptamer sequence was chosen according to previous research [44]; it had high affinity and specificity, and was modified at the 3-terminus with a thiol group for the purpose of immobilization on the gold surface. The CEA-sensitive aptamer sequence was as follows: 3′-SH-ATACC AGCTT ATTCA ATT-5′. The aptamers were dissolved in a PBS buffer (pH 7.4) to certain concentrations for further experiments.

2.2. Preparation of FTO-ZnO-Au Structure

ZnO NRs were prepared by a hydrothermal method onto the FTO substrate. A thin film of ZnO was deposited onto the surface of FTO by radio frequency (RF) spurting as a seeding layer. The hydrothermal deposition was carried out in a solution of an equal volume of the same concentration of 0.02 M $Zn(NO_3)_2$ and methenamine in a sealed beaker. The surface onto which the arrays were expected to grow was put downward, with temperature for the growth at 95 °C for 2 h. After growth, the substrate was removed from the container and rinsed with deionized water thoroughly and blown-dry with N_2 for the deposition of a layer of Au by evaporation. Then, the prepared FTO-ZnO-Au structure could be used as an electrochemical electrode for further experiments.

2.3. Immobilization of Aptamer

To covalently immobilize a CEA-sensitive aptamer onto the gold surface of an electrochemical electrode, TCEP offers a good linkage between the gilded groups and the aptamer. To achieve this, a CEA-sensitive aptamer (10 μM, 18 μL), TCEP solution (10 mM, 3 μL), and NaAc solution (500 mM, 2 μL) were mixed, and the solution was stirred for 1 h at room temperature, allowing for the full reaction.

Prior to aptasensor preparation, the bare FTO-ZnO-Au electrodes (specs of 1 cm × 1 cm) were pretreated for cleaning. Firstly, the electrodes were ultrasonically washed in a 1 M NaOH solution for 30 min. Then, the electrodes were rinsed with ultrapure water and dried in a stream of nitrogen gas. Afterwards, the electrodes were ultrasonically washed in a 1 M HCl solution for 5 min and rinsed with ultrapure water and dried in a stream of nitrogen gas. Next, the electrodes were ultrasonically washed in a piranha solution ($H_2O_2:H_2SO_4$ = 1:3, 1 mL/3 mL) for 8 min. Finally, the electrodes were rinsed with ultrapure water and dried in a stream of nitrogen gas.

The preparation process of the aptasensor is schematically shown in Figure 1. First, 20 μL of the previously prepared CEA-sensitive aptamer solution was added dropwise onto the surface of pretreated electrodes with the FTO-ZnO-Au structure, which was incubated for 12 h at room temperature. Then, the electrode surface was rinsed with the PBS buffer (PH 7.4) to remove the aptamers that were not bound to the gold surface, and dried naturally at room temperature. After that, the previously prepared MCH solution (1 mM, 10 μL) was added dropwise onto the modified electrode surface, which was incubated for

30 min at room temperature to seal the non-specific binding sites. Then, the excess MCH on the electrode surface was removed with the PBS buffer (PH 7.4) to obtain an electrochemical aptasensor suitable for the detection of CEA. The prepared aptasensors were stored in a refrigerator at 4 °C for future use.

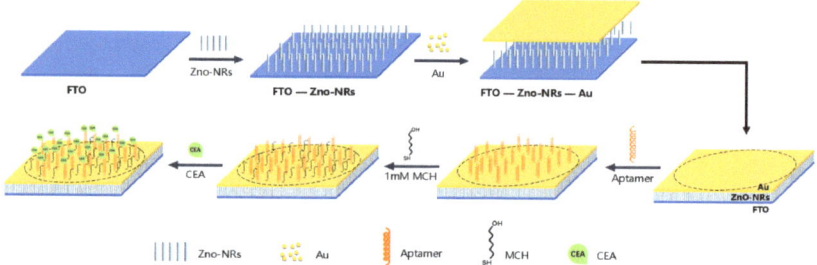

Figure 1. Schematic diagram showing preparation process of electrochemical aptasensors based on FTO-ZnO-Au structure and CEA-sensitive aptamer.

2.4. Electrical Measurement

The three-electrode system is a commonly used system in electrochemical experiments, especially in electroanalytical chemistry. In this study, the electrochemical measurement circuit adopted a three-electrode system, including a zinc oxide nanorod gold electrode as the working electrode, a silver/silver chloride electrode as the reference electrode, and a Pt wire as the counter-electrode. The electrochemical measurements were performed on an electrochemical workstation (CHI600E, Chenhua, Shanghai, China), with a signal generator update rate of 10 MHz. When the scanning speed of cyclic voltammetry was 1000 V/s, the potential increment was only 0.1 mV. When the scanning speed was 5000 V/s, the potential increment was 1 mV. The data acquisition used was two synchronous 16-bit high-resolution, low-noise analog-to-digital converters, and the maximum speed for dual-channel simultaneous sampling was 1 MHz. The scanning voltage of the CV parameter was set to -0.4~0.8 V and the scanning speed was 0.1 V/s. The electrochemical impedance signal acquisition system used a 32-bit high-precision, high-resolution analog/digital signal converter with high-dynamic-range technology and a scanning voltage range of 14 V. The frequency range of the EIS parameter was set to 100 kHz~100 mHz.

Electrical measurements were carried out in electrolytes containing 5 mM $K_3[Fe(CN)_6]$, 5 mM $K_4[Fe(CN)_6]$, and 0.1 M KCl. Firstly, the electrochemical characterization of bare FTO-ZnO-Au electrodes, adapter-modified electrodes, and sealed electrodes were measured separately. Next, the electrode surface was rinsed with the PBS buffer (PH 7.4) and dried naturally at room temperature. Then, a certain concentration gradient of the CEA solution (1 ng/mL, 5 ng/mL, 20 ng/mL, 50 ng/mL, 60 ng/mL, 80 ng/mL) was added dropwise onto the electrode surface, which was incubated for 1 h at room temperature. The electrochemical measurement of different CEA concentrations were performed separately. After that was the electrochemical characterization of the electrodes, which were hydrolyzed with trypsin for 5 min at room temperature. Finally, we performed the electrochemical characterization of the electrodes, which were incubated with 20 ng/mL CRP, BSA, and Areg-Ag to test the specificity of this aptasensor.

2.5. Saliva Sample Collection

All subjects brushed their teeth after dinner one day before sampling, then abstained from eating and drinking and any oral cleaning measures until saliva samples were collected in the morning of the second day. Saliva sample: Ask the subject not to cough up sputum before sample collection and use non-irritating methods to obtain the saliva sample, and ask the patient to hold the saliva collection container, sit still, lower their head, open their mouth slightly, open their eyes, and tilt their head slightly forward; swallowing is avoided and the position is held for 5 min, so that saliva naturally flows into the 15 mL sterile centrifuge tube, and the collected amount is at least 5 mL (saliva does not contain impurities such as blood, food residue, and sputum); otherwise, the above steps should be repeated. All the collected samples were grouped and numbered, frozen in liquid nitrogen, and sent to a laboratory within 2 h, and stored in a refrigerator at −80 °C.

2.6. Signal Analysis

Accurate resistance values on the surface of the working electrode chip under different testing environments were analyzed using ZView 3.1 software to fit and merge the measured electrochemical impedance spectra. We performed linear fitting and created plot fitting curves for the resistance of electrodes incubated with different concentrations of CEA using SPSS 13.0. The electrochemical impedance map and cyclic voltammetry map were all plotted using Origin 8.0.

3. Results and Discussion

3.1. Characterization of Prepared FTO-ZnO-Au Structure

The FTO-ZnO-Au structure was utilized as a transducer for the development of the electrochemical aptasensor towards CEA detection. The surface with ZnO-NRs could improve the coupling efficiency between the transducer surface and aptamers. This could consequently improve the sensing performance for target detection. The surface gold layer was employed to facilitate the immobilization of thiol-modified aptamers onto the sensor surface via the strong Au-S bond. As a result, it is crucial to characterize the surface morphology of the FTO-ZnO-Au structure. In this study, an atomic force microscope (AFM) and scanning electron microscope (SEM) were employed to characterize the electrode surface morphology. The characterization resulting images are shown in detail in Figure 2. As shown in Figure 2A, the SEM result indicated that ZnO-NRs are uniformly and densely distributed on the electrode surface. ZnO-NRs grow approximately vertically, with a hexagonal cross-section and a diameter of ~75 nm. From the AFM result shown in Figure 2B, it is indicated that ZnO-NRs grow uniformly, which further demonstrate the nanostructure on the sensor surface. Figure 2C shows SEM images of the surface of ZnO-NRs after the deposition of a thin layer of chromium and gold. We can see that the chromium and gold layers are evenly distributed on the surface of ZnO-NRs by magnetron sputtering, and the morphology of ZnO-NRs can still be observed, which indicate that the three-dimensional structure of ZnO NRs is still retained on the electrode surface after gold deposition. All the results demonstrate that the FTO-ZnO-Au structure has been successfully prepared and is suitable for further experiments.

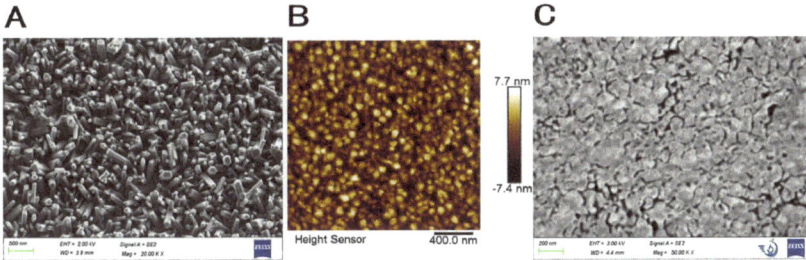

Figure 2. (**A**) The SEM characterization of ZnO−NRs; (**B**) the AFM characterization of FTO−ZnO−Au; (**C**) an SEM image shows the surface of ZnO−NRs after the deposition of a thin layer of gold.

3.2. Electrochemical Characterization of Aptasensor Preparation

The electrochemical aptasensors were constructed by assembling CEA-sensitive aptamers on the electrode surface, and then treated with 1 mM MCH to block the non-specific sites on the surface of the aptamer-modified electrode (Figure 1). Electrochemical impedance spectroscopy (EIS) and cyclic voltammetry (CV) were performed to describe its electrochemical characterization. For the Nyquist diagram, the semicircle diameter of the electrochemical impedance spectroscopy equals the surface electron transfer resistance (R_{ct}). As shown in Figure 3A, the bare FTO-ZnO-Au electrode showed a very small semicircle domain, which has almost no effect on electron transfer, exhibiting a very fast electron transfer process of $[Fe(CN)_6]^{3-/4-}$. When the aptamer was immobilized onto the electrode surface via the Au-S bond, the R_{ct} value increased significantly. This is mainly due to the negatively charged aptamer acting as an electrostatic barrier in the $[Fe(CN)_6]^{3-/4-}$ system, hindering the electron transfer on the chip surface, which leads to an increase in the R_{ct} value. After MCH blocks the aptamer-modified electrode surface, the electron surface defects are repaired, which makes the electron surface layer more compact and further hinders the electron transfer, with a consequent further increase in the R_{ct} value.

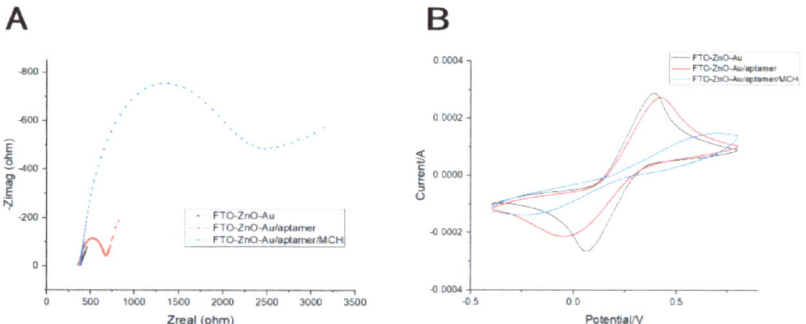

Figure 3. (**A**) The Nyquist plots and (**B**) CVs of the preparation steps of the electrochemical aptasensor based on the CEA-sensitive aptamer.

Cyclic voltammograms (CVs) were also carried out to describe the electrochemical characterization of the surface modification process. As shown in Figure 3B, the peak current decreased with the connections of aptamers and the closure of MCH and increase in the peak-to-peak separation. This is mainly caused by the electrostatic repulsion between negative charges of the DNA aptamer and the anionic redox probe and the formation of the compact layer by the modification of MCH, which together results in the hampering of the interfacial electron transfer. The CV results were consistent to the EIS measurement results, further confirming the successful fabrication of the sensing interface.

3.3. Optimization of Measurement Conditions

To achieve the best detection performance, we optimized the relevant experimental parameters such as CEA-sensitive aptamer concentrations and incubation time of the CEA-sensitive aptamer. We measure the adhesion of the aptamer to the electrode surface by measuring the electrode surface resistance with EIS. The more CEA aptamers attached onto the electrode surface, the greater R_{ct} value that could be detected.

Optimizing the time of the incubate CEA-sensitive aptamer was an important step in building the electrochemical aptasensors. As we can see from the curves shown in Figure 4A, when we use the CEA-sensitive aptamer with a concentration of 5 μM for incubation, the peak R_{ct} value first increased and then decreased with the CEA-sensitive aptamer incubation time increasing from 1 h to 20 h. Meanwhile, an excessive amount of incubation time might have a negative impact on the aptamer attachment onto the electrode surface. Therefore, in this study, 12 h was selected as the appropriate incubation time. Observing the trends depicted in Figure 4B, it was evident that the peak R_{ct} value also initially increased and later leveled off with the increase in CEA-sensitive aptamer concentrations from 1 μM to 40 μM. This indicated that the number of binding sites for the CEA-sensitive aptamer on the electrode surface had reached saturation at the concentration of 10 µM. Therefore, a concentration of 10 µM for the CEA-sensitive aptamer was deemed suitable for optimal performance.

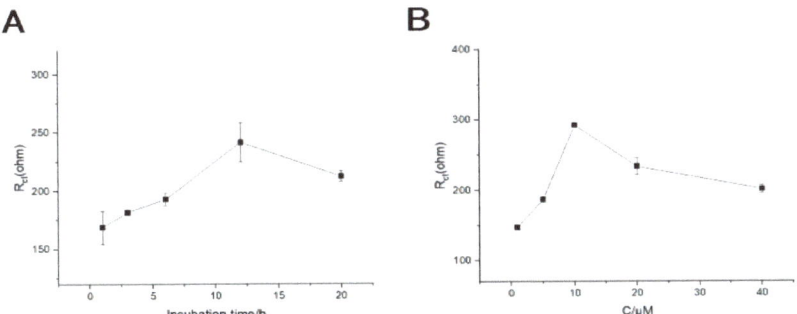

Figure 4. Optimization of aptasensor preparation conditions. (**A**) Influence of incubation time of CEA-sensitive aptamer (5 µM); incubation time of aptamer is 1 h, 3 h, 6 h, 12 h, and 20 h. (**B**) Influence of concentrations of CEA-sensitive aptamer (incubating for 12 h); concentrations of CEA-sensitive aptamer are 1 µM, 5 µM, 10 µM, 20 µM, and 40 µM.

3.4. Analytical Performance Testing of Electrochemical Aptasensor

When different concentrations of CEA were added under optimal conditions, as shown in Figure 5A, the electrode impedance increased with increasing CEA concentrations. The linear regression equation was $R_{ct}(\text{ohm}) = 27.68 \times C[\text{CEA}] + 1335.69$ (ng/mL) ($R^2 = 0.9986$). The linear range between the electrode impedance and CEA concentrations was from 1 to 80 ng/mL (Figure 5B), with a limit of detection (LOD) estimated to be 0.75 ng/mL (S/N = 3). Typically, the salivary CEA concentration in healthy individuals is found to be minimal (0~3 ng/mL), whereas in cancer patients, the CEA concentration is significantly elevated (>5 ng/mL), correlating with the presence of malignancies [45]. A prior investigation employing the enzyme-linked immunosorbent assay (ELISA) has reported salivary CEA levels of 42.6 ± 21.1 ng/mL in OSCC patients, contrasting with the levels of 22.6 ± 22.1 ng/mL observed in the control cohort [46]. Then, Table 1 summarizes the detection performance of the developed electrochemical aptasensor and the previously reported CEA biosensors. The developed electrochemical aptasensor using the FTO-ZnO-Au nanostructure showed exceptional analytical performance, and was characterized by a wide linear range, a lower LOD, and an exemplary linear correlation. In addition, compared with the anti-CEA antibody-based biosensors, the developed biosensor adopted an aptamer-based strategy to effectively improve the accuracy and reliability of detection.

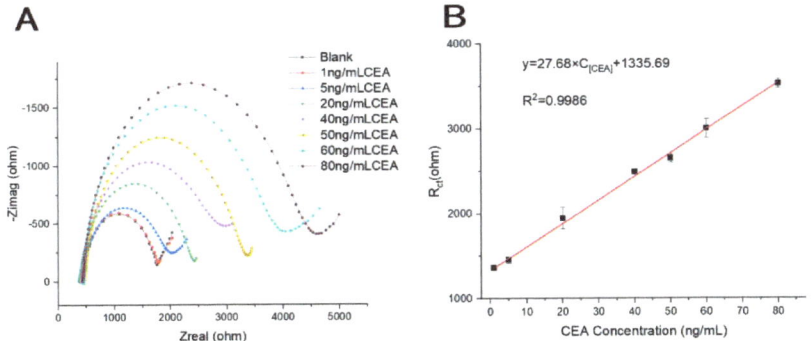

Figure 5. (**A**) EIS responses of the electrochemical aptasensor for the detection of different concentrations of CEA. (**B**) The linear calibration plot between the electrode impedance and CEA concentrations.

Table 1. Comparison of analytical methods for the detection of CEA.

Analytical Method	Sensitive Elements	Analytes	Linear Ranges (ng/mL)	LOD (ng/mL)	R^2	Reference	Real Sample Test
Fluoroimmunosensor	Anti-CEA antibody	CEA	1×10^{-1}~2.5 2.5~20	1.0×10^{-1}	0.988 0.996	[17]	Yes
Aptasensor	Aptamer	CEA	2×10^{-2}~6.0	3.25×10^{-1}	0.962	[47]	No
Aptasensor	Aptamer	CEA	1~50	1.0	0.990	[48]	Yes
Fluoroimmunosensor	Anti-CEA antibody	CEA	7×10^{-1}~80	5.5×10^{-3}	0.993	[49]	Yes
Aptasensor	Aptamer	CEA	1~80	7.5×10^{-1}	0.999	This work	Yes

3.5. Selectivity, Reproducibility, Stability, and Regeneration of the Aptasensor

To examine the selectivity of this electrochemical aptasensor, we measured other proteins present in human saliva, such as amphiregulin (Areg), C-reactive protein (CRP), Interleukin-8 (IL-8), and bovine serum albumin (BSA), which is commonly used for blocking. As shown in Figure 6A, only CEA showed significant responses, and other interferers only had subtle signal changes. These similar signal changes indicated that the interfering substances had little effect on the sensor, which proved that the developed aptasensor has high specificity, which can be used for the specific detection of CEA. Secondly, we prepared two batches of six electrodes under the same conditions to detect 5 and 50 ng/mL CEA. The results were basically consistent, and the relative standard deviations (RSDs) were 2.44 and 1.83%, respectively. The detection results showed that this aptasensor had good reproducibility. At the same time, the stability of this biosensor was also investigated (Figure 6B). The results showed that after storage of 12 days at 4 °C, compared with the current test signal in the initial experiment, the signal change rate was 94.52%. Therefore, this aptasensor could be stored stably for at least 12 days. The reasons that make the sensor stable are as follows. Firstly, the gold plating on the electrode surface has good biocompatibility and does not damage the aptamer activity. Secondly, the aptamer is firmly bound to the electrode surface through the Au-SH bond and falling off is not easy. Finally, the steric hindrance effect provided by the three-dimensional structure of ZnO-NRs will also prevent the aptamer from detaching from the electrode surface to a certain extent.

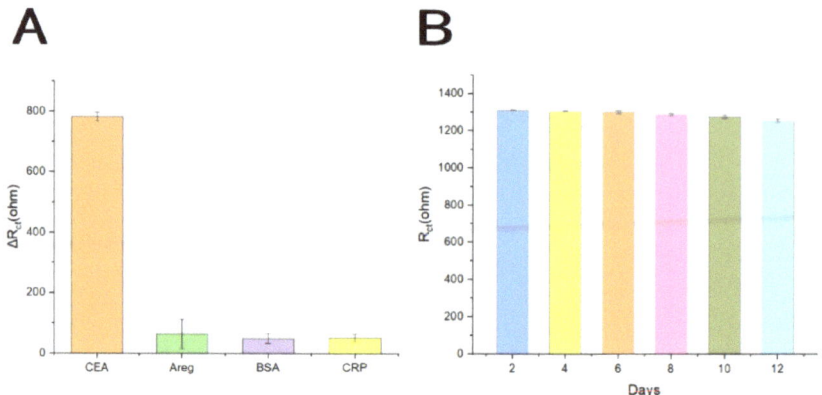

Figure 6. (**A**) Selectivity of the aptasensor against CEA (20 ng/mL), Areg (50 ng/mL), BSA (50 ng/mL), and CRP (20 ng/mL). (**B**) Stability of the aptasensor for 12 days of storage.

The regeneration of the electrode surface is of great significance for reusable biosensors. In this study, we selected trypsin to digest CEA specifically bound to the aptamer, changing its structure and thus separating it from the aptamer to achieve the purpose of sensor regeneration, and achieved good results. The resistance of the sensor chip incubated with 5 and 60 ng/mL CEA for 5 min by trypsin digestion decreased to 67.04% of the original resistance, and the resistance increased to 89.81% of the original resistance after the same concentration of CEA was incubated again, indicating that the state of the chip surface does not return to the state of unbound protein after digestion, and the activity of the aptamer is affected to a certain extent.

3.6. Anti-Interference Performance and Preliminary Analysis of Human Saliva Samples

In order to explore the electrode detection performance in the complex saliva environment, we studied the saliva samples of healthy people. We measured the impedance values of CEA concentrations of 5 ng/mL, 20 ng/mL, 50 ng/mL, and 60 ng/mL in 10% healthy human saliva, and compared them with those of corresponding concentrations measured in 0.1 M PBS. As shown in Figure 7, when the CEA concentration is 5 ng/mL, the error of the impedance value measured in the saliva sample is 0.73%; when the CEA concentration is 20 ng/mL, the error of the impedance value measured in the saliva sample is 1.15%; when the CEA concentration is 50ng/mL, the error of the impedance value measured in the saliva sample is 2.07%. When the CEA concentration was 60ng/mL, the impedance error was 0.41%. The above results indicate that the electrode has good anti-interference performance in a complex salivary environment, and has a significant signal for CEA detection.

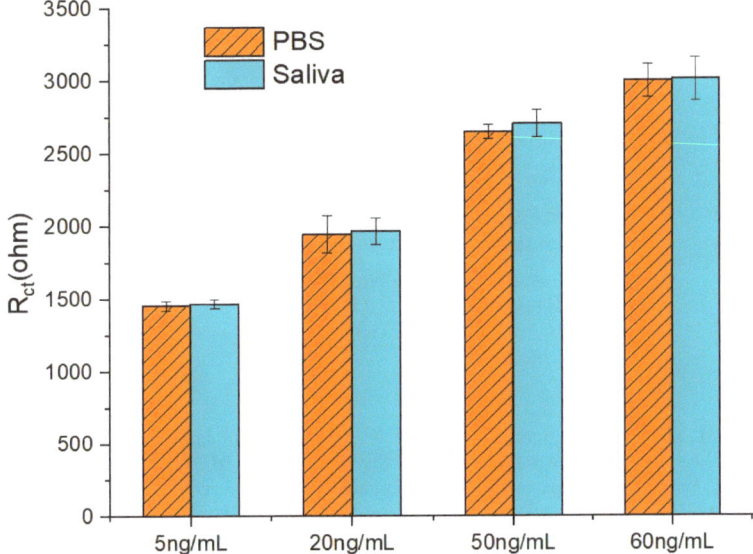

Figure 7. The impedance values of CEA concentrations of 5 ng/mL, 20 ng/mL, 50 ng/mL, and 60 ng/mL were detected in 0.1 M PBS and 10% saliva.

4. Conclusions

This study developed a simple and highly sensitive aptasensor for detecting the salivary biomarker CEA in OSCC. This aptasensor is based on an FTO-ZnO-Au composite structure, where the corresponding ligands are immobilized on the electrode surface for target protein capture and signal detection. The combination of ZnO-NRs' electrode and ligands offers multiple advantages, including a high sensitivity, specificity, stability, and ease of fabrication. EIS and CV electrochemical characterization confirmed that the aptasensor was successfully constructed and could effectively capture the target protein CEA. The results showed good linearity in the range of 1~80 ng/mL, with a LOD of 0.75 ng/mL. Experimentally determined tiny responses of AREG, CRP, and BSA to the aptasensor ensure that the aptamer sensor achieves the specific detection of CEA in complex environments. Additionally, consistent detection results were obtained from multiple batches of sensors, confirming their reproducibility. Stability testing over a 12-day period showed excellent performance of the aptasensor. In addition, we found that trypsin had a good digestion effect on the aptamer-bound CEA, but at the same time, it also had a certain effect on the activity of the aptamer, making it less effective for electrode regeneration. This indicated that this aptasensor has great potential for application in the detection of a tumor marker in saliva, and a class of non-invasive, portable, and highly sensitive OSCC detection devices is expected to be developed based on this aptasensor.

Author Contributions: J.L. (Junrong Li): Conceptualization, Investigation, Methodology, Project administration, Writing—Original draft preparation, Writing—Review and editing; Y.D.: Conceptualization, Data curation, Investigation, Methodology, Software, Visualization, Writing—Review and editing; Y.S.: Conceptualization, Data curation, Formal analysis, Methodology, Software, Visualization; Z.L.: Conceptualization, Formal analysis, Methodology, Validation, Visualization; J.L. (Jun Lin): Formal analysis, Investigation, Methodology, Software, Writing—Review and editing; R.C.: Data curation, Investigation, Methodology, Software, Validation; M.W.: Investigation, Methodology, Validation, Visualization; Y.T.: Data curation, Investigation, Methodology, Software; X.Z.: Investigation, Methodology, Resources; Z.Q.: Conceptualization, Data curation, Investigation, Methodology, Software, Validation, Writing—Review and editing; L.D.: Conceptualization, Funding acquisition, Resources, Supervision, Validation, Writing—Review and editing; C.W.: Conceptualization, Funding acquisition, Project administration, Resources, Supervision, Writing—Review and editing. All authors have read and agreed to the published version of the manuscript.

Funding: This research was funded by the National Natural Science Foundation of China (Grant Nos. 32071370, 32271427, and 32471433), and the S&T Program of Hebei (Grant No.: 22321902D).

Institutional Review Board Statement: Not applicable.

Informed Consent Statement: Not applicable.

Data Availability Statement: The data presented in this study are available on request from the corresponding author due to privacy.

Conflicts of Interest: Author Xiaolin Zong was employed by the company Jiashan JunYuan New Material Sci&Tech Co., Ltd. The remaining authors declare that the research was conducted in the absence of any commercial or financial relationships that could be construed as a potential conflict of interest.

References

1. Badwelan, M.; Muaddi, H.; Ahmed, A.; Lee, K.T.; Tran, S.D. Oral Squamous Cell Carcinoma and Concomitant Primary Tumors, What Do We Know? A Review of the Literature. *Curr. Oncol.* **2023**, *30*, 3721–3734. [CrossRef] [PubMed]
2. Bugshan, A.; Farooq, I. Oral squamous cell carcinoma: Metastasis, potentially associated malignant disorders, etiology and recent advancements in diagnosis. *F1000Research* **2020**, *9*, 229. [CrossRef] [PubMed]
3. Li, Q.; Hu, Y.; Zhou, X.; Liu, S.; Han, Q.; Cheng, L. Role of Oral Bacteria in the Development of Oral Squamous Cell Carcinoma. *Cancers* **2020**, *12*, 2797. [CrossRef]
4. Al Feghali, K.A.; Ghanem, A.I.; Burmeister, C.; Chang, S.S.; Ghanem, T.; Keller, C.; Siddiqui, F. Impact of smoking on pathological features in oral cavity squamous cell carcinoma. *J. Cancer Res. Ther.* **2019**, *15*, 582–588. [CrossRef]
5. Imbesi Bellantoni, M.; Picciolo, G.; Pirrotta, I.; Irrera, N.; Vaccaro, M.; Vaccaro, F.; Squadrito, F.; Pallio, G. Oral Cavity Squamous Cell Carcinoma: An Update of the Pharmacological Treatment. *Biomedicines* **2023**, *11*, 1112. [CrossRef]
6. Chakraborty, D.; Natarajan, C.; Mukherjee, A. Advances in oral cancer detection. *Adv. Clin. Chem.* **2019**, *91*, 181–200. [CrossRef]
7. Khurshid, Z.; Zafar, M.S.; Khan, R.S.; Najeeb, S.; Slowey, P.D.; Rehman, I.U. Role of Salivary Biomarkers in Oral Cancer Detection. *Adv. Clin. Chem.* **2018**, *86*, 23–70. [CrossRef]
8. Abati, S.; Bramati, C.; Bondi, S.; Lissoni, A.; Trimarchi, M. Oral Cancer and Precancer: A Narrative Review on the Relevance of Early Diagnosis. *Int. J. Environ. Res. Public Health* **2020**, *17*, 9160. [CrossRef]
9. Goldoni, R.; Scolaro, A.; Boccalari, E.; Dolci, C.; Scarano, A.; Inchingolo, F.; Ravazzani, P.; Muti, P.; Tartaglia, G. Malignancies and Biosensors: A Focus on Oral Cancer Detection through Salivary Biomarkers. *Biosensors* **2021**, *11*, 396. [CrossRef]
10. Jayanthi, V.; Das, A.B.; Saxena, U. Recent advances in biosensor development for the detection of cancer biomarkers. *Biosens. Bioelectron.* **2017**, *91*, 15–23. [CrossRef]

11. Cho, S.; Yang, H.C.; Rhee, W.J. Simultaneous multiplexed detection of exosomal microRNAs and surface proteins for prostate cancer diagnosis. *Biosens. Bioelectron.* **2019**, *146*, 111749. [CrossRef] [PubMed]
12. Wu, L.L.; Zhang, Z.L.; Tang, M.; Zhu, D.L.; Dong, X.J.; Hu, J.; Qi, C.B.; Tang, H.W.; Pang, D.W. Spectrally Combined Encoding for Profiling Heterogeneous Circulating Tumor Cells Using a Multifunctional Nanosphere-Mediated Microfluidic Platform. *Angew. Chem. Int. Ed. Engl.* **2020**, *59*, 11240–11244. [CrossRef] [PubMed]
13. Zheng, J.; Sun, L.; Yuan, W.; Xu, J.; Yu, X.; Wang, F.; Sun, L.; Zeng, Y. Clinical value of Naa10p and CEA levels in saliva and serum for diagnosis of oral squamous cell carcinoma. *J. Oral Pathol. Med.* **2018**, *47*, 830–835. [CrossRef]
14. Rajguru, J.P.; Mouneshkumar, C.D.; Radhakrishnan, I.C.; Negi, B.S.; Maya, D.; Hajibabaei, S.; Rana, V. Tumor markers in oral cancer: A review. *J. Fam. Med. Prim. Care* **2020**, *9*, 492–496. [CrossRef]
15. Zarrin, P.S.; Jamal, F.I.; Roeckendorf, N.; Wenger, C. Development of a Portable Dielectric Biosensor for Rapid Detection of Viscosity Variations and Its In Vitro Evaluations Using Saliva Samples of COPD Patients and Healthy Control. *Healthcare* **2019**, *7*, 11. [CrossRef]
16. Senf, B.; Yeo, W.-H.; Kim, J.-H. Recent Advances in Portable Biosensors for Biomarker Detection in Body Fluids. *Biosensors* **2020**, *10*, 127. [CrossRef]
17. Li, Y.; Hu, S.; Chen, C.; Alifu, N.; Zhang, X.; Du, J.; Li, C.; Xu, L.; Wang, L.; Dong, B. Opal photonic crystal-enhanced upconversion turn-off fluorescent immunoassay for salivary CEA with oral cancer. *Talanta* **2023**, *258*, 124435. [CrossRef]
18. Qin, W.; Wang, K.; Xiao, K.; Hou, Y.; Lu, W.; Xu, H.; Wo, Y.; Feng, S.; Cui, D. Carcinoembryonic antigen detection with "Handing"-controlled fluorescence spectroscopy using a color matrix for point-of-care applications. *Biosens. Bioelectron.* **2017**, *90*, 508–515. [CrossRef]
19. Rouhi, J.; Kakooei, S.; Sadeghzadeh, S.M.; Rouhi, O.; Karimzadeh, R. Highly efficient photocatalytic performance of dye-sensitized K-doped ZnO nanotapers synthesized by a facile one-step electrochemical method for quantitative hydrogen generation. *J. Solid. State Electrochem.* **2020**, *24*, 1599–1606. [CrossRef]
20. Wang, K.; Yang, J.; Xu, H.; Cao, B.; Qin, Q.; Liao, X.; Wo, Y.; Jin, Q.; Cui, D. Smartphone-imaged multilayered paper-based analytical device for colorimetric analysis of carcinoembryonic antigen. *Anal. Bioanal. Chem.* **2020**, *412*, 2517–2528. [CrossRef]
21. Wang, L.; Xiong, Q.; Xiao, F.; Duan, H. 2D nanomaterials based electrochemical biosensors for cancer diagnosis. *Biosens. Bioelectron.* **2017**, *89*, 136–151. [CrossRef] [PubMed]
22. Zhang, C.; Zhang, S.; Jia, Y.; Li, Y.; Wang, P.; Liu, Q.; Xu, Z.; Li, X.; Dong, Y. Sandwich-type electrochemical immunosensor for sensitive detection of CEA based on the enhanced effects of Ag NPs@CS spaced Hemin/rGO. *Biosens. Bioelectron.* **2019**, *126*, 785–791. [CrossRef] [PubMed]
23. Paniagua, G.; Villalonga, A.; Eguílaz, M.; Vegas, B.; Parrado, C.; Rivas, G.; Díez, P.; Villalonga, R. Amperometric aptasensor for carcinoembryonic antigen based on the use of bifunctionalized Janus nanoparticles as biorecognition-signaling element. *Anal. Chim. Acta* **2019**, *1061*, 84–91. [CrossRef] [PubMed]
24. Shamsuddin, S.H.; Gibson, T.D.; Tomlinson, D.C.; McPherson, M.J.; Jayne, D.G.; Millner, P.A. Reagentless Affimer- and antibody-based impedimetric biosensors for CEA-detection using a novel non-conducting polymer. *Biosens. Bioelectron.* **2021**, *178*, 113013. [CrossRef]
25. Song, Y.; Chen, K.; Li, S.; He, L.; Wang, M.; Zhou, N.; Du, M. Impedimetric aptasensor based on zirconium-cobalt metal-organic framework for detection of carcinoembryonic antigen. *Mikrochim. Acta* **2022**, *189*, 338. [CrossRef]
26. Hashkavayi, A.B.; Raoof, J.B.; Ojani, R. Preparation of Epirubicin Aptasensor Using Curcumin as Hybridization Indicator: Competitive Binding Assay between Complementary Strand of Aptamer and Epirubicin. *Electroanalysis* **2017**, *30*, 378–385. [CrossRef]
27. Byun, J. Recent Progress and Opportunities for Nucleic Acid Aptamers. *Life* **2021**, *11*, 193. [CrossRef]
28. Zhao, H.; Ming, T.; Tang, S.; Ren, S.; Yang, H.; Liu, M.; Tao, Q.; Xu, H. Wnt signaling in colorectal cancer: Pathogenic role and therapeutic target. *Mol. Cancer* **2022**, *21*, 123. [CrossRef]
29. Li, J.; Yang, F.; Chen, X.; Fang, H.; Zha, C.; Huang, J.; Sun, X.; Mohamed Ahmed, M.B.; Guo, Y.; Liu, Y. Dual-ratiometric aptasensor for simultaneous detection of malathion and profenofos based on hairpin tetrahedral DNA nanostructures. *Biosens. Bioelectron.* **2023**, *227*, 114853. [CrossRef]
30. Maral, M.; Erdem, A. Carbon Nanofiber-Ionic Liquid Nanocomposite Modified Aptasensors Developed for Electrochemical Investigation of Interaction of Aptamer/Aptamer–Antisense Pair with Activated Protein C. *Biosensors* **2023**, *13*, 458. [CrossRef]
31. Zhang, X.-W.; Du, L.; Liu, M.-X.; Wang, J.-H.; Chen, S.; Yu, Y.-L. All-in-one nanoflare biosensor combined with catalyzed hairpin assembly amplification for in situ and sensitive exosomal miRNA detection and cancer classification. *Talanta* **2024**, *266*, 125145. [CrossRef]
32. Lv, S.; Zhang, K.; Zhu, L.; Tang, D. ZIF-8-Assisted NaYF4:Yb,Tm@ZnO Converter with Exonuclease III-Powered DNA Walker for Near-Infrared Light Responsive Biosensor. *Anal. Chem.* **2020**, *92*, 1470–1476. [CrossRef] [PubMed]
33. Que, M.; Lin, C.; Sun, J.; Chen, L.; Sun, X.; Sun, Y. Progress in ZnO Nanosensors. *Sensors* **2021**, *21*, 5502. [CrossRef] [PubMed]
34. Shetti, N.P.; Bukkitgar, S.D.; Reddy, K.R.; Reddy, C.V.; Aminabhavi, T.M. ZnO-based nanostructured electrodes for electrochemical sensors and biosensors in biomedical applications. *Biosens. Bioelectron.* **2019**, *141*, 111417. [CrossRef] [PubMed]
35. Xia, Y.; Chen, Y.; Tang, Y.; Cheng, G.; Yu, X.; He, H.; Cao, G.; Lu, H.; Liu, Z.; Zheng, S.Y. Smartphone-Based Point-of-Care Microfluidic Platform Fabricated with a ZnO Nanorod Template for Colorimetric Virus Detection. *ACS Sens.* **2019**, *4*, 3298–3307. [CrossRef] [PubMed]

36. Sinha, K.; Chakraborty, B.; Chaudhury, S.S.; Chaudhuri, C.R.; Chattopadhyay, S.K.; Das Mukhopadhyay, C. Selective, Ultra-Sensitive, and Rapid Detection of Serotonin by Optimized ZnO Nanorod FET Biosensor. *IEEE Trans. NanoBiosci.* **2022**, *21*, 65–74. [CrossRef]
37. Wang, Z.L. From nanogenerators to piezotronics—A decade-long study of ZnO nanostructures. *MRS Bull.* **2012**, *37*, 814–827. [CrossRef]
38. Khan, M.; Nagal, V.; Masrat, S.; Tuba, T.; Alam, S.; Bhat, K.S.; Wahid, I.; Ahmad, R. Vertically Oriented Zinc Oxide Nanorod-Based Electrolyte-Gated Field-Effect Transistor for High-Performance Glucose Sensing. *Anal. Chem.* **2022**, *94*, 8867–8873. [CrossRef]
39. Lv, M.; Zhou, W.; Tavakoli, H.; Bautista, C.; Xia, J.; Wang, Z.; Li, X. Aptamer-functionalized metal-organic frameworks (MOFs) for biosensing. *Biosens. Bioelectron.* **2021**, *176*, 112947. [CrossRef]
40. Wang, Z.; Dai, L.; Yao, J.; Guo, T.; Hrynsphan, D.; Tatsiana, S.; Chen, J. Enhanced adsorption and reduction performance of nitrate by Fe–Pd–Fe3O4 embedded multi-walled carbon nanotubes. *Chemosphere* **2021**, *281*, 130718. [CrossRef]
41. Chen, Y.; Wang, A.-J.; Yuan, P.-X.; Luo, X.; Xue, Y.; Feng, J.-J. Three dimensional sea-urchin-like PdAuCu nanocrystals/ferrocene-grafted-polylysine as an efficient probe to amplify the electrochemical signals for ultrasensitive immunoassay of carcinoembryonic antigen. *Biosens. Bioelectron.* **2019**, *132*, 294–301. [CrossRef] [PubMed]
42. Jouyandeh, M.; Sajadi, S.M.; Seidi, F.; Habibzadeh, S.; Munir, M.T.; Abida, O.; Ahmadi, S.; Kowalkowska-Zedler, D.; Rabiee, N.; Rabiee, M.; et al. Metal nanoparticles-assisted early diagnosis of diseases. *OpenNano* **2022**, *8*, 100104. [CrossRef]
43. Kim, H.-M.; Park, J.-H.; Lee, S.-K. Fiber optic sensor based on ZnO nanowires decorated by Au nanoparticles for improved plasmonic biosensor. *Sci. Rep.* **2019**, *9*, 17371. [CrossRef] [PubMed]
44. Luo, C.; Wen, W.; Lin, F.G.; Zhang, X.H.; Gu, H.S.; Wang, S.F. Simplified aptamer-based colorimetric method using unmodified gold nanoparticles for the detection of carcinoma embryonic antigen. *RSC Adv.* **2015**, *5*, 10994–10999. [CrossRef]
45. Joshi, S.; Kallappa, S.; Kumar, P.; Shukla, T.; Ghosh, R. Simple diagnosis of cancer by detecting CEA and CYFRA 21-1 in saliva using electronic sensors. *Sci. Rep.* **2022**, *12*, 15421. [CrossRef]
46. Honarmand, M.H.; Farhad-Mollashahi, L.; Nakhaee, A.; Nehi, M. Salivary Levels of ErbB2 and CEA in Oral Squamous Cell Carcinoma Patients. *Asian Pac. J. Cancer Prev.* **2016**, *17*, 77–80. [CrossRef]
47. Xu, Z.; Wang, C.; Ma, R.; Sha, Z.; Liang, F.; Sun, S. Aptamer-based biosensing through the mapping of encoding upconversion nanoparticles for sensitive CEA detection. *Analyst* **2022**, *147*, 3350–3359. [CrossRef]
48. Shahbazi, N.; Hosseinkhani, S.; Ranjbar, B. A facile and rapid aptasensor based on split peroxidase DNAzyme for visual detection of carcinoembryonic antigen in saliva. *Sens. Actuators B Chem.* **2017**, *253*, 794–803. [CrossRef]
49. Yu, Q.; Wang, X.; Duan, Y. Capillary-Based Three-Dimensional Immunosensor Assembly for High-Performance Detection of Carcinoembryonic Antigen Using Laser-Induced Fluorescence Spectrometry. *Anal. Chem.* **2014**, *86*, 1518–1524. [CrossRef]

Disclaimer/Publisher's Note: The statements, opinions and data contained in all publications are solely those of the individual author(s) and contributor(s) and not of MDPI and/or the editor(s). MDPI and/or the editor(s) disclaim responsibility for any injury to people or property resulting from any ideas, methods, instructions or products referred to in the content.

Article

Paper-Based Analytical Devices Based on Amino-MOFs (MIL-125, UiO-66, and MIL-101) as Platforms towards Fluorescence Biodetection Applications

Sofía V. Piguillem [1], Germán E. Gomez [2,*], Gonzalo R. Tortella [3,4], Amedea B. Seabra [5], Matías D. Regiart [6], Germán A. Messina [6] and Martín A. Fernández-Baldo [6,*]

1. Instituto de Investigaciones en Físico-Química de Córdoba (INFIQC), Departamento de Físico-Química, Universidad Nacional de Córdoba, CONICET, Haya de la Torre Esquina Medina Allende S/N, Córdoba 5000, Argentina; svpiguillem@unsl.edu.ar
2. Instituto de Investigaciones en Tecnología Química (INTEQUI), Departamento de Química, Universidad Nacional de San Luis (UNSL), CONICET, Ejército de los Andes 950, San Luis D5700BWS, Argentina
3. Centro de Excelencia en Investigación Biotecnológica Aplicada al Medio Ambiente (CIBAMA), Facultad de Ingeniería y Ciencias, Universidad de La Frontera, Av. Francisco Salazar 01145, Temuco 4811230, Chile; gonzalo.tortella@ufrontera.cl
4. Departamento de Ingeniería Química, Facultad de Ingeniería y Ciencias, Universidad de La Frontera, Av. Francisco Salazar 01145, Temuco 4811230, Chile
5. Center for Natural and Human Sciences, Federal University of ABC (UFABC), Avenida dos Estados, Saint Andrew 09210-580, Brazil; amedea.seabra@ufabc.edu.br
6. Instituto de Química de San Luis, Facultad de Química, Universidad Nacional de San Luis, INQUISAL (UNSL—CONICET), Ejército de los Andes 950, San Luis D5700BWS, Argentina; mregiart@unsl.edu.ar (M.D.R.); messina@unsl.edu.ar (G.A.M.)
* Correspondence: gegomez@unsl.edu.ar (G.E.G.); mbaldo@unsl.edu.ar (M.A.F.-B.)

Abstract: In this study, we designed three promising platforms based on metal–organic frameworks (MOFs) to develop paper-based analytical devices (PADs) for biosensing applications. PADs have become increasingly popular in field sensing in recent years due to their portability, low cost, simplicity, efficiency, fast detection capability, excellent sensitivity, and selectivity. In addition, MOFs are excellent choices for developing highly sensitive and selective sensors due their versatility for functionalizing, structural stability, and capability to adsorb and desorb specific molecules by reversible interactions. These materials also offer the possibility to modify their structure and properties, making them highly versatile and adaptable to different environments and sensing needs. In this research, we synthesized and characterized three different amino-functionalized MOFs: UiO-66-NH$_2$ (Zr), MIL-125-NH$_2$ (Ti), and MIL-101-NH$_2$ (Fe). These MOFs were used to fabricate PADs capable of sensitive and portable monitoring of alkaline phosphatase (ALP) enzyme activity by laser-induced fluorescence (LIF). Overall, amino-derived MOF platforms demonstrate significant potential for integration into biosensor PADs, offering key properties that enhance their performance and applicability in analytical chemistry and diagnostics.

Keywords: analytical device; paper biosensor; MOFs platform; biomolecules; alkaline phosphatase enzyme

1. Introduction

Metal–organic frameworks (MOFs) are attractive crystalline materials made of metal ions or clusters coordinated by multifunctional organic ligands via robust chemical bonds and supramolecular interactions. These assemblies create periodic polymeric networks with organic and inorganic components, spanning two or three dimensions [1]. Moreover, MOF functionalization without altering their structure could be achieved by post-synthesis modifications, allowing unique properties by improving their reactivities [2]. In recent years, MOFs have experienced exponential growth, showing potential applications in diverse fields such as gas adsorption/separation, controlled drug release, catalysis, and luminescence. This progress allows the rational design of materials with tailored electrochemical and optical properties based on carefully selecting and combining metals and ligands. In general, the properties of MOFs are intimately linked to their structural features, encompassing compositions and architectures [3].

Biochemical analysis often involves lengthy procedures and large, expensive equipment that must be operated by qualified professionals in a safe environment. To overcome these difficulties, the lab-on-a-chip concept was introduced [4], which aims to reduce the costs of analysis using platforms made of miniaturized active materials supported on quartz, glass, or polymers [5]. Recently, paper-based analytical devices (PADs) have gained attention as an innovative alternative, taking advantage of the accessibility and affordability of paper, which is renewable and recyclable [6–8]. Furthermore, PADs are easy to store, transport, and handle, and could be printed or coated for many applications [9]. Their high surface area, compatibility with biomolecules, and excellent filtration properties make them suitable for lateral flow assays, chromatographic separations, and microfluidic devices. In addition, most PADs are biodegradable and can be disposed by incineration, making them environmentally friendly, biocompatible, and efficient tools that require minimal use of samples and reagents, while ensuring easy operation [9].

In recent decades, biosensors have attracted considerable interest in both the scientific and industrial sectors [10–15]. These devices cleverly integrate biology, physics, chemistry, and engineering knowledge, offering relevant and practical solutions to contemporary analytical challenges [10–15]. However, several studies have reported on biosensors based on MOF platforms for biomolecule immobilization, but only a few examples based on PADs have been explored [16–25]. Ortiz-Gómez et al. (2017) [18] developed a microfluidic paper-based analytical device (µPAD) for glucose determination using a supported MIL-101 (Fe) MOF acting as a peroxidase mimic, and under optimal conditions, the value for the S coordinate increases linearly for glucose concentrations up to 150 µmol·L^{-1}, with a 2.5 µmol·L^{-1} detection limit. In that same sense, Wei et al. (2021) [19] reported an electrochemical sensor based on Co-MOF modified carbon cloth/paper applied to quantitative glucose detection, effectively increasing the specific area and catalytic sites more than a traditional plane electrode. Also, Hassanzadeh et al. (2019) [20] presented a luminescent PAD to estimate total phenolic content based on a H_2O_2—rhodamine b—Co-MOF. It was discovered that the reaction of H_2O_2 with rhodamine b molecules, loaded into the Co-MOF nanopores, can enhance their emission. Recently, Catalá-Icardo et al. (2024) [21] fabricated MIL-53 (Al)-modified PADs for the efficient extraction of neonicotinoid insecticides. The whole method showed satisfactory analytical performance with recoveries between 86 and 114%, suitable precision (with RSD lower than 14%), and detection limits ranging from 1 to 1.6 µg·L^{-1}. Moreover, Chang et al. (2024) [22] proposed colorimetric PADs based on a two-dimensional MOF nanozyme for a dichlorophen assay. On the other hand, Feng et al. (2024) [23] reported a paper-based electrochemiluminescence biosensor for detecting pathogenic bacteria *Staphylococcus aureus*. This biosensor was constructed using porous Zn-MOFs to form [Ru (bpy)$_3$]$^{+2}$ functionalized MOF nanoflowers (MOF-5 (Ru) NFs) [23]. In addition, Guan et al. (2024) [24] fabricated a hydrogel-based colorimetric PADs platform for the visual colorimetric assay of creatinine using CdTe@UiO-66-PC-Cu MOF. Definitely, PAD biosensors based on MOF platforms for biomolecule immobilization open exciting doors for the development of analytical devices.

In the present research, we synthesized and characterized three amino-functionalized MOF platforms (UiO-66-NH_2 (Zr), MIL-125-NH_2 (Ti), and MIL-101-NH_2 (Fe)) and employed them to develop sensitive, portable PADs for monitoring ALP enzyme activity by laser-induced fluorescence.

2. Materials and Methods

2.1. Reagents and Materials

All of the used reagents were analytical grade and were purchased without further purification treatment. 2-aminoterephthalic acid (H_2ATA), methanol (MeOH), zirconium tetrachloride ($ZrCl_4$), titanium isopropoxide (Ti $(OiPr)_4$), N,N'-dimethylformamide (DMF), and 2-amino-1,4-benzenedicarboxylic acid were purchased from Merck (Darmstadt, Germany). Dopamine hydrochloride, Tris-HCl, and Iron (III) chloride hexahydrate ($FeCl_3·6H_2O$, 97%) were purchased from J&K Scientific Ltd. (San José, CA, USA); 4-methylumbelliferyl phosphate (4-MUP) from Fluka and alkaline phosphatase (ALP) enzyme were purchased from Sigma-Aldrich (Buenos Aires, Argentina). Hydrogen peroxide (H_2O_2) 30% (v/v), glutaraldehyde (GLU, 5% aqueous solution), and acetone were purchased from Merck (Darmstadt, Germany). Filter paper N° 1 was provided by Whatman (Maidstone, UK). All buffer solutions were prepared with milli-Q water.

2.2. Equipment

An Orion Research Inc. model EA 940 equipped with a glass combination electrode (Orion Research 95 Inc., Cambridge, MA, USA) was employed for pH determinations. In addition, MOF platforms were characterized using an LEO 1450VP scanning electron microscope (SEM). A Xerox Color-Qube 8870 printer was used to mark the paper microzones. The fluorescent signal was monitored under LED excitation at 430 nm. All the samples were placed at 45 ° with respect to the LED beamline. The fluorescent radiation was detected with the assembly's optical axis perpendicular to the device's plane. A microscope objective (10:1, PZO, Poland) mounted on a BIOLAR L-PZO microscope was used for light collection. A fiber-optic collection bundle was mounted on a sealed housing at the end of the microscope lens, connected to an Ocean Optics model QE65000-FL spectrometer. In order to avoid spurious light, the entire assembly was covered with a large black box. The powder X-ray diffraction (PXRD) plots were recorded with a Rigaku—Ultima IV type II diffractometer. A scanning step of 0.05° between 5 and 50 2-theta Bragg angles with an exposure time of 5 s per step was used to obtain the best counting statistics. Fourier transform infrared (FTIR) spectroscopic measurements were obtained in a Perkin Elmer Spectrum 65 FTIR spectrometer in a region from 400 to 4000 cm^{-1}.

2.3. Synthesis of Amino-MOFs

The synthesis of MIL-125-NH_2 (Ti) was carried out following a previously reported procedure [10] with some modifications in the amounts of the components: 0.54 g of H_2ATA ($0.294·10^{-3}$ mol·L^{-1}) and 0.315 mL of Ti$(OiPr)_4$ ($1.05·10^{-3}$ mol·L^{-1}) were mixed in 24 mL of DMF solution and 6 mL of MeOH solution. The mixture was stirred for 30 min, then transferred to a 45 mL Parr reactor coated internally with Teflon and heated to 150 °C for 72 h. After that, a pale yellow powder was obtained. The solid underwent a two-step washing process with continuous stirring at 500 rpm: initially with 45 mL of a DMF solution, followed by 20 mL of an MeOH solution. Afterward, the solid was dried at 75 °C in an oven for 48 h (yield 48%). The resulting powder was reduced in particle size through manual grinding for 90 min in an agate mortar. Then, a 0.5% w/v suspension of MIL-125-NH_2 (Ti) in 0.1 mol·L^{-1} phosphate buffer solution (PBS) pH 7.2 was prepared and sonicated for 60 min. The UiO-66-NH_2 (Zr) synthesis was carried out following previously described steps [11] with some modifications in the amounts of the components: 0.288 g of anhydrous $ZrCl_4$ (1.23×10^{-3} mol·L^{-1}) and 0.209 g (1.92×10^{-3} mol·L^{-1}) of H_2ATA were dissolved in 60 mL of DMF and stirred for 30 min. Subsequently, the solution was transferred to a 45 mL Teflon-lined Parr reactor and brought to 120 °C for 24 h (yield 55%). After that, it was cooled to room temperature, and the resulting light brown powder was washed three times with DMF solution and finally dried at 70 °C for 24 h. A 0.1 mol·L^{-1} pH 7.2 PBS buffer solution containing 0.5% w/v of UiO-66-NH_2 (Zr) was prepared for analytical measurements. The synthesis of MIL-101-NH_2 (Fe) followed steps previously described [12], with some modifications. Masses of 0.504 g of anhydrous $FeCl_3·6H_2O$

(2×10^{-3} mol·L^{-1}) and 0.181 g (1×10^{-3} mol·L^{-1}) of H$_2$ATA were mixed in 15 mL of DMF under constant stirring for 30 min. The mixture was transferred into a 120 mL stainless steel Parr autoclave reactor internally lined with Teflon, which was hermetically sealed and heated at 120 °C for 24 h. After cooling to room temperature, the resulting dark brown powder was washed with constant agitation using 20 mL of DMF for 24 h. Finally, the product was dried at 75 °C for 48 h (yield 19%). Before employing the product, a 0.1 mol·L^{-1} PBS buffer solution pH 7.2 containing 0.5% w/v of the MIL-101-NH$_2$ (Fe) was prepared.

2.4. Biosensor Design

Whatman No.1 filter paper was used to design and construct the immunosensor. A 6 mm diameter hydrophobic containment barrier was delimited circularly on a microzone of this paper by wax stamping, using Corel Draw 9 graphic software and the Xerox ColorQube 8870 wax printer. The cut papers were then placed on a hot plate at 80 °C to homogenize the walls forming the hydrophobic wax barrier.

The papers were subjected to a constant flow of oxygen plasma for 3 min to generate as many aldehyde groups as possible on their surface by cellulose oxidation, in order to have a more significant surface reaction.

2.5. Paper Modification

Suspensions of UiO-66-NH$_2$ (Zr), MIL-125-NH$_2$ (Ti), and MIL-101-NH$_2$ (Fe) were prepared at a concentration of 0.5% w/v in 0.1 mol·L^{-1} PBS buffer solution pH 7.2 and dispersed under sonication for 15 min. When the suspensions were completely homogenized, 20 μL was placed onto the pre-treated papers using a drop casting method and left in a humid chamber for 30 min. During this time, the aldehyde groups on the cellulose surface linked to the amino groups of the mentioned MOFs. After that, the papers were washed three times with 0.1 mol·L^{-1} PBS buffer solution pH 7.2 and dried under N$_2$ flow.

2.6. Enzyme Immobilization

To achieve covalent bonds between the amino groups of the immobilized MOFs on the paper with the amino groups present in the ALP, a cross-linking agent solution of glutaraldehyde (GLU) was employed. This procedure consisted of the following steps: First, the papers were immersed for 30 min in a 0.5% w/v GLU acetone solution at pH 8. After being washed three times with 0.1 mol·L^{-1} pH 7.2 PBS buffer solution and dried, 20 μL of 0.1% w/v ALP solution in 0.1 mol·L^{-1} PBS buffer solution pH 7.2 was added and incubated at 37 °C for 10 min. Subsequently, the papers were washed three times with PBS buffer solution to remove excess non-bounded protein (Figure 1).

2.7. Study of the Enzymatic Response by Laser-Induced Fluorescence

An optical system was constructed using a single frequency 430 nm DPSS laser (Cobolt Zouk^{TM}) operated at 10 mW that served as the fluorescence excitation source to detect enzyme activity and test the paper biosensor. The laser was focused on the detection channel at 90° geometry with respect to the surface.

The relative fluorescence signal, which corresponded to the ALP, was measured in situ on the modified paper device. The fluorescence of 4-MUP was measured using excitation at 430 nm and emission at 452 nm. Diethanolamine (DEA) buffer (0.1 mol·L^{-1} diethanolamine, 5×10^{-2} mol·L^{-1} KCl, 1×10^{-3} mol·L^{-1} MgCl$_2$, pH 9.6) was used to prepare the 4-MUP solution. Finally, 5 μL of substrate solution (2.5×10^{-3} mol·L^{-1} 4-MUP in DEA buffer solution, pH 9.6) was injected into the modified paper, and LIF was used to measure the enzyme product.

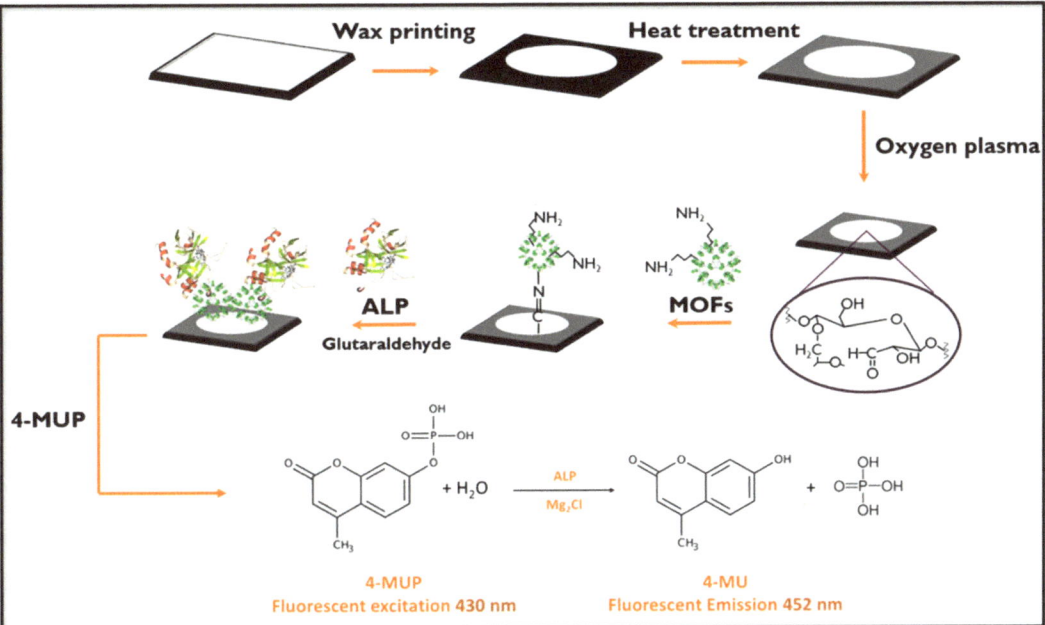

Figure 1. Schematic representation of fluorescent paper-based biosensor construction showing the modification of the paper surface and the alkaline phosphatase determination procedure.

The paths of the reflected beams were meticulously arranged to prevent collisions with other components of the device cell, thereby minimizing the risk of photobleaching. Fluorescent radiation was detected along the optical axis of the array. An optical fiber collection system was installed within a sealed housing linked to the QE65000-FL scientific-grade spectrophotometer (Ocean Optics, Inc., New York, NY, USA). The entire system was enclosed within a black box during measurements to eliminate spurious light interference. These assessments were conducted for both a blank assay where the enzyme was directly immobilized onto unmodified paper, and assays where the enzyme was immobilized onto papers modified with three amino-functionalized MOFs.

3. Results and Discussion

3.1. Elemental Characterization of MOFs
SEM–EDS

The synthesized MOFs were characterized by SEM and EDS analysis to obtain information regarding the particle size and semi-quantitative elemental percentage. As shown in Figure 2a, MIL-125-NH_2 (Ti) particles exhibit a spherical morphology with an average size of approximately 0.6 ± 0.2 µm. Moreover, the EDS data indicate the presence of high lines from titanium metal ions and signals from C and O belonging to the organic backbone (Figure 3a).

In Figure 2b, the particles of UiO-66-NH_2 (Zr) showed a flat morphology with an average size of 0.53 ± 0.2 µm, confirming the elemental composition of the metal Zr and the ligand components C and O (Figure 3b). Finally, besides Figure 2c, the MIL-101-NH_2 (Fe) particles exhibited a spherical shape with a uniform average size of 0.26 ± 0.06 µm. Also, EDS analysis revealed characteristic X-ray lines corresponding to the iron along with O and C from the ligands.

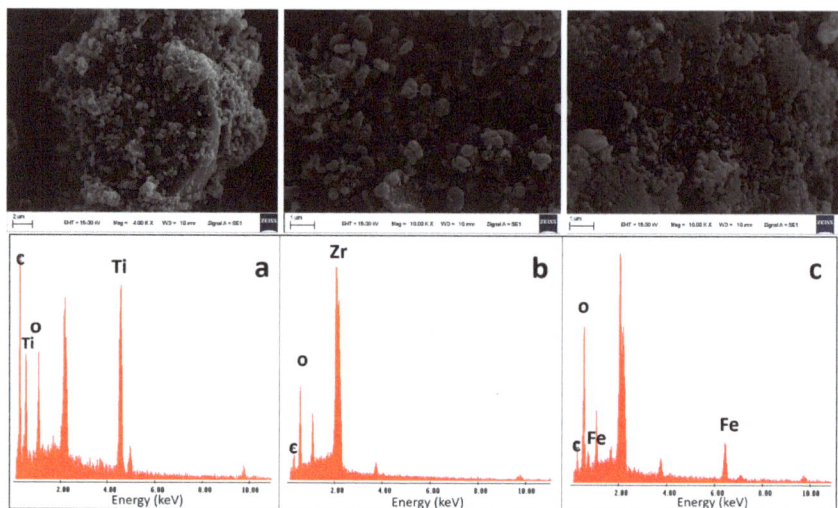

Figure 2. SEM image characterization: (**a**) MIL-125-NH$_2$ (Ti), (**b**) UiO-66-NH$_2$ (Zr), and (**c**) MIL-101-NH$_2$ (Fe).

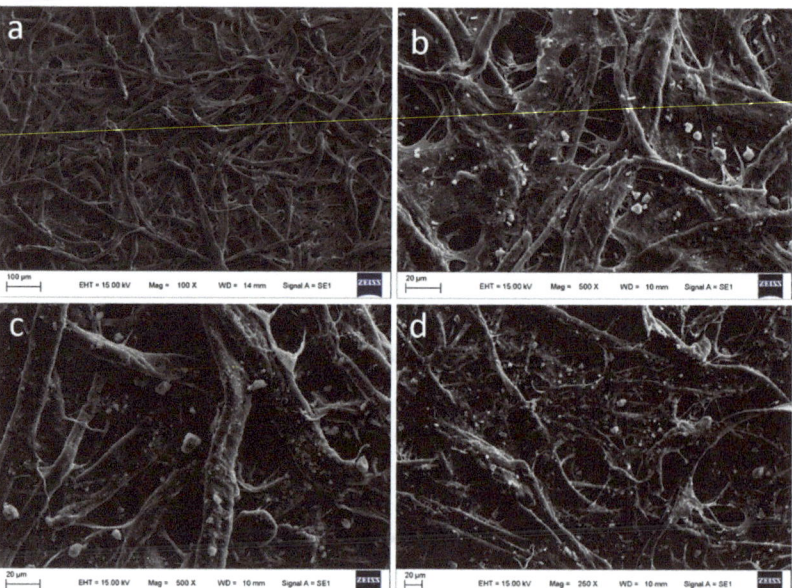

Figure 3. SEM image characterization: (**a**) unmodified paper surface, (**b**) paper surface modified with amino-functionalized MIL-101-NH$_2$ (Fe), (**c**) paper surface modified with amino-functionalized UiO-66-NH$_2$ (Zr), and (**d**) paper surface modified with amino-functionalized MIL-125-NH$_2$ (Ti).

The paper supports, with and without modifications, were also characterized by SEM and EDS techniques, as illustrated in Figures 3 and 4. In the unmodified paper, the cellulose fibers are shown in their original state provided by the manufacturer, with characteristic EDS peaks confirming their C and O composition (see Figure 4a). Figure 3b shows the impregnated paper with MIL-101-NH$_2$ (Fe), revealing visible particles deposited on the support and confirming this observation through an X-ray line spectrum (see Figure 4b).

In Figure 3c, the MOF UiO-66-NH$_2$ (Zr) is observed impregnating the cellulose support, with visible particles deposited and elemental composition that confirms the presence of cellulose and mesoporous elements (see Figure 4c). Finally, Figure 3d shows cellulose paper impregnated with MIL-125-NH$_2$ (Ti), demonstrating the presence of elements through composition analysis (see Figure 4c) and a uniform dispersion of particles throughout the paper.

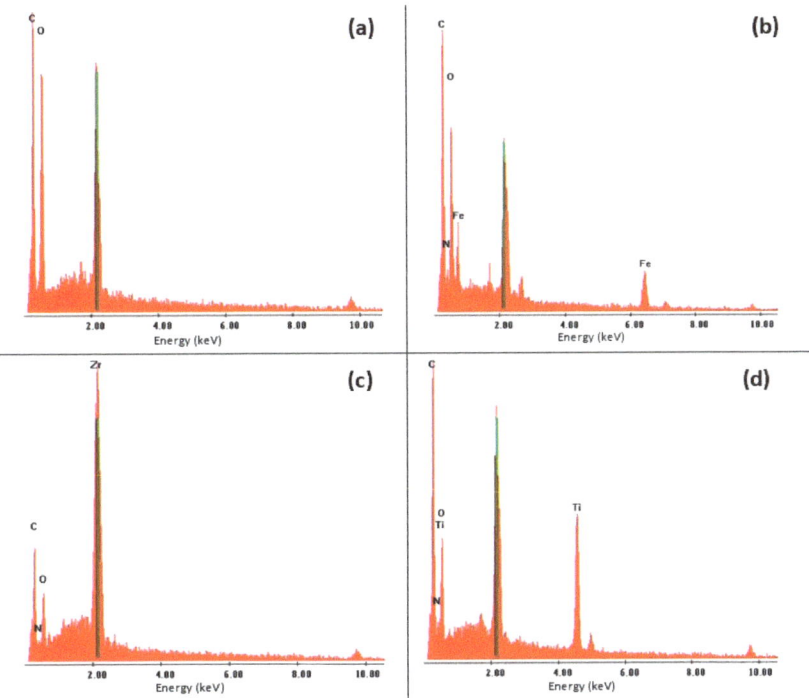

Figure 4. EDS spectra characterization: (**a**) unmodified paper surface, (**b**) paper surface modified with amino-functionalized MIL-101-NH$_2$ (Fe), (**c**) paper surface modified with amino-functionalized UiO-66-NH$_2$ (Zr), and (**d**) paper surface modified with amino-functionalized MIL-125-NH$_2$ (Ti). Note: the peak centered at 2.12 keV corresponds to the sputtered gold employed for the SEM technique.

3.2. Powder X-ray Diffraction (PXRD)

Using powder PXRD, experimental diffractograms were compared with the corresponding simulated powder plots (the latter were obtained using cif files extracted from the literature [10–12]). In MIL-101-NH$_2$ (Fe), the presence of a single phase could be verified. However, lower intensities are observed in the diffraction peaks, possibly due to a lower crystallinity related to the synthesis method (see Figure 5a). This MOF crystallized in the cubic space group Fd-3 m (No. 227). Performing the same analysis for the UiO-66-NH$_2$ (Zr) phase, a coincidence is also observed between the experimental pattern and its corresponding theoretical one (see Figure 5b). The compound with the formula crystallized in the cubic space group Fm-3 m. Figure 5c shows the comparison between the theoretical and experimental diffractograms of MIL-125-NH$_2$ (Ti), concluding in a broad coincidence between the diffraction peaks, confirming the presence of a single pure phase. This MOF was indeed found to be crystallized in the orthorhombic space group I4/mmm (N° 139).

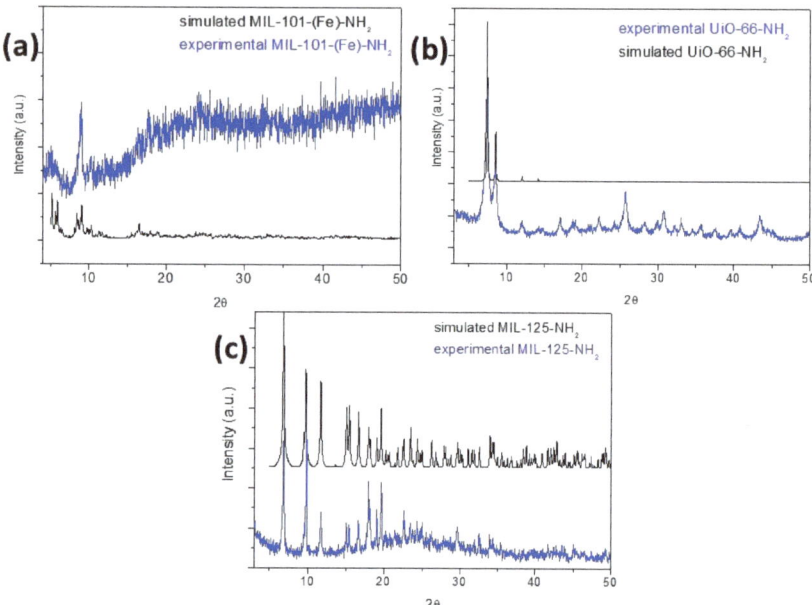

Figure 5. Experimental PXRD patterns of (**a**) MIL-101-NH$_2$ (Fe), (**b**) UiO-66-NH$_2$ (Zr), and (**c**) MIL-125-NH$_2$ (Ti) (blue lines) compared to the simulated diffractograms.

MIL-101-NH$_2$-Fe, with composition {[Fe$_3$ (μ_3-O) (μ4-2-ATA)$_3$ (H$_2$O)$_2$Cl]}, belongs to the family of materials known as MIL-101 (where MIL stands for Matériaux de l'Institut Lavoisier) [25]. The crystal structure of MIL-101 (Fe)-NH$_2$ is composed of 2-ATA linkers and Fe (III) ions, which serve as metallic nodes. The Fe (III) ions form a trigonal planar cluster and a trigonal prismatic secondary building unit (see Figure 6a) [26]. Four of these building units are connected through six 2-ATA linkers, forming a supertetrahedral unit. Figure 6b shows the final crystal structure of MIL-101-NH$_2$ (Fe), which contains different cages and sizes: a microporous cage with entrance windows approximately 8.6 Å in free diameter, and two mesoporous cages with diameters of approximately 29 Å and 34 Å.

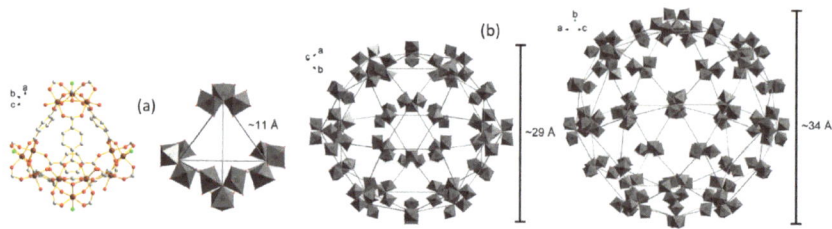

Figure 6. (**a**) Secondary building unit (color codes: gray: carbon; red: oxygen; green: chloride, brown: iron) and (**b**) the formed cages in MIL-101-NH$_2$ (Fe).

Figure 7 shows the most stable form of UiO-66-NH$_2$ (Zr), which consists of a face-centered-cubic structure containing an fm−3m symmetry with a lattice parameter of 20.7 Å. Moreover, it contains two separate cages, a tetrahedron cage of 7.5 Å, another octahedron cage of 12 Å, and a pore aperture of 6 Å [27].

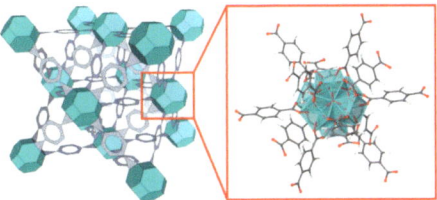

Figure 7. A schematic structure of UiO-66-NH$_2$ (Zr). The face-center-cubic UiO-66 structure type is composed of the metal node (aqua) and ligand (gray) with an atomic representation of the node and 12-connected 2-ATA linkers.

MIL-125-NH$_2$ (Ti) comprises [Ti$_8$O$_8$(OH)$_4$-(2-ATA)$_6$] building units, and its crystal structure is established from the cyclic octamers constructed from octahedral titanium units present at the edge or corner [28]. Also, these octamers are associated with 12 other cyclic octamers, giving rise to a porous 3-D quasi-cubic tetragonal structure (see Figure 8). Also, the structure has two types of cages, an octahedral (12.5 Å) and a tetrahedral (6 Å) cage, accessible through narrow triangular windows of ca. 6 Å.

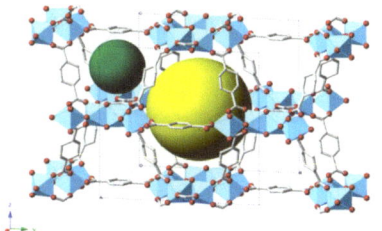

Figure 8. A schematic structure of MIL-125-NH$_2$-(Ti) (color codes: gray bars: carbon backnone; red: oxygen; light-blue polyhedra: titanium centers) shows the two cage types (green and yellow spheres).

3.3. Characterization of ALP Immobilization on the Support

FTIR

Another method employed to confirm the immobilization processes conducted on the paper substrate was Fourier transform infrared (FTIR) characterization.

The spectra provided below (see Figure 9) depict a comparison between the unaltered paper (shown in blue spectra) and the same paper modified with MIL-125-NH$_2$ (Ti), UiO-66-NH$_2$ (Zr), and MIL-101-NH$_2$ (Fe)—(black traces), followed by the subsequent stage where glutaraldehyde molecules were attached, leaving exposed terminal aldehyde groups (depicted in red spectra) for all three cases.

By the FTIR technique, it is possible to analyze the principal vibrational modes of the cellulose fibers and their interactions with the MOF particles. The paper exhibits the C-H stretching modes of cellulose located at 2850 cm^{-1} and a prominent band attributed to the vibrations of the C-O-C groups within the β-glucopyranose ring of cellulose at 1066 cm^{-1}. This band decreases significantly upon modifying the paper with MOFs, suggesting a robust covalent interaction between these bonds and the amino groups functionalizing the materials [9]. The following chemical reaction illustrates the bonding between the amino groups of MOFs and the aldehyde groups of cellulose.

$$\text{MOF-NH}_2 + \text{H-(CO)-C}_3\text{H}_6\text{-(CO)-H} \rightarrow \text{MOF-N=CH-C}_3\text{H}_6\text{-(CO)-H}$$

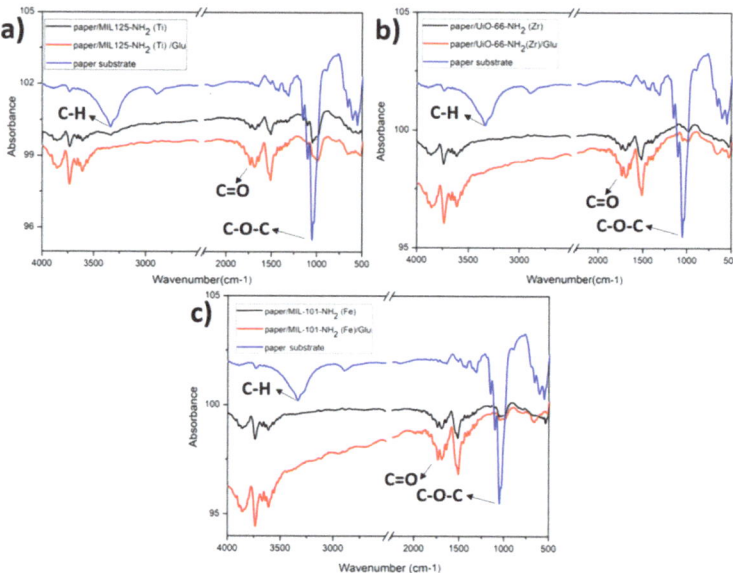

Figure 9. FTIR study of paper modification with (**a**) MIL-125-NH$_2$ (Ti) and subsequent stage with GLU; (**b**) UiO-66-NH$_2$ (Zr) and subsequent stage with GLU; and (**c**) MIL-101-NH$_2$ (Fe) and subsequent stage with GLU.

In all cases where the paper support is modified with different MOFs, the presence of solvent trapped in its pores is evident at 3625 cm^{-1}. Finally, there is an increase in the intensity of the carbonyl band (C=O stretching) centered at 1745 cm^{-1} when glutaraldehyde is bound to the MOFs.

3.4. Optimization of Experimental Parameters

To carry out the elemental parameter optimization tests, the following considerations were taken into account: adjust the MOF concentration for the paper support modification, adjust the ALP enzyme concentration, optimize the 4-MUP substrate concentration, determine the reaction time, and identify the optimum pH for enzymatic activity.

For the ALP enzyme, the substrate used was 4-MUP. After enzymatic action, the substrate produces 4-methylumbelliferone (4-MU), a fluorescent compound emitting at 452 nm.

In addition, to optimize the concentrations of the different MOFs used to modify the paper support, dispersions of the three MOFs were prepared in 1 mol·L^{-1} PBS buffer solution pH 7.2, in a concentration range of 0 to 0.8% w/v. The ALP enzyme was immobilized at 0.1% w/v in 0.1 mol·L^{-1} PBS buffer pH 7.2 for each modified paper with the different MOF concentrations. Finally, the ALP response was monitored using 5 µL of fluorescent substrate solution (2.5 × 10^{-3} mol·L^{-1} 4-MUP in DEA buffer solution (0.1 mol·L^{-1} diethanolamine, 5 × 10^{-2} mol·L^{-1} KCl, 1 × 10^{-3} mol·L^{-1} MgCl$_2$) pH 9.6) in each paper microzone. Subsequently, the enzyme product was measured using LIF.

The obtained signal from the enzymatic response was greater when the microzones of the paper were modified with different concentrations of MIL-125-NH$_2$ (Ti), which demonstrates that there is an increase in the immobilization surface due to the incorporation of MIL-125-NH$_2$ (Ti) in comparison to UiO-66-NH$_2$ (Zr) and MIL-101-NH$_2$ (Fe). On the other hand, when higher concentrations of MIL-125-NH$_2$ (Ti) than 0.5% w/v were used, a system saturation was observed, so this was selected as the concentration limit for all subsequent tests (see Figure 10).

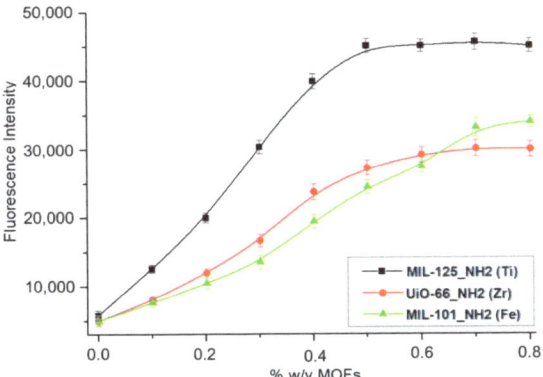

Figure 10. Optimization of impregnated MOF concentrations for paper support modification.

In light of these results highlighting the differential response of the modified papers with MIL-125-NH$_2$ (Ti)/ALP compared to those with UiO-66-NH$_2$ (Zr)/ALP, MIL-101-NH$_2$ (Fe)/ALP, it was decided to continue with the optimizations of experimental variables using the MOF MIL-125-NH$_2$ (Ti) as a proposal for the best surface to apply to the development of biosensors.

According to the literature, MIL-101-NH$_2$ (Fe) exhibits a higher BET surface area of 3528 m$^2 \cdot$g^{-1} and a pore volume of 1.48 cm$^3 \cdot$g^{-1} [26] with respect to those values from MIL-125-NH$_2$ (Ti) (1469 m$^2 \cdot$g^{-1} and a pore volume of 0.6 cm$^3 \cdot$g^{-1}) [28] and UiO-66-NH$_2$ (Zr) (822 m$^2 \cdot$g^{-1} and a pore volume of 0.23 cm$^3 \cdot$g^{-1}) [29]. Nevertheless, MIL-125-NH$_2$ (Ti) exhibits better immobilization performance, which could be reasonably attributed to a lower pore volume and, consequently, lower water occupancy.

Optimizing the concentration of ALP employed for immobilization was a critical parameter to evaluate. Therefore, a comprehensive study was conducted within the 0.1 to 2.5 mg·mL^{-1} concentration range. Figure 11a illustrates that the fluorescence intensity increased proportionally with the concentration of immobilized ALP, reaching a peak at 1 mg·mL^{-1}. Beyond this concentration, the intensity plateaued, indicating saturation. Consequently, 1 mg·mL^{-1} of ALP emerged as the optimal concentration for immobilization on modified paper with 0.5% w/v of MIL-125-NH$_2$ (Ti) suspension.

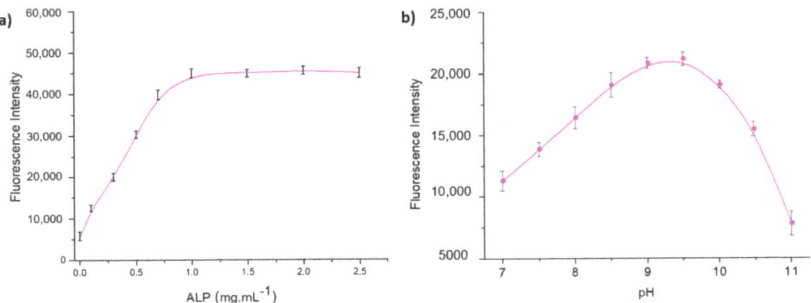

Figure 11. (**a**) Optimization of the ALP concentration used for immobilization; (**b**) optimization of the optimal pH range of the enzymatic activity.

The impact of pH on the enzymatic response was also evaluated by adjusting the pH of the substrate solution to a range of 7 to 11. The enzyme activity exhibited its maximum value when the DEA buffer solution was at a pH of 9.6, as shown in Figure 11b. Therefore, pH 9.6 was selected as the optimal pH for all subsequent experimental measurements.

The impact of 4-MUP concentration on the enzymatic response of ALP was investigated within the range of 0.1×10^{-3}–5.0×10^{-3} mol·L^{-1} in DEA buffer solution at pH 9.6. The optimal concentration of 4-MUP, which produces the highest enzymatic response, was determined to be 2.5×10^{-3} mol·L^{-1}, as shown in Figure 12a. As a consequence, this concentration was adopted for all subsequent experimental measurements.

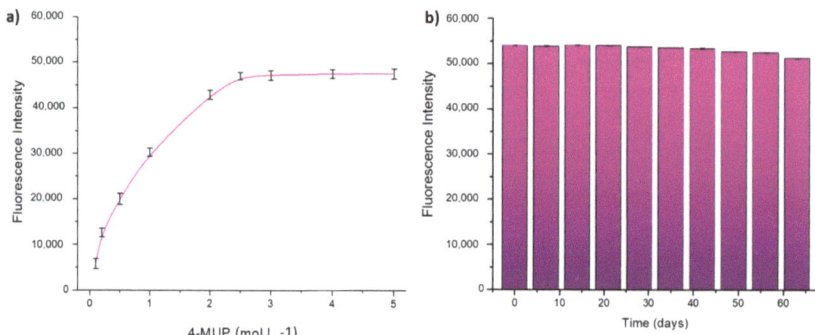

Figure 12. (**a**) Optimization of 4-MUP concentration; (**b**) stability tests of the modified paper support.

Long-term stability testing of the system was performed by freeze-drying the MIL-125-NH$_2$ (Ti)/ALP-modified paper devices and storing them at 4 °C, protected from light. Measurements were taken every 7 days, as illustrated in Figure 12b. In particular, the fluorescent signal response remained practically constant after 45 days, with only a 5% decrease observed after 60 days. This result shows the reproducibility of the sensor signal.

Reaction time is also a crucial parameter to define, especially in applications that demand fast and reliable analysis. This study used MIL-125-NH$_2$ (Ti)/ALP/4-MUP-modified papers and analyzed them at various incubation times ranging from 20 to 240 s. The fluorescence intensity progressively increased with longer incubation periods until reaching 180 s. Beyond this point, a saturation effect became evident. Consequently, the optimal reaction time was determined to be 180 s (see Figure 13). Finally, a summary of the experimental working experiments is depicted in Table 1.

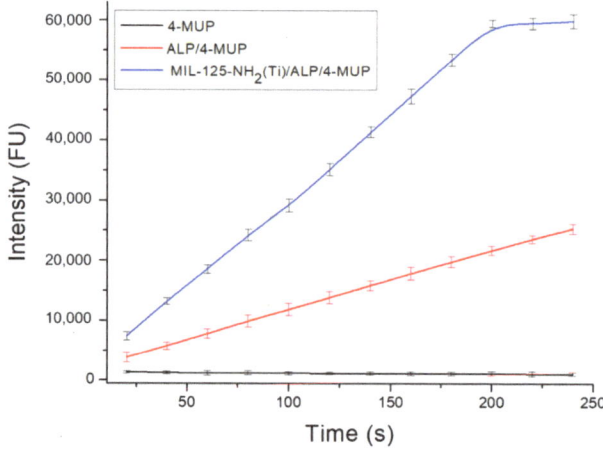

Figure 13. Fluorescence response at different incubation times of MIL-125-NH$_2$ (Ti)/ALP/4-MUP by modifying the paper support.

Table 1. Summary of optimal working conditions.

Sequence	Condition	Time (min)/Temperature (°C)
Cellulose oxidation	Oxygen plasma	3/room temperature
Modification of paper	0.5% w/v MOFs suspensions in PBS buffer solution, pH 7.2	30/room temperature in wet chamber
Washing buffer	PBS buffer solution, pH 7.2	1 min at a time
Cross-linking	0.5% w/v GLA in acetone pH 8	30/room temperature
Washing buffer	PBS buffer solution, pH 7.2	1 min at a time
Enzyme	1 mg.mL^{-1} ALP in PBS buffer solution pH 7.2	10/37 °C
Washing buffer	PBS buffer solution, pH 7.2	1 min at a time
Substrate	2.5×10^{-3} mol.L^{-1} 4-MUP in DEA buffer solution, pH 9.6	2

4. Conclusions

In our research, MIL-125-NH$_2$ (Ti), UiO-66-NH$_2$ (Zr), and MIL-101-NH$_2$ (Fe) MOFs were synthesized, characterized, and successfully used for the modification of paper surfaces towards development of PADs biosensors. It could be concluded that the MIL-125-NH$_2$ (Ti) MOF significantly increased the effective surface area for the immobilization of biomolecules in comparison to UiO-66-NH$_2$ (Zr) and MIL-101-NH$_2$ (Fe). This feature was confirmed by immobilizing the ALP enzyme and monitoring its activity by laser-induced fluorescence, giving rise to a higher signal intensity for the same concentrations of the three dispersions and modifying the paper support. In the light of these features, we chose MIL-125-NH$_2$ (Ti) for further optimization trials to conclude a reliable and reproducible platform suitable for biomolecule detection. Consequently, the MIL-125-NH$_2$(Ti)/ALP-modified papers showed excellent stability and reproducibility, demonstrating the potential use of this new modified platform for developing biosensors for the rapid detection of various bioanalytes of interest.

Author Contributions: Conceptualization, S.V.P., G.E.G. and M.A.F.-B.; methodology, S.V.P., G.E.G., and M.A.F.-B.; validation, S.V.P., G.R.T., A.B.S., M.D.R. and G.A.M.; formal analysis, S.V.P., G.E.G., M.D.R. and M.A.F.-B.; resources, S.V.P., G.R.T., A.B.S., M.D.R. and G.A.M.; writing—original draft preparation, S.V.P., G.E.G. and M.A.F.-B.; writing—review and editing, S.V.P., G.E.G., G.R.T., A.B.S., G.A.M. and M.A.F.-B.; supervision, G.E.G. and M.A.F.-B.; project administration, G.E.G., G.A.M. and M.A.F.-B.; funding acquisition, G.A.M. and M.A.F.-B. All authors have read and agreed to the published version of the manuscript.

Funding: Funding is acknowledged from the Universidad Nacional de San Luis (PROICO 02-2220), from Agencia Nacional de Promoción Científica y Tecnológica (AGENCIA, FONCyT, Argentina, PICT-2021-GRF-TII-0390, PICT-2021-GRF-TI-0136, PICT-2020-2347, and PICT-2020-2369), from Consejo Nacional de Investigaciones Científicas y Técnicas (CONICET, Argentina, PIP11220200100033CO), from Ministerio de Ciencia, Tecnología e Innovación, Argentina, Proyectos de Redes Federales de Alto Impacto: NANOQUIMISENS (IF-2023-85161983-APN-DNOYPI#MCT), and from Agencia Nacional de Investigación y Desarrollo (ANID) (ANID/FOVI220003).

Institutional Review Board Statement: Not applicable.

Informed Consent Statement: Not applicable.

Data Availability Statement: The complete data of this study can be obtained upon request from the corresponding author.

Acknowledgments: This research was supported by Universidad Nacional de San Luis, Instituto de Química de San Luis (INQUISAL); Consejo Nacional de Investigaciones Científicas y Técnicas (CONICET); and Agencia Nacional de Promoción Científica y Tecnológica (FONCYT) (PICT-BID).

Conflicts of Interest: The authors declare no conflicts of interest.

References

1. Batten, S.R.; Champness, N.R.; Chen, X.M.; Garcia-Martinez, J.; Kitagawa, S.; Öhrström, L.; O'Keeffe, M.; Suh, M.P.; Reedijk, J. Terminology of Metal-Organic Frameworks and Coordination Polymers (IUPAC Recommendations 2013). *Pure Appl. Chem.* **2013**, *85*, 1715–1724. [CrossRef]
2. Long, J.R.; Yaghi, O.M. The Pervasive Chemistry of Metal-Organic Frameworks. *Chem. Soc. Rev.* **2009**, *38*, 1213–1214. [CrossRef]
3. Xu, Q.; Kitagawa, H. MOFs: New Useful Materials—A Special Issue in Honor of Prof. Susumu Kitagawa. *Adv. Mater.* **2018**, *30*, 1803613. [CrossRef]
4. Figeys, D.; Pinto, D. Lab-on-a-Chip: A Revolution in Biological and Medical Sciences. *Anal. Chem.* **2000**, *72*, 330–335. [CrossRef]
5. Lu, R.; Shi, W.; Jiang, L.; Qin, J.; Lin, B. Rapid Prototyping of Paper-Based Microfluidics with Wax for Low-Cost, Portable Bioassay. *Electrophoresis* **2009**, *30*, 1497–1500. [CrossRef] [PubMed]
6. Cate, D.M.; Adkins, J.A.; Mettakoonpitak, J.; Henry, C.S. Recent Developments in Paper-Based Microfluidic Devices. *Anal. Chem.* **2015**, *87*, 19–41. [CrossRef]
7. Martinez, A.W.; Phillips, S.T.; Whitesides, G.M.; Carrilho, E. Diagnostics for the Developing World: Microfluidic Paper-Based Analytical Devices. *Anal. Chem.* **2010**, *82*, 3–10. [CrossRef] [PubMed]
8. Wang, S.; Ge, L.; Song, X.; Yan, M.; Ge, S.; Yu, J.; Zeng, F. Simple and Covalent Fabrication of a Paper Device and Its Application in Sensitive Chemiluminescence Immunoassay. *Analyst* **2012**, *137*, 3821–3827. [CrossRef]
9. Pelton, R. Bioactive Paper Provides a Low-Cost Platform for Diagnostics. *TrAC Trends Anal. Chem.* **2009**, *28*, 925–942. [CrossRef]
10. Hendon, C.H.; Tiana, D.; Fontecave, M.; Sanchez, C.; D'Arras, L.; Sassoye, C.; Rozes, L.; Mellot-Draznieks, C.; Walsh, A. Engineering the Optical Response of the Titanium-MIL-125 Metal-Organic Framework through Ligand Functionalization. *J. Am. Chem. Soc.* **2013**, *135*, 10942–10945. [CrossRef]
11. Ding, L.; Shao, P.; Luo, Y.; Yin, X.; Yu, S.; Fang, L.; Yang, L.; Yang, J.; Luo, X. Functionalization of UiO-66-NH$_2$ with Rhodanine via Amidation: Towarding a Robust Adsorbent with Dual Coordination Sites for Selective Capture of Ag (I) from Wastewater. *Chem. Eng. J.* **2020**, *382*, 123009. [CrossRef]
12. Shan, Y.; Xu, C.; Zhang, H.; Chen, H.; Bilal, M.; Niu, S.; Cao, L.; Huang, Q. Polydopamine-Modified Metal–Organic Frameworks, NH$_2$-Fe-MIL-101, as PH-Sensitive Nanocarriers for Controlled Pesticide Release. *Nanomaterials* **2020**, *10*, 2000. [CrossRef] [PubMed]
13. Pashazadeh-Panahi, P.; Belali, S.; Sohrabi, H.; Oroojalian, F.; Hashemzaei, M.; Mokhtarzadeh, A.; de la Guardia, M. Metal-Organic Frameworks Conjugated with Biomolecules as Efficient Platforms for Development of Biosensors. *TrAC Trends Anal. Chem.* **2021**, *141*, 116285. [CrossRef]
14. Sohrabi, H.; Salahshour Sani, P.; Orooji, Y.; Majidi, M.R.; Yoon, Y.; Khataee, A. MOF-Based Sensor Platforms for Rapid Detection of Pesticides to Maintain Food Quality and Safety. *Food Chem. Toxicol.* **2022**, *165*, 113176. [CrossRef]
15. Osman, D.I.; El-Sheikh, S.M.; Sheta, S.M.; Ali, O.I.; Salem, A.M.; Shousha, W.G.; EL-Khamisy, S.F.; Shawky, S.M. Nucleic Acids Biosensors Based on Metal-Organic Framework (MOF): Paving the Way to Clinical Laboratory Diagnosis. *Biosens. Bioelectron.* **2019**, *141*, 111451. [CrossRef]
16. Dai, C.; Gan, Y.; Qin, J.; Ma, L.; Liu, Q.; Huang, L.; Yang, Z.; Zang, G.; Zhu, S. An Ultrasensitive Solid-State ECL Biosensor Based on Synergistic Effect between Zn-NGQDs and Porphyrin-Based MOF as "on-off-on" platform. *Colloids Surf. B Biointerfaces* **2023**, *226*, 113322. [CrossRef] [PubMed]
17. Li, M.; Zhang, G.; Boakye, A.; Chai, H.; Qu, L.; Zhang, X. Recent Advances in Metal-Organic Framework-Based Electrochemical Biosensing Applications. *Front. Bioeng. Biotechnol.* **2021**, *9*, 797067. [CrossRef]
18. Ortiz-Gómez, I.; Salinas-Castillo, A.; García, A.G.; Álvarez-Bermejo, J.A.; de Orbe-Payá, I.; Rodríguez-Diéguez, A.; Capitán-Vallvey, L.F. Microfluidic Paper-Based Device for Colorimetric Determination of Glucose Based on a Metal-Organic Framework Acting as Peroxidase Mimetic. *Microchim. Acta* **2018**, *185*, 47. [CrossRef]
19. Wei, X.; Guo, J.; Lian, H.; Sun, X.; Liu, B. Cobalt Metal-Organic Framework Modified Carbon Cloth/Paper Hybrid Electrochemical Button-Sensor for Nonenzymatic Glucose Diagnostics. *Sens. Actuators B Chem.* **2021**, *329*, 129205. [CrossRef]
20. Hassanzadeh, J.; Al Lawati, H.A.J.; Al Lawati, I. Metal-Organic Framework Loaded by Rhodamine b as a Novel Chemiluminescence System for the Paper-Based Analytical Devices and Its Application for Total Phenolic Content Determination in Food Samples. *Anal. Chem.* **2019**, *91*, 10631–10639. [CrossRef]
21. Catalá-Icardo, M.; Gómez-Benito, C.; Martínez-Pérez-Cejuela, H.; Simó-Alfonso, E.F.; Herrero-Martínez, J.M. Green Synthesis of MIL53 (Al)-Modified Paper-Based Analytical Device for Efficient Extraction of Neonicotinoid Insecticides from Environmental Water Samples. *Anal. Chim. Acta* **2024**, *1316*, 342841. [CrossRef] [PubMed]
22. Chang, J.; Hu, R.; Zhang, J.; Hou, T.; Li, F. Two-dimensional metal-organic framework nanozyme-mediated portable paper-based analytical device for dichlorophen assay. *Biosens. Bioelectron.* **2024**, *255*, 116271. [CrossRef] [PubMed]
23. Feng, Q.; Wang, C.; Miao, X.; Wu, M. A novel paper-based electrochemiluminescence biosensor for non-destructive detection of pathogenic bacteria in real samples. *Talanta* **2024**, *267*, 125224. [CrossRef] [PubMed]
24. Guan, J.; Xiong, Y.; Wang, M.; Liu, Q.; Chen, X. A novel functionalized CdTe@MOFs based fluorometric and colorimetric biosensor for dual-readout assay of creatinine. *Sens. Actuators B Chem.* **2024**, *399*, 134842. [CrossRef]
25. Lebedev, O.I.; Millange, F.; Serre, C.; Van Tendeloo, G.; Férey, G. First direct imaging of giant pores of the metal−organic framework MIL-101. *Chem. Mater.* **2005**, *17*, 6525–6527. [CrossRef]

26. Capková, D.; Almáši, M.; Kazda, T.; Čech, O.; Király, N.; Čudek, P.; Fedorková, A.S.; Hornebecq, V. Metal-organic framework MIL-101 (Fe)–NH$_2$ as an efficient host for sulphur storage in long-cycle Li–S batteries. *Electrochim. Acta* **2020**, *354*, 136640. [CrossRef]
27. Winarta, J.; Shan, B.; Mcintyre, S.M.; Ye, L.; Wang, C.; Liu, J.; Mu, B. A decade of UiO-66 research: A historic review of dynamic structure, synthesis mechanisms, and characterization techniques of an archetypal metal–organic framework. *Cryst. Growth Des.* **2019**, *20*, 1347–1362. [CrossRef]
28. Kim, S.-N.; Kim, J.; Kim, H.-Y.; Cho, H.-Y.; Ahn, W.-S. Adsorption/catalytic properties of MIL-125 and NH$_2$-MIL-125. *Catal. Today* **2013**, *204*, 85–93. [CrossRef]
29. Cao, Y.; Zhang, H.; Song, F.; Huang, T.; Ji, J.; Zhong, Q.; Chu, W.; Xu, Q. UiO-66-NH$_2$/GO composite: Synthesis, characterization and CO$_2$ adsorption performance. *Materials* **2018**, *11*, 589. [CrossRef]

Disclaimer/Publisher's Note: The statements, opinions and data contained in all publications are solely those of the individual author(s) and contributor(s) and not of MDPI and/or the editor(s). MDPI and/or the editor(s) disclaim responsibility for any injury to people or property resulting from any ideas, methods, instructions or products referred to in the content.

Article

Metal–Organic Framework-Derived CeO₂/Gold Nanospheres in a Highly Sensitive Electrochemical Sensor for Uric Acid Quantification in Milk

Miloš Ognjanović [1], Milena Marković [2], Vladimír Girman [3,4], Vladimir Nikolić [5], Sanja Vranješ-Đurić [1], Dalibor M. Stanković [1,5,*] and Branka B. Petković [2,*]

1. VINČA Institute of Nuclear Sciences, National Institute of the Republic of Serbia, University of Belgrade, 11000 Belgrade, Serbia; miloso@vin.bg.ac.rs (M.O.); sanjav@vin.bg.ac.rs (S.V.-Đ.)
2. Faculty of Sciences and Mathematics, University of Priština in Kosovska Mitrovica, Lole Ribara 29, 38220 Kosovska Mitrovica, Serbia; milenatasic4@gmail.com
3. Institute of Physics, Faculty of Science, P. J. Šafárik University in Košice, Park Angelinum 9, 041 54 Košice, Slovakia; vladimir.girman@upjs.sk
4. Institute of Materials Research, Slovak Academy of Sciences, Watsonova 47, 040 01 Košice, Slovakia
5. Faculty of Chemistry, University of Belgrade, Studentski trg 12-16, 11000 Belgrade, Serbia; nikolicv@chem.bg.ac.rs
* Correspondence: dalibors@chem.bg.ac.rs (D.M.S.); branka.petkovic@pr.ac.rs (B.B.P.)

Abstract: In this work, CeBTC (a cerium(III) 1,3,5-benzene-tricarboxylate), was used as a precursor for obtaining CeO₂ nanoparticles (nanoceria) with better sensor performances than CeO₂ nanoparticles synthesized by the solvothermal method. Metal–organic framework-derived nanoceria (MOFdNC) were functionalized with spheric gold nanoparticles (AuNPs) to further improve non-enzymatic electrode material for highly sensitive detection of prominent biocompound uric acid (UA) at this modified carbon paste electrode (MOFdNC/AuNPs&CPE). X-ray powder diffraction (XRPD) and transmission electron microscopy (TEM) analysis were used for morphological structure characterization of the obtained nanostructures. Cyclic voltammetry and electrochemical impedance spectroscopy, both in an $[Fe(CN)_6]^{3-/4-}$ redox system and uric acid standard solutions, were used for the characterization of material electrocatalytic performances, the selection of optimal electrode modifier, and the estimation of nature and kinetic parameters of the electrode process. Square-wave voltammetry (SWV) was chosen, and the optimal parameters of technique and experimental conditions were established for determining uric acid over MOFdNC/AuNPs&CPE. Together with the development of the sensor, the detection procedure was optimized with the following analytical parameters: linear operating ranges of 0.05 to 1 µM and 1 to 50 µM and a detection limit of 0.011 µM, with outstanding repeatability, reproducibility, and stability of the sensor surface. Anti-interference experiments yielded a stable and nearly unchanged current response with negligible or no change in peak potential. After minor sample pretreatment, the proposed electrode was successfully applied for the quantification of UA in milk.

Keywords: MOF; CeO₂; uric acid; electrochemical sensor; milk samples

Citation: Ognjanović, M.; Marković, M.; Girman, V.; Nikolić, V.; Vranješ-Đurić, S.; Stanković, D.M.; Petković, B.B. Metal–Organic Framework-Derived CeO₂/Gold Nanospheres in a Highly Sensitive Electrochemical Sensor for Uric Acid Quantification in Milk. *Chemosensors* **2024**, *12*, 231. https://doi.org/10.3390/chemosensors12110231

Received: 30 September 2024
Revised: 31 October 2024
Accepted: 1 November 2024
Published: 3 November 2024

Copyright: © 2024 by the authors. Licensee MDPI, Basel, Switzerland. This article is an open access article distributed under the terms and conditions of the Creative Commons Attribution (CC BY) license (https://creativecommons.org/licenses/by/4.0/).

1. Introduction

Uric acid (UA) is a weak organic acid, hardly soluble in water or ethanol, and it forms ions and salts known as urates and acid urates. Scheele separated monosodium urate crystals from kidney gout nodules and connected this compound to gout, a common, complex, and painful form of arthritis [1]. After that, Fourcroy discovered uric acid in human urine. Uric acid is a final oxidation product of purine metabolism, extracted in urine. The concentration of UA in the blood can increase, reaching 408 µM in serum (6.8 mg/dL), which can be diagnosed as pathological hyper-uricemia [2], if the body does not contain enough uricase, an enzyme for the catalysis of UA oxidation when UA is converted in water-soluble allantoin. Besides gout, this condition may be a precursor to the development of metabolic syndrome, diabetes, hypertension, and kidney and cardiovascular diseases, but some genetic diseases are related to small concentrations of UA [3]. As the body converts purines into uric acid, the consumption of food rich in purines can contribute to hyper-uricemia-developed diseases [4]. So, keeping uric acid at optimum levels, in a range between 5.0 and 7.0 mg/dL, is important for the normal functioning of the organism, and food intake could be a major contributor to this. It is important to mention that uric acid also plays an active, vital role in the body's function in terms of initiating the inflammatory process that is necessary for tissue repair, mobilizing the progenitor of endothelial cells, and providing an antioxidant defense from oxygen free radicals that cause aging and cancer [5]. In food technology, a recent study confirmed that the antioxidant activity of uric acid can protect milk from rapid microbiological spoilage [6]. Zuo et al. determined uric acid by Hydrophilic Interaction HPLC [7], and they found 24.1 ± 0.05 to 36.8 ± 0.04 mg mL^{-1} of UA in bovine milk. Motshakeri et al. applied cyclic voltammetry to different milk samples and found 52.3 ± 2.1 to 92.9 ± 2.3 µM of UA, with no significant difference in uric acid content between fresh milk, UHT, and other processed milk samples [8]. Khamzina and co-workers found a high content of uric acid in fresh cow's milk (208 ± 1 µM), while in pasteurized milk samples, they found much lower concentrations of 61 ± 6 and 77 ± 3 µM of uric acid [6]. Bearing in mind all of this, researching uric acid effects in the human body tissue and food and developing new, reliable, and accurate analytical procedures and methods for determining low UA concentrations are of high importance for medical sciences and industrial technologies.

Besides the spectral, fluorescence, capillary electrophoresis detections and various HPLC methods [9], the significance of determining uric acid has also conditioned the development of electrochemical methods for this purpose [10,11]. Among other analytical tools, electrochemical sensors proved themselves to be one of the most promising, bearing in mind low-cost instrumentation, facile fabrication, easy operation, fast response, good sensitivity and selectivity, and minor sample preparation [12,13]. CeO_2 NPs (nanoceria) belong to high-quality electrode modification materials because of their unique physicochemical properties like switching between Ce^{+3} and Ce^{+4} oxidation states, ionic conductivity, oxygen storing capacity, chemically inertness, and thermal stability. For this reason, CeO_2-based materials are suitable for diverse applications, such as catalysts, sensors, actuators, transducers, and supercapacitors [11]. Employed like uric acid sensor electrode material (often simultaneously determining other biocompounds like ascorbic acid, dopamine, and others), nanoceria were reported in different structures like CeO_2 nanocubes [14], sponge-like CeO_2 [15], and ZnO-CeO_2 hollow nanospheres [16] or doped with cobalt [17] or indium nanoparticles [18]. As can be seen, for synergic improvement in sensor properties, CeO_2 nanoparticles can be easily combined or functionalized with other nanomaterials, such as noble and other metals and metal oxide nanoparticles, and often with carbon nanomaterials. Au nanostructures were reported in the literature in different sizes, shapes, and compositions and applied in electrochemical sensors because of their biocompatibility, good thermal and electrical conductivity, chemical stability, and high volume/surface ratio [19]. Gold nanoparticles can be synthesized by physical, chemical, biological, and electrochemical methods [20], and some of them are commercially available. Nanoparticles stabilized in active CeO_2 support are among the most popular gold-based systems for CO oxidation and other oxidation reactions [21]. Different AuNPs/CeO_2 nanocomposites were successfully fabricated: Huang and coauthors applied atomically precise Au_{25} superatoms with electron-deficient Au_{12} shells and electron-rich Au_{13} cores immobilized on the surface of CeO_2 nanorods as catalysts for styrene oxidation [22]. Gold nanoclusters supported on mesoporous CeO_2 nanospheres were used for the reduction of nitrobenzene; Jiao et al. obtained strawberry-like Au@CeO_2 nanoparticles by the assembly of block copolymer composite micelles [23], while Tang and coauthors employed Au-CeO_2 composite aerogels with tunable Au nanoparticle sizes as plasmonic photocatalysts for CO_2 reduction [24]. Also, Palanisamy et al. reported electrochemical sensing and the simultaneous detection of dopamine and uric acid at electrochemically deposited gold 3D nanoclusters on nanoceria of biomolecule adsorption and mass transport [25]. Considering the catalytic performance and selectivity of nanoceria/Au nanocomposite materials, it can be concluded that the structure properties and morphologies of metal and metal oxide nanoparticles play a key role in the tailoring of new functional materials.

Metal–organic frameworks (MOFs) present a prestigious group of materials with tunable pore sizes and high surface area, suitable for diverse applications [26,27]. On the other hand, the insufficient electronic conductivity and relatively poor chemical stability of these materials are a significant disadvantage, which can be overcome in MOF-derived functional materials: metal oxides and hierarchical carbon-based hybrid structures [28], in this form applicable for electrochemical sensing [29]. The approach through template-assisted nanomaterials offers promising protocols for obtaining nanoparticles with improved performances [30]. Then, MOF-derived metal oxide nanostructures can be combined with other functional materials in order to design further improved electronic structure of the sensor interface and enhance catalytic performance, which was a goal of this work.

In light of this, it is worthwhile to explore and compare the structural and electrochemical properties of CeO_2 NPs, prepared in two different ways: by one of the most preferred methods—solvothermal synthesis—and nanoceria, easily prepared from CeBTC MOF, as a precursor for obtaining a functional CeO_2 nanostructure. Electrode modification with CeO_2 NPs can contribute to resistance to fouling from the oxidation of interferences and oxidation products of uric acid [16,31]. As MOF-derived CeO_2 NPs can promote electron transfer between uric acid and the electrode, adding AuNPS in the electrode modifier can further improve the electrocatalytic activity of CeO_2. Together, they can form a stable composite material that enhances sensitivity and response time for the electrochemical detection of uric acid. Hence, the selected favorable composite electrode modifier for a uric acid sensor with improved activity was prepared by combining two nanomaterials: 1—nanoceria, obtained by thermolysis of CeBTC MOF, synthesized at room temperature, and 2—gold nanoparticles, commercially available. Morphological and electrochemical characterization was used for structure analysis and to determine the sensor properties of the proposed binary nanocomposite. A sensitive and fast analytical SWV method at CeBTC-derived nanoceria/AuNP-modified CPE was developed by changing and optimizing experimental and technique parameters. Selectivity over possible interferences was estimated, and MOFdNC/AuNPs&CPE was successfully applied to determine uric acid in milk samples.

2. Materials and Methods

2.1. Chemicals

Cerium(III)-chloride heptahydrate and benzene-1,3,5-tricarboxylic acid (BTC) were purchased from Alfa Aesar (Haverhill, MA, USA). Ethylene glycol ($\geq$99.5%) was acquired by Fluka, while poly(vinylpyrrolidone) (Mw 40,000) and sodium acetate trihydrate ($\geq$99%) were obtained from Sigma Aldrich (St. Louis, MO, USA). Gold nanoparticles, 10 nm diameter, OD 1, stabilized suspension in citrate buffer, paraffin oil, and glassy carbon powder were all products of the Sigma Aldrich company (USA). All other chemicals used to perform the experiments in this work were reagent grade. Britton–Roinson buffer (pH 2 to 9) was prepared by adjusting the pH of a 0.05 M solution of acetic, boric, and phosphoric acid with a 0.2 M NaOH solution. Ultrapure water was produced from the Millipore Milli Q system.

2.2. Methods and Instrumentation

The crystal structures of the synthesized nanoceria were determined through X-ray powder diffraction (XRPD) analysis. Dried powder samples were examined using a high-resolution Smart Lab® diffractometer (Rigaku, Japan) equipped with a CuKα radiation source working at 40 kV accelerating voltage and 30 mA current. Diffraction patterns were collected within the 2θ range of 10–70°.

Transmission electron microscopy (TEM) was performed using a JEOL JEM 2100F UHR microscope operated at 200 kV to analyze the nanoparticle morphology. Diluted aqueous dispersions of the samples were deposited onto commercial copper support TEM grids and left to dry overnight at room temperature. TEM images were acquired in bright field mode in a STEM regime.

All electrochemical measurements were performed at an Autolab 302 N with corresponding NOVA 2.0.2 software. The three-electrode electrochemical working station consisted of an Ag/AgCl (saturated KCl) as the reference electrode, a Pt-wire as the counter electrode, and a modified carbon paste electrode as a working electrode. For the electrochemical characterization of electrode modifiers, cyclic voltammetric (CV) and electrochemical impedance spectroscopy (EIS) measurements were performed in a redox test solution of 5 mM $K_3[Fe(CN)_6]/K_4[Fe(CN)_6]$ (1:1) in 0.1 M KCl. EIS measurements were performed with the frequency changed from $1·10^5$ Hz to $1·10^2$ Hz and a signal amplitude of 5 mV at the potential of 0.05 V. For analytical measurements, the SWV technique was used. The optimal parameters were found by varying the pulse amplitude from 10 to 50 mV, the frequency from 10 to 90 Hz, and the potential step from 2 to 8 mV/s. Measurements of uric acid standard solutions for the calibration curve and analysis of milk samples were performed by applying optimized parameters of the SWV technique: a pulse amplitude of 40 mV, a frequency of 80 Hz, and a potential step of 4 mV/s.

2.3. Synthesizing CeBTC MOF and Obtaining CeO$_2$ NPs

CeO_2 nanoparticles were synthesized by the solvothermal method. Cerium oxide nanoparticles (CeO_2) were synthesized following a modified literature procedure [32]. Specifically, the concentrations of $CeCl_3·7H_2O$ and poly(vinylpyrrolidone) (PVP40) were varied, and the experimental method was adjusted. A typical synthesis involved dissolving 4 mmol of $CeCl_3·7H_2O$ in 60 mL of ethylene glycol (EG) using ultrasound. Then, 80 mmol of PVP40 was slowly added under vigorous magnetic stirring and mild heating until a homogeneous, colorless solution was formed. Finally, 20 mg of sodium acetate trihydrate (NaAc) was added to the solution. The mixture was transferred to a 100 mL Teflon-lined stainless-steel autoclave, sealed, and heated at 190 °C for 4 h for hydrothermal treatment. After cooling to room temperature, the precipitate was centrifuged and washed several times with deionized water and ethanol to remove excess EG, PVP, and NaAc. The product was dried overnight at 80 °C and afterward calcined at 400 °C for 4 h for better crystallization.

During the synthesis of MOF-derived CeO_2 nanoparticles, a CeBTC metal–organic framework was easily synthesized by a complex reaction between Ce^{3+} and trimesic acid as a ligand at room temperature, according to the previously reported procedure [33]. In the first solution, 0.25 mmol of $CeCl_3·7H_2O$ was added to 50 mL of ultrapure water, and the second solution was prepared when 0.25 mmol of benzene-1,3,5-tricarboxylic acid was dissolved in 50 mL of an ethanol/water solution (v/v = 1:1) under vigorous stirring. These two solutions were mixed, and after a continuous stirring for 1.5 h, a white precipitate of CeBTC was collected by centrifugation. This precipitate was washed several times with ethanol/water (v/v, 1:1) and then dried at 60 °C. CeO_2 NPs were obtained by thermolysis of the as-prepared CeBTC. The calcination was performed at a temperature of 450 °C at 10 °C/min for 3 h in air conditions and cooled down to room temperature naturally.

2.4. Preparation of MOFdNC/AuNP-Modified CPE

First, 10 mg of MOF-derived CeO_2 NPs was mixed with 1 mL of the gold nanoparticle suspension and ultrasonicated for 1 h. Liquid from the suspension of CeO_2 NPs/Au NPs was left to evaporate at 40 °C. Then, 10 mg of light orange powder was mixed with paraffin oil and glassy carbon powder in a mortar, at a ratio of 1:4, up to a total amount of material of 100 mg. After standing overnight, the resulting paste was used as CeO_2 NPs/Au NP-modified CPE.

2.5. Preparation of Milk Samples

Before analysis, two samples of pasteurized milk were prepared by removing proteins with the following procedure: a certain amount of pasteurized milk samples was fully mixed with trichloroacetic acid (10%) and then centrifuged at 5000 rpm for 20 min to remove the protein fraction, and the supernatant was collected. Then, 100 µM of this milk supernatant was added into an electrochemical cell with BR buffer pH 6 and spiked with a certain amount of UA before each SWV measurement.

3. Results and Discussion

3.1. Structural Analysis and Morphological Characterization

The crystal structure of the synthesized cerium oxide nanoparticles, prepared both hydrothermally and by the pyrolysis of the cerium-based metal–organic framework (MOF), were analyzed using X-ray powder diffraction (XRPD). The obtained diffractograms indicated that both samples crystallized in the pure cubic CeO_2 phase (JCPDS card #43-1002) without any detectable impurities or other crystalline phases (Figure 1a) [34]. The presence of intense and broadened diffraction peaks in the XRPD patterns indicated the crystalline nature of the cerium oxide particles, as well as their nanoscale dimensions. The average crystallite size of the hydrothermally prepared CeO_2 was determined to be 8 ± 1 nm using the Scherrer equation applied to the most intense diffraction peaks. Similarly, the crystallite size of the MOF-derived CeO_2 was found to be 7 ± 1 nm. This analysis indicates that both synthesis methods successfully yielded pure cerium oxide nanoparticles with comparable crystallite sizes within the experimental error range. The XRD diffraction pattern of the parent compound (CeBTC) was also checked (Figure S1a). The observed diffraction peaks in the CeBTC particles closely matched those reported in the previous literature, which were indexed to $Ce(1,3,5-BTC)(H_2O)_6$ [33]. In addition, FT-IR spectra of CeBTC were analyzed (Figure S1b). The FTIR spectrum of mesoporous CeBTC exhibits prominent absorption bands at 1367, 1431, and 1608 cm^{-1}, characteristic of carboxylate group vibrations. These bands correspond to the asymmetric, symmetric C=O stretching modes and the C–O stretching mode associated with the binding of carboxylate (COOH) groups to Ce ions. The analyzed spectrum also clearly shows most of the typical bands associated with the COO$^-$ groups of BTC$_3^-$ ligands in the CeBTC metal–organic framework.

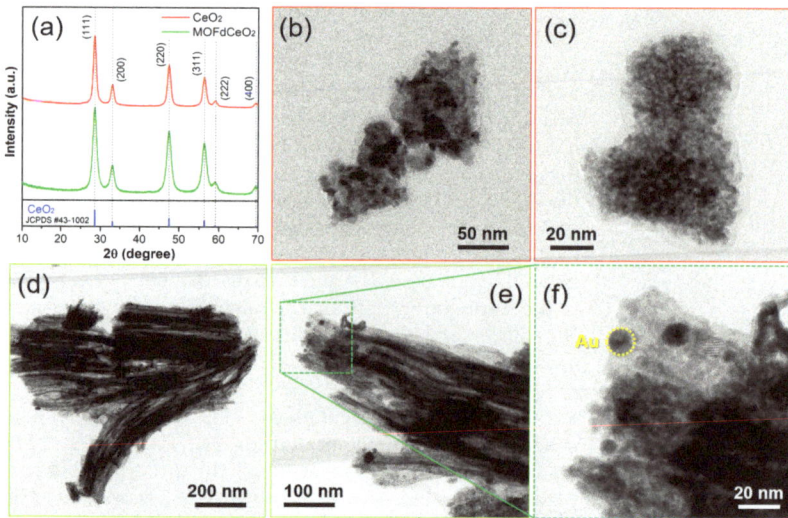

Figure 1. (a) XRPD diffractograms of solvothermal-prepared CeO_2 and MOFdCeO$_2$. The standard diffraction pattern of CeO_2 JCPDS #43-1002 is given as a reference. (b,c) TEM micrographs of solvothermal-prepared CeO_2. (d–f) TEM micrographs of MOFdCeO$_2$/AuNP nanocomposites.

The morphology of the synthesized samples was first characterized by transmission electron microscopy (TEM) measurements (Figure 1b–f). In contrast to XRPD measurements, the morphology of nanoceria samples prepared in different ways differs significantly. Nanoceria prepared hydrothermally consist of small spherical nanoparticles of an average diameter of 8.7 ± 1.2 nm, which are heavily agglomerated into cluster-like structures (Figure 1b,c). On the other hand, MOFdNC consists of layered nanotube-like structures that are stacked on top of each other. The length of each tube is between half and one micron, while the thickness of each layer is extremely small and amounts to 6.5 ± 1.5 nm, which is comparable to the average crystallite size from XRPD measurements (Figure 1d,e). Additionally, we prepared a MOFdNC/AuNP nanocomposite, which is best seen in Figure 1f. Spherical gold nanoparticles were decorated on the surface of MOFdNC tubes, potentially enhancing the surface area of the material and facilitating improved electron transfer and electrocatalytic properties. The morphology of the CeBTC particles was investigated using scanning electron microscopy (SEM). The pristine Ce-MOF exhibited an urchin-like structure composed of closely packed nanorods, with an overall diameter of about 4–5 μm (as displayed in Figure S1c–e). This structure resulted from the self-assembly of nanorods through a spilling growth mechanism, leading to the formation of larger hierarchical architectures, a phenomenon often observed in anisotropic crystal structures.

Liu et al. demonstrated a significant influence of the precursor morphology on the catalytic activity of CuCeO in their research [35]. They found that an urchin-like morphology exhibited superior catalytic activity compared to rod-like or strawsheaf-like morphologies, attributed to the presence of highly dispersed species and oxygen vacancies. The morphology of nanoceria plays a key role in the catalytic activity of modified electrodes as well, especially because of its surface reactivity and ability to store and release oxygen through changes in the oxidation state between Ce^{3+} and Ce^{4+}. These factors are crucial both for electrochemical and catalytic applications. A thorough electrochemical characterization is necessary to establish the correlation between particle morphology and catalytic activity on the electrode modifier.

3.2. Determination of the Electrochemical Properties of the Selected Catalysts in the $[Fe(CN)_6]^{3-/4-}$ Redox Probe

The results of our recent study confirmed the suitability of CeBTC-derived nanoceria as a potential sensor modifier. Based on those results, comparisons of the electrocatalytic behavior of pristine, solvothermal-prepared CeO_2 NP-, Au NP-, and MOF-derived nanoceria and gold NP-functionalized MOF-derived nanoceria-modified carbon paste electrodes were performed by cyclic voltammetry. Studies were performed in 5 mM ferro-ferri-cyanide/KCl solution within the potential window of −0.6 to +1 V at a scanning rate of 50 mV/s. In Figure 2, it is obvious that MOF-derived nanoceria show better electrochemical response compared with nanoceria obtained by solvothermal synthesis, and further functionalization of this material with gold nanospheres rapidly enhanced peak current. Slight changes in the potential of the peaks, increasing the potential of the oxidation peak towards more positive values and the reduction peak towards more negative values, are attributed to the heterogeneous surface of the electrode due to the homemade preparation of the electrode, which is expected for such measurements. The final electrode modification shows excellent cathodic and anodic peak current values of −34 μA and +36 μA, respectively, with a total increase of over 100% compared with the unmodified electrode. This effect can be attributed to the extraordinary properties of the selected materials and the collective effect of the prepared bimetallic nanocomposite material.

This current improvement can be attributed to the synergetic effect of gold nanoparticles and nanoceria, where the final composite demonstrates fast electron transfer kinetic and mass/charge transfer properties.

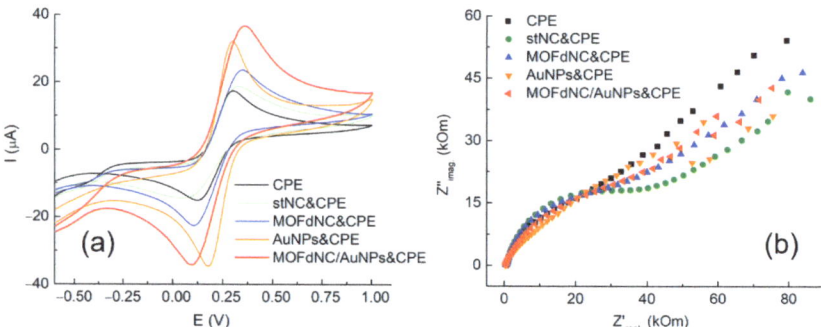

Figure 2. (a) CV responses of pristine CPE, solvothermal-synthesized nanoceria and MOF-derived nanoceria-modified CPEs, and CPE modified with binary nanocomposite MOF-derived nanoceria/AuNPs. (b) EIS responses of the pristine and modified electrodes.

Electrochemical impedance spectroscopy measurements were conducted to study the interface properties of electrode surfaces. The electron transfer kinetics of $[Fe(CN)_6]^{3-/4-}$ at pristine and different modified electrodes revealed a typical Nyquist plot, as can be seen in Figure 2b. All obtained EIS curves include a semicircle part at high frequencies that corresponds to the electron transfer limited process consistent with charge transfer resistance. The results demonstrated that modified electrodes possess smaller semicircles compared with pristine CPE, and the best electrochemical properties and promotion of the electron transfer process were spotted at MOF-derived nanoceria functionalized with gold nanospheres, corresponding to previously considered CV data.

3.3. Electrochemical Behavior and Kinetics of UA over MOFdNC/AuNPs&CPE

Previous electrochemical studies confirmed that MOF-derived nanoceria functionalized with gold nanospheres facilitated the charge transfer and made the electrode area highly conductive, while further preliminary experiments showed that CPE modified with this binary composite is extremely sensitive to the presence of uric acid. As can be seen in Figure 3a, the peak current for determining 20 µM of UA at MOFdNC/AuNPs&CPE in BR buffer, pH 6, is about 3.7-fold higher, 0.32 µA, than it was recorded at pristine CPE, 9 µA. On both electrodes, the peak originating from UA oxidation is evidenced at about 0.39 V, and the electrode reaction can be considered irreversible because of the absence of a reduction peak. By changing the pH value of the BR buffer as a supporting electrolyte from 2 to 9, it was found that the peak current highly depends on pH, and protons are directly involved in the oxidation reaction of UA. The peak current is the highest at pH 6, and this pH value was chosen and kept constant in further experiments. Figure 3b shows that the peak potential shifts toward less positive potentials with increasing pH, and the linear relationship between Ep and pH was found with the following regression Equation (1):

$$E_{pa} (V) = 0.75 - 0.058 \text{ pH} \quad (R = 0.993) \qquad (1)$$

The slope of 58 mV/pH showed that the electrochemical oxidation of UA followed the Nernst equation. This suggests that the electrode process involved an equal number of protons and electrons included in the electrode reaction process (two protons and two electrons in the first step of the overall electrochemical reaction, which is in agreement with previous reports [15,36–38]). According to [39,40], the electro-oxidation pathway of uric acid occurred through the exchange of two protons and electrons and forming diamine in the transition state, which further successively absorbed two molecules of water to form imine-alcohol and uric acid-4,5-diol. In natural pH, this compound can be decomposed on allantoin and CO_2. The proposed electrochemical reaction mechanism of UA oxidation is given in Figure 4.

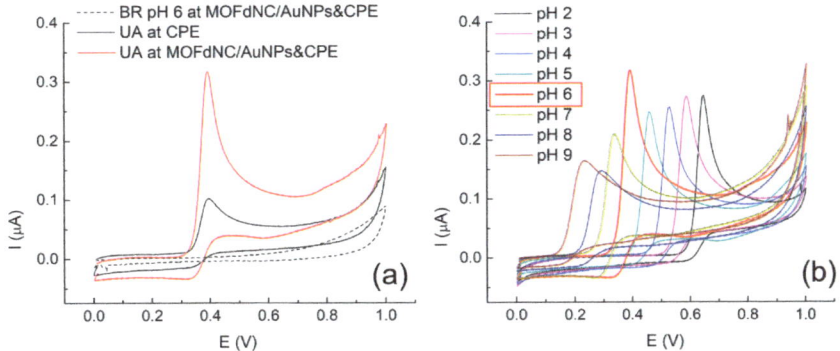

Figure 3. (a) CV profiles of 20 µM of UA at pristine and MOFdNC/AuNP-modified CPE in BR buffer at pH 6. (b) Effect of pH on CV profiles at MOFdNC/AuNPs&CPE (scan rate of 100 mV/s).

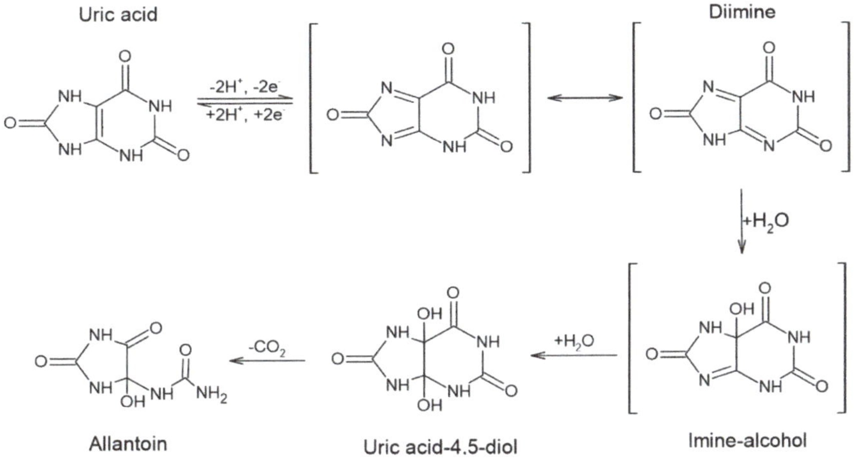

Figure 4. The suggested electro-oxidation mechanism of UA at MOFdNC/AuNPs&CPE.

3.4. Effect of the Potential Scan Rate and Study of the Electrode Process

The oxidation mechanism at MOFdNC/AuNPs&CPE was examined by recording cyclic voltammograms of 20 µM of uric acid in BR buffer, pH 6, applying scan rates of 10, 15, 25, 50, 75, 100, and 125 mV/s (Figure 5a). In this figure, it is evident that the peak current increased with an increasing scan rate and the peak potential (E_p) was slightly shifted to positive potentials. The irreversibility of the reaction was confirmed by the linear plot of E_p vs. the logarithm of the scan rate (Figure 5b), described by regression equation E_p (V) = 0.336 + 0.025 log v (mV/s) (R = 0.990). The linear relationship between the peak current and scan rate is shown in Figure 5c, with the corresponding fitting equation I (A) = $-4.43 \times 10^{-8} + 2.76 \times 10^{-9} \times$ v (mV/s); (R = 0.992). This linearity between I and v indicated that an absorption-controlled process occurred at the electrode surface. The linear plot log I vs. log v, fitted by the formula log I (A) = $-8.01 + 0.81 \times$ log v (V/s); (R = 0.991), is given in Figure 5d. The slope of 0.81 indicates the predominance of absorption over diffusion at the electrode surface, bearing in mind that the theoretical value for the fully absorption-controlled process is 1.0 [41].

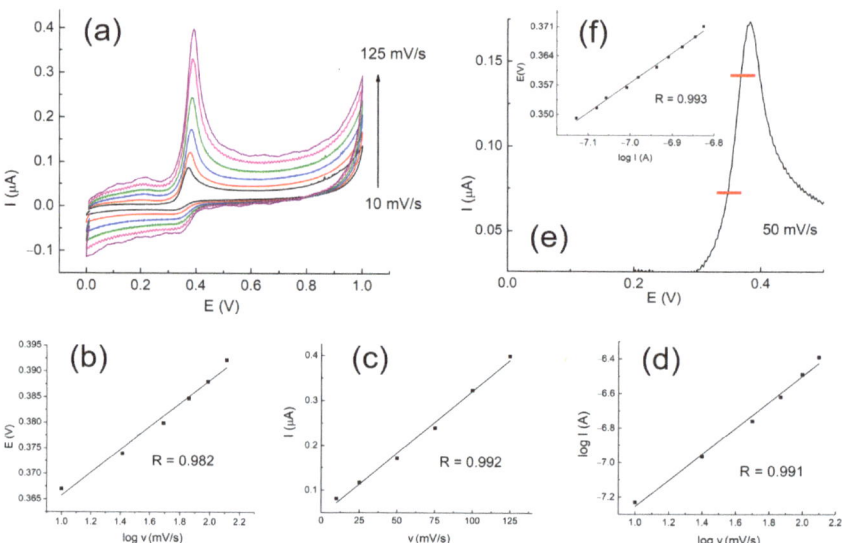

Figure 5. (**a**) Effect of the scan rate (20 µM of UA at various scan rates from 10 to 125 mV/s at MOFdNC/AuNPs&CPE in BR buffer at pH 6). (**b**) Plot of peak potential vs. log of the scan rate, derived from the graph in a. (**c**) Plot of the peak current vs. the scan rate. (**d**) Plot of log of the peak current vs. log of the scan rate. (**f**) Taffel plot derived from CV recorded at 50 mV/s, as presented in graph (**e**).

The Tafel plot (Figure 5e) derived from the rising part points (between red marks on the graph) of the cyclic voltammogram of 20 µM of UA recorded at a scan rate of 50 mV/s (Figure 5d) was fitted by the equation E (V) = 0.071 log I (A) + 0.856 (R = 0.993). The Tafel slope of 71 mV is close to the theoretical value of 60 mV for two electrons involved in charge transfer in the reaction process of UA at MOFdNC/AuNPs&CPE, which is in correlation with the proposed electro-oxidation mechanism (Figure 4). The electron transfer coefficient (α) was calculated to be 0.61, and it was found when the slope of E vs. log I was equated to $2.3RT/Fn(1-\alpha)$ [38] for the anodic reaction. In the given relation, R and F are the Gas and Faraday constant, T is absolute temperature, and n is the number of electrons. According to the Laviron theory for an adsorption-controlled and fully irreversible electrode process [42], Ep is defined by the following Equation (2):

$$E_p = E^0 + \left(\frac{2.303RT}{\alpha nF}\right)\log\left(\frac{RTk^0}{\alpha nF}\right) + \left(\frac{2.303RT}{\alpha nF}\right)\log v \qquad (2)$$

The value of the formal potential E_0 of 0.370 V was obtained from the intercept of E_p vs. v by extrapolation of the vertical axis to v = 0 mV/s. Thus, using the Laviron equation, the standard heterogeneous rate constant of the reaction, k_0, was calculated to be 1.39×10^3 s^{-1}. These results may be attributed to the good electrical conductivity and large electrode surface area of the used MOF-derived CeO$_2$ nanoparticles functionalized with gold nanospheres, which synergistically accelerate the electron transfer process between the UA molecules and proposed modified CPE.

3.5. Quantification of UA at MOFdNC/AuNPs&CPE

The development of an analytical procedure for the determination of uric acid at MOFdNC/AuNPs&CPE in optimal pH 6 was the next step. The comparison between two highly sensitive analytical electrochemical pulse techniques (differential pulse and square wave voltammetry in Figure S2) showed that the SWV proved to be a more suitable technique for the quantification of UA because of the almost 3-fold higher electrochemical response in the presence of this compound. The optimization of SWV parameters was completed by varying each of the examined experimental parameters systematically while the others were kept constant. The pulse amplitude was changed in the range from 10 to 50 mV, the square-wave frequency was varied from 10 to 90 Hz, and the influence of the potential step was examined from 2 to 8 mV. In further experiments, an amplitude of 40 mV, a frequency of 80 Hz, and a potential step of 4 mV were chosen and used in the quantification of UA.

Under the optimized experimental conditions and parameters of the technique (Figure S3), the SW voltammograms of different concentrations (0, 0.05, 0.1, 0.3, 0.5, 0.7, 1, 3, 5, 7, 10, 20, 30, 40, 50 μM) of uric acid were recorded at the proposed modified electrode and voltammetric profiles of whole investigated range (Figure 6a,b). The linearity of the data was observed for the range of higher concentrations from 1 to 50 μM and for the range of lower concentrations from 0,05 to 1 μM. The corresponding calibration curves are given below for SWVs (Figure 6c,d), while in Figure 6e, the lower concentration range is presented in better resolution. The distinct oxidation peak was recorded at 0.39 V with a proportional increase in the current response as the concentration of uric acid increased, and it was applied for quantitative determination. At the same time, no changes in the peak potentials were observed, additionally confirming excellent electrocatalytic performances of the proposed sensor. The equations that describe linearity in calibration curves are as follows:

$I\ (\mu A) = 0.688 + 0.321 \times C_{UA}\ (\mu M); R = 0.986$, for concentration range from 0.05 to 1 μM;

$I\ (\mu A) = 0.805 + 0.180 \times C_{UA}\ (\mu M); R = 0.997$, for concentration range from 1 to 50 μM.

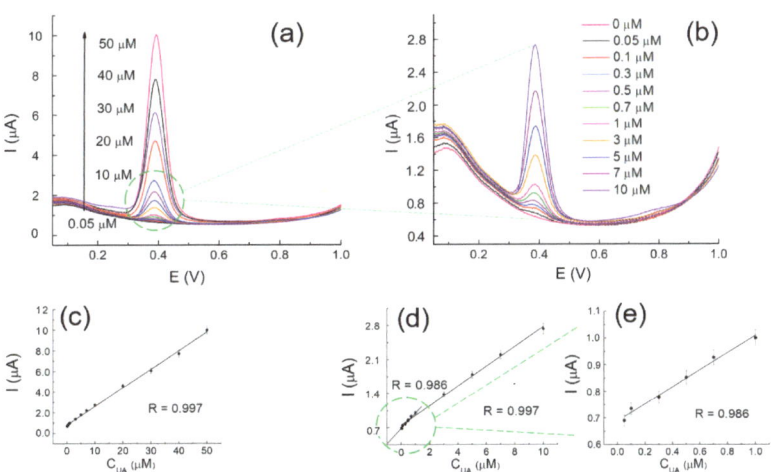

Figure 6. SW voltammograms of different concentrations of UA in BR buffer at pH 6 at MOFdNC/AuNPs&CPE in (**a**) the whole investigated concentration range and (**b**) for a range of lower concentrations. The corresponding calibration curves (**c,d**) are given below each voltammetric profile. (**e**) The calibration curve for low concentrations ranging from 0.05 to 1 μM.

The correlation coefficient R shows a high correlation between the corresponding linear relationship and the data for both concentration ranges.

The limit of the detection (LOD) was calculated from the formula LOD = 3 s/m, where s is the standard deviation of the blank solution (S/N = 3) and m is the slope of the calibration curve (for a range of lower concentrations). The LOD for the proposed UA determination at MOFdNC/AuNPs&CPE amounts to 0.011 μM, and based on this result and the width of the working range, the developed analytical procedure on the proposed sensor represents one of the electrochemical methods with the highest sensitivity for uric acid. In Table 1, previously reported electrochemical sensors are compared to this one in terms of concentration range and LOD. When considering the upper limit of the working range of concentrations, the low solubility of uric acid in water of 6 mg/100 mL should be kept in mind, which corresponds to 360 μM.

Table 1. Comparison of UA electrochemical sensors in terms of analytical parameters.

Electrochemical Sensor	Applied Technique	Linear Range (μM)	LOD (μM)	Ref.
CeO_2 nanocubes/GCE	DPV	10–700	4.3	[14]
CPE/CeO_2 sponge-lake porous	DPV	0.25–10; 10–300	0.06	[15]
ZnO-CeO_2/GCE	DPV	10–100	0.49	[16]
Co-CeO_2/GCE	SWV	1–2200	0.12	[17]
In–CeO_2/GCPE	SWV	0.079–148	0.0074	[18]
ITO-rGO-AuNPs	LSV	10–500	2.26	[43]
PVP-GR/GCE	SDLSV	0.04–100	0.02	[38]
UOx/Fc/Cu_2O/GCE	DPV	0.1–1000	0.0596	[44]
Cu-BTC/CPE	DPV	0.5–600	0.2	[45]
Ce@Zn-MOF/GCE	DPV	0–1.78 (0–300 ng/mL) *	0.003 (0.51 ng) *	[46]
PEDOT/GCE	CV	6–100	7	[47]
MOFdNC/AuNPs&CPE	SWV	0.05–1; 1–50	0.011	This work

* Original data on the UA concentration in the article.

Electrochemical sensors based on CeO_2 NPs have been commonly used to determine uric acid, as seen in Table 1. It is interesting to mention that even Ce salt was used to dope Zn-MOF to fabricate a highly sensitive sensor for UA [46]. This sensor deposited uric acid on the surface of the electrode modifier, and contrary to other electrochemical sensors, the addition of UA led to a decrease in the peak current. The absorption of UA at the electrode surface of our electrode was also found to be dominant in the electro-oxidation process. In a previous study, CeO_2 NPs in combination with gold were also investigated by depositing 3D nanoclusters (Au NCs) on a CeO_2 NP-modified Ti substrate, which was used for the simultaneous electrochemical detection of dopamine and uric acid [25]. The authors suggested the application of this electrode for DPV and the amperometric detection of DA in the presence of UA because the application of UA makes the proposed determination much more sensitive. Nevertheless, amperometric detection at this electrode did not show any signal for UA in the range of 1–100 μM. In this work, we suggest a combination of nanoceria and commercially available Au NPs, which makes it much easier to prepare sensors with improved analytical performances in terms of the working range and detection limit compared with nanoceria-based and other reported electrodes.

From the perspective of the application, the reproducibility of the proposed sensor was examined by repeating seven SWV measurements of 0.3 and 1 μM UA under the same conditions. The value of the relative standard deviation (RSD) of the response currents was 3.4% and 4.2%, respectively. Five MOFdNC/AuNPs&CPE were prepared in the same way, and the same conditions demonstrated an RSD of 4.4% for the peak current obtained by recording 1.5 μM of UA. The stability of MOFdNC/AuNPs&CPE was also studied by measuring the electrochemical response after 10 and 30 days. The results showed excellent stability of this sensor as there was no significant change in the peak current after 10 days of storage, while after 30 days of storage, the current remained at 79.2% compared with the initial value. Based on the obtained analytical and statistical parameters, CeO_2 nanoparticles prepared by calcination of CeBTC MOF and functionalized with gold nanospheres incorporated in a carbon paste matrix provided a highly sensitive, repeatable, and reproducible platform for UA quantification.

3.6. Interference Studies and Milk Sample Analysis

Next, measurements were performed to determine how the presence of various inorganic ions and bioactive compounds influenced the results of uric acid analysis at MOFdNC/AuNPs&CPE. The tolerance limit for interfering species was considered as the maximum concentration of foreign species that caused a relative error of less than ±5% for the determination of 1 µM of UA under the optimized conditions. First, selectivity over antioxidants and compounds that can usually be found with uric acid in body fluids (vitamins C, B6, B12, and dopamine-DOP) was studied. Figure 7a reveals that the electrochemical response to dopamine at this sensor is higher than the responses to the investigated vitamins, but in a ratio of 1:1 with UA, there is no peak for DOP. There is no interfering effect on the current response of a 1 µM concentration of UA when much higher concentrations (50-fold higher) of all these biocompounds were added simultaneously in the UA test solution.

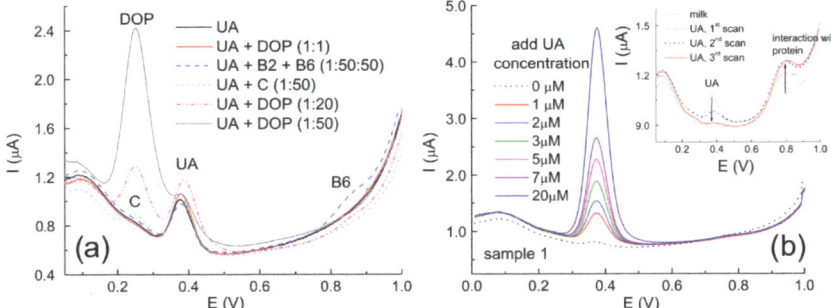

Figure 7. (**a**) The interfering effect of selected biocompounds on 1 µM of UA at MOFdNC/AuNPs&CPE. (**b**) Determination of UA in treated milk sample 1 (insert graph—milk protein interfering effect).

Although this selectivity study suggested the possibility of the successful determination of UA in body fluids by the proposed sensor, we decided to investigate the possibility of applying this sensor in the milk sample because electrochemical UA sensors are rarely used in these samples [6,8,47,48]. Uric acid is one of the main antioxidant biomarkers found in milk [49], and the concentration of this compound was found to be connected with milk spoilage [6]. First, the interfering effect of some ions and biocompounds that can be found in milk was investigated. According to the results obtained, fructose, glucose, lactose, citric acid, folic acid, and Mg^{2+}, Na^+, Ca^{2+}, $NH4^+$, F^-, Cl^-, SO_4^{2-}, and NO_3^- in 100-fold excess also did not interfere with the determination of uric acid, suggesting the possibility of successful analysis of milk samples. Two samples of pasteurized milk were investigated by the developed analytical procedure and sensor. Figure 7b shows the SWV of prepared milk sample 1 (S1) in BR buffer at pH 6 and the SWVs of the samples prepared with spiking. As can be seen in the inset graph, determining UA concentration by spiking an unprepared milk sample was impossible because of the interaction between the added UA and the proteins present in milk, as they can reduce the concentration of uric acid [50]. The peak current proportional to the concentration of UA decreased with the addition of UA and repeated measurements with the same concentration, while the peak at 0.8 V increased rapidly. After removing the protein fraction (Section 2.5), the determination of UA in the treated milk samples was successfully performed and correlated with comparative spectrophotometric measurements (Table 2).

Table 2. Determination of uric acid in milk samples.

S1 (Working Solution)			S2 (Working Solution)		
Added (μM)	Found (μM) by SWV *	Recovery (%)	Added (μM)	Found (μM) by SWV *	Recovery (%)
0	0.41 ± 0.03	-	0	0.44 ± 0.04	-
1	1.45 ± 0.06	102.8	5	5.63 ± 0.09	103.5
2	2.42 ± 0.04	100.4	10	10.33 ± 0.19	98.9
3	3.39 ± 0.04	99.4	15	16.80 ± 0.59	102.2
5	5.57 ± 0.07	102.9	20	20.54 ± 01.59	100.5
7	7.24 ± 0.18	98.2	The final UA Concentration (μM) in Milk Samples SWV * S1: 82 ± 6.23 S2: 88 ± 2.06		
20	20.23 ± 0.31	99.1			

* average of three measurements ± SD.

Finally, 100 μM of prepared milk samples were analyzed by the SWV technique in BR buffer at pH 6, and the results indicated the low effect of the tested solution matrix in this condition. The unknown UA concentration in milk was calculated from the calibration curve. The final analyte concentration in the pasteurized milk samples, calculated taking into account a 200-fold dilution, is 82 μM and 88 μM, respectively. The obtained results are summarized in Table 2 (the determination of UA in treated milk sample 2 is shown in Figure S4).

4. Conclusions

In this study, functional nanostructured CeO_2 obtained by calcination of MOF in air conditions and CeO_2 NPs prepared by the solvothermal method were compared based on their structural, morphological, and electrochemical properties. MOF-derived CeO_2 nanoparticles obtained by thermolysis proved themselves to be a suitable electrochemical sensor modifier, especially when they are functionalized with commercially available gold nanospheres. This binary nanocomposite, incorporated in a carbon paste electrode, represents an easily prepared, eco-friendly electrochemical platform with excellent electrocatalytic activity for the highly sensitive determination of uric acid. The developed analytical method based on this electrode material is distinguished by a wide linear range, higher sensitivity, and one of the lowest LODs for the detection of uric acid among the electrochemical sensors that have been reported until now. This combination of high conductivity, catalytic activity, redox mediation, and stability makes the proposed MOF-derived CeO_2/AuNPs sensor an effective electrode modifier for the sensitive and selective electrochemical detection of uric acid. Moreover, good results were obtained in the analysis of milk samples, proving the great practical applicability potential of this modified electrode.

Supplementary Materials: The following supporting information can be downloaded at https://www.mdpi.com/article/10.3390/chemosensors12110231/s1: Figure S1: (a) X-ray powder diffraction pattern of CeBTC; (b) FT-IR spectra of CeBTC and c-e) SEM micrographs of CeBTC at different magnifications; Figure S2: A comparison of UA responses differential pulse voltammetry (DPV) and square wave voltammetry (SWV); Figure S3: Optimization of SWV method: (a) selection of the best modulation amplitude; (b) Frequency selection and (c) Scan rate selection; Figure S4: Determination of UA in treated milk sample 2.

Author Contributions: Conceptualization, M.O., V.G., M.M. and B.B.P.; methodology, M.O., M.M., D.M.S. and B.B.P.; validation, V.N.; formal analysis, M.O., M.M.; V.G., D.M.S. and B.B.P.; investigation, M.O., M.M., V.G., D.M.S. and B.B.P.; resources, S.V.-Đ.; writing—original draft preparation, M.O., D.M.S. and B.B.P.; writing—review and editing, M.O., M.M., D.M.S. and B.B.P.; supervision, M.O. and B.B.P. All authors have read and agreed to the published version of the manuscript.

Funding: Financial support for this study was granted by the Ministry of Education, Science and Technological Development of the Republic of Serbia, Grant Nos. 451-03-65/2024-03/200123 and 451-03-66/2024-03/200168, and the Faculty of Sciences and Mathematics, University of Priština in Kosovska Mitrovica, Project Number IJ-2301.

Institutional Review Board Statement: Not applicable.

Informed Consent Statement: Not applicable.

Data Availability Statement: The data presented in this study are available on request from the corresponding author.

Acknowledgments: The authors thank Djordje Veljović for the SEM measurements.

Conflicts of Interest: The authors declare no conflicts of interest.

References

1. Scheele, K.W. Examen Chemicum Calculi Urinarii. *Opuscula* **1776**, *2*, 1776.
2. Dalbeth, N.; Merriman, T.R.; Stamp, L.K. Gout. *Lancet* **2016**, *388*, 2039–2052. [CrossRef] [PubMed]
3. Jin, M. Uric Acid, Hyperuricemia and Vascular Diseases. *Front. Biosci.* **2012**, *17*, 656. [CrossRef] [PubMed]
4. Aihemaitijiang, S.; Zhang, Y.; Zhang, L.; Yang, J.; Ye, C.; Halimulati, M.; Zhang, W.; Zhang, Z. The Association between Purine-Rich Food Intake and Hyperuricemia: A Cross-Sectional Study in Chinese Adult Residents. *Nutrients* **2020**, *12*, 3835. [CrossRef]
5. El Ridi, R.; Tallima, H. Physiological Functions and Pathogenic Potential of Uric Acid: A Review. *J. Adv. Res.* **2017**, *8*, 487–493. [CrossRef]
6. Khamzina, E.; Bukharinova, M.; Stozhko, N. Uric Acid as a Marker of Milk Microbiological Spoilage. *BIO Web Conf.* **2023**, *76*, 02001. [CrossRef]
7. Zuo, R.; Zhou, S.; Zuo, Y.; Deng, Y. Determination of Creatinine, Uric and Ascorbic Acid in Bovine Milk and Orange Juice by Hydrophilic Interaction HPLC. *Food Chem.* **2015**, *182*, 242–245. [CrossRef]
8. Motshakeri, M.; Phillips, A.R.J.; Kilmartin, P.A. Application of Cyclic Voltammetry to Analyse Uric Acid and Reducing Agents in Commercial Milks. *Food Chem.* **2019**, *293*, 23–31. [CrossRef]
9. Wang, Q.; Wen, X.; Kong, J. Recent Progress on Uric Acid Detection: A Review. *Crit. Rev. Anal. Chem.* **2020**, *50*, 359–375. [CrossRef]
10. Chen, X.; Wu, G.; Cai, Z.; Oyama, M.; Chen, X. Advances in Enzyme-Free Electrochemical Sensors for Hydrogen Peroxide, Glucose, and Uric Acid. *Microchim. Acta* **2014**, *181*, 689–705. [CrossRef]
11. Sun, M.; Cui, C.; Chen, H.; Wang, D.; Zhang, W.; Guo, W. Enzymatic and Non-Enzymatic Uric Acid Electrochemical Biosensors: A Review. *ChemPlusChem* **2023**, *88*, e202300262. [CrossRef] [PubMed]
12. Knežević, S.; Ognjanović, M.; Nedić, N.; Mariano, J.F.M.L.; Milanović, Z.; Petković, B.; Antić, B.; Djurić, S.V.; Stanković, D. A Single Drop Histamine Sensor Based on AuNPs/MnO_2 Modified Screen-Printed Electrode. *Microchem. J.* **2020**, *155*, 104778. [CrossRef]
13. Stanković, D.M.; Ognjanović, M.; Fabián, M.; Avdin, V.V.; Manojlović, D.D.; Đurić, S.V.; Petković, B.B. CeO_2-Doped—Domestic Carbon Material Decorated with MWCNT as an Efficient Green Sensing Platform for Electrooxidation of Dopamine. *Surf. Interfaces* **2021**, *25*, 101211. [CrossRef]
14. Tharani, D.S.; Sivasubramanian, R. CeO_2 Nanocubes-Based Electrochemical Sensor for the Selective and Simultaneous Determination of Dopamine in the Presence of Uric Acid and Ascorbic Acid. *J. Chem. Sci.* **2023**, *135*, 93. [CrossRef]
15. Hashemzaei, Z.; Saravani, H.; Sharifitabar, M.; Shahbakhsh, M. Combustion Synthesis of Sponge-like CeO_2 Powder for Selective Determination of Uric Acid in Biological Fluids. *J. Part. Sci. Technol.* **2021**, *7*, 73–82. [CrossRef]
16. Zhang, Y.; Yan, X.; Chen, Y.; Deng, D.; He, H.; Lei, Y.; Luo, L. ZnO-CeO_2 Hollow Nanospheres for Selective Determination of Dopamine and Uric Acid. *Molecules* **2024**, *29*, 1786. [CrossRef]
17. Lavanya, N.; Sekar, C.; Murugan, R.; Ravi, G. An Ultrasensitive Electrochemical Sensor for Simultaneous Determination of Xanthine, Hypoxanthine and Uric Acid Based on Co Doped CeO_2 Nanoparticles. *Mater. Sci. Eng. C* **2016**, *65*, 278–286. [CrossRef]
18. Temerk, Y.; Ibrahim, H. A New Sensor Based on In Doped CeO_2 Nanoparticles Modified Glassy Carbon Paste Electrode for Sensitive Determination of Uric Acid in Biological Fluids. *Sens. Actuators B Chem.* **2016**, *224*, 868–877. [CrossRef]
19. Petrucci, R.; Bortolami, M.; Di Matteo, P.; Curulli, A. Gold Nanomaterials-Based Electrochemical Sensors and Biosensors for Phenolic Antioxidants Detection: Recent Advances. *Nanomaterials* **2022**, *12*, 959. [CrossRef]
20. Xiao, T.; Huang, J.; Wang, D.; Meng, T.; Yang, X. Au and Au-Based Nanomaterials: Synthesis and Recent Progress in Electrochemical Sensor Applications. *Talanta* **2020**, *206*, 120210. [CrossRef]
21. Centeno, M.; Ramírez Reina, T.; Ivanova, S.; Laguna, O.; Odriozola, J. Au/CeO_2 Catalysts: Structure and CO Oxidation Activity. *Catalysts* **2016**, *6*, 158. [CrossRef]
22. Huang, P.; Chen, G.; Jiang, Z.; Jin, R.; Zhu, Y.; Sun, Y. Atomically Precise Au25 Superatoms Immobilized on CeO_2 Nanorods for Styrene Oxidation. *Nanoscale* **2013**, *5*, 3668. [CrossRef] [PubMed]

23. Jiao, Y.; Li, N.; Yu, H.; Li, W.; Zhao, J.; Li, X.; Zhang, X. Fabrication of Strawberry-like Au@CeO$_2$ Nanoparticles with Enhanced Catalytic Activity by Assembly of Block Copolymer Composite Micelles. *RSC Adv.* **2017**, *7*, 662–668. [CrossRef]
24. Tang, H.; Ang Chen, Z.; Wu, M.; Li, S.; Ye, Z.; Zhi, M. Au-CeO$_2$ Composite Aerogels with Tunable Au Nanoparticle Sizes as Plasmonic Photocatalysts for CO$_2$ Reduction. *J. Colloid Interface Sci.* **2024**, *653*, 316–326. [CrossRef]
25. Palanisamy, S. Simultaneous Electrochemical Detection of Dopamine and Uric Acid over Ceria Supported Three Dimensional Gold Nanoclusters. *Mater. Res. Express* **2014**, *1*, 045020. [CrossRef]
26. Yusuf, V.F.; Malek, N.I.; Kailasa, S.K. Review on Metal–Organic Framework Classification, Synthetic Approaches, and Influencing Factors: Applications in Energy, Drug Delivery, and Wastewater Treatment. *ACS Omega* **2022**, *7*, 44507–44531. [CrossRef]
27. Ma, J.; Fan, L.; Wang, X.; Li, L.; Zhang, Y.; Zhao, G.; Chai, B.; Gao, J. Defect Engineering on Metal-Organic Frameworks for Enhanced Photocatalytic Reduction of Cr(VI). *Surf. Interfaces* **2024**, *54*, 105174. [CrossRef]
28. Lu, X.F.; Fang, Y.; Luan, D.; Lou, X.W.D. Metal–Organic Frameworks Derived Functional Materials for Electrochemical Energy Storage and Conversion: A Mini Review. *Nano Lett.* **2021**, *21*, 1555–1565. [CrossRef]
29. Hammad, S.F.; Abdallah, I.A.; Bedair, A.; Abdelhameed, R.M.; Locatelli, M.; Mansour, F.R. Metal Organic Framework-Derived Carbon Nanomaterials and MOF Hybrids for Chemical Sensing. *TrAC Trends Anal. Chem.* **2024**, *170*, 117425. [CrossRef]
30. Alhalili, Z. Metal Oxides Nanoparticles: General Structural Description, Chemical, Physical, and Biological Synthesis Methods, Role in Pesticides and Heavy Metal Removal through Wastewater Treatment. *Molecules* **2023**, *28*, 3086. [CrossRef]
31. Hartati, Y.W.; Topkaya, S.N.; Gaffar, S.; Bahti, H.H.; Cetin, A.E. Synthesis and Characterization of Nanoceria for Electrochemical Sensing Applications. *RSC Adv.* **2021**, *11*, 16216–16235. [CrossRef] [PubMed]
32. Ho, C.; Yu, J.C.; Kwong, T.; Mak, A.C.; Lai, S. Morphology-Controllable Synthesis of Mesoporous CeO$_2$ Nano- and Microstructures. *Chem. Mater.* **2005**, *17*, 4514–4522. [CrossRef]
33. He, J.; Xu, Y.; Wang, W.; Hu, B.; Wang, Z.; Yang, X.; Wang, Y.; Yang, L. Ce(III) Nanocomposites by Partial Thermal Decomposition of Ce-MOF for Effective Phosphate Adsorption in a Wide pH Range. *Chem. Eng. J.* **2020**, *379*, 122431. [CrossRef]
34. Liu, X.-M.; Gao, W.-L.; Zhang, J. Facile Synthesis of Monodispersed CeO$_2$ Nanostructures. *J. Phys. Chem. Solids* **2011**, *72*, 1472–1476. [CrossRef]
35. Liu, Y.; Jie, W.; Liu, F.; Liu, Q.; Qiu, M.; Gong, X.; Hu, J.; Gong, L. Effect of Ce-BTC Precursor Morphology on CuO/CeO$_2$ Catalysts for CO Preferential Oxidation in H2-Rich Gas. *Solid State Sci.* **2023**, *139*, 107182. [CrossRef]
36. Hu, G.; Ma, Y.; Guo, Y.; Shao, S. Electrocatalytic Oxidation and Simultaneous Determination of Uric Acid and Ascorbic Acid on the Gold Nanoparticles-Modified Glassy Carbon Electrode. *Electrochim. Acta* **2008**, *53*, 6610–6615. [CrossRef]
37. Zhao, Y.; Gao, Y.; Zhan, D.; Liu, H.; Zhao, Q.; Kou, Y.; Shao, Y.; Li, M.; Zhuang, Q.; Zhu, Z. Selective Detection of Dopamine in the Presence of Ascorbic Acid and Uric Acid by a Carbon Nanotubes-Ionic Liquid Gel Modified Electrode. *Talanta* **2005**, *66*, 51–57. [CrossRef]
38. Wu, Y.; Deng, P.; Tian, Y.; Feng, J.; Xiao, J.; Li, J.; Liu, J.; Li, G.; He, Q. Simultaneous and Sensitive Determination of Ascorbic Acid, Dopamine and Uric Acid via an Electrochemical Sensor Based on PVP-Graphene Composite. *J. Nanobiotechnol.* **2020**, *18*, 112. [CrossRef]
39. Chelmea, L.; Badea, M.; Scarneciu, I.; Moga, M.A.; Dima, L.; Restani, P.; Murdaca, C.; Ciurescu, D.; Gaman, L.E. New Trends in Uric Acid Electroanalysis. *Chemosensors* **2023**, *11*, 341. [CrossRef]
40. Knežević, S.; Ognjanović, M.; Stanković, V.; Zlatanova, M.; Nešić, A.; Gavrović-Jankulović, M.; Stanković, D. La(OH)$_3$ Multi-Walled Carbon Nanotube/Carbon Paste-Based Sensing Approach for the Detection of Uric Acid—A Product of Environmentally Stressed Cells. *Biosensors* **2022**, *12*, 705. [CrossRef]
41. Bard, A.J.; Faulkner, L.R. *Electrochemical Methods: Fundamentals and Applications*, 2nd ed.; Wiley: New York, NY, USA, 2001; ISBN 978-0-471-04372-0.
42. Laviron, E. General Expression of the Linear Potential Sweep Voltammogram in the Case of Diffusionless Electrochemical Systems. *J. Electroanal. Chem. Interfacial Electrochem.* **1979**, *101*, 19–28. [CrossRef]
43. Mazzara, F.; Patella, B.; Aiello, G.; O'Riordan, A.; Torino, C.; Vilasi, A.; Inguanta, R. Electrochemical Detection of Uric Acid and Ascorbic Acid Using R-GO/NPs Based Sensors. *Electrochim. Acta* **2021**, *388*, 138652. [CrossRef]
44. Yan, Q.; Zhi, N.; Yang, L.; Xu, G.; Feng, Q.; Zhang, Q.; Sun, S. A Highly Sensitive Uric Acid Electrochemical Biosensor Based on a Nano-Cube Cuprous Oxide/Ferrocene/Uricase Modified Glassy Carbon Electrode. *Sci. Rep.* **2020**, *10*, 10607. [CrossRef]
45. Azizpour Moallem, Q.; Beitollahi, H. Electrochemical Sensor for Simultaneous Detection of Dopamine and Uric Acid Based on a Carbon Paste Electrode Modified with Nanostructured Cu-Based Metal-Organic Frameworks. *Microchem. J.* **2022**, *177*, 107261. [CrossRef]
46. Zhang, J.; Gao, L.; Zhang, Y.; Guo, R.; Hu, T. A Heterometallic Sensor Based on Ce@Zn-MOF for Electrochemical Recognition of Uric Acid. *Microporous Mesoporous Mater.* **2021**, *322*, 111126. [CrossRef]
47. Motshakeri, M.; Travas-Sejdic, J.; Phillips, A.R.J.; Kilmartin, P.A. Rapid Electroanalysis of Uric Acid and Ascorbic Acid Using a Poly(3,4-Ethylenedioxythiophene)-Modified Sensor with Application to Milk. *Electrochim. Acta* **2018**, *265*, 184–193. [CrossRef]
48. Jawad, M.A.; Dorie, J.; El Murr, N. Electrochemical Quantitative Analysis of Uric Acid in Milk. *J. Food Sci.* **1991**, *56*, 594–595. [CrossRef]

49. Lindmark-Månsson, H.; Åkesson, B. Antioxidative Factors in Milk. *Br. J. Nutr.* **2000**, *84*, 103–110. [CrossRef]
50. Garrel, D.; Verdy, M.; PetitClerc, C.; Martin, C.; Brulé, D.; Hamet, P. Milk- and Soy-Protein Ingestion: Acute Effect on Serum Uric Acid Concentration. *Am. J. Clin. Nutr.* **1991**, *53*, 665–669. [CrossRef]

Disclaimer/Publisher's Note: The statements, opinions and data contained in all publications are solely those of the individual author(s) and contributor(s) and not of MDPI and/or the editor(s). MDPI and/or the editor(s) disclaim responsibility for any injury to people or property resulting from any ideas, methods, instructions or products referred to in the content.

MDPI AG
Grosspeteranlage 5
4052 Basel
Switzerland
Tel.: +41 61 683 77 34

Chemosensors Editorial Office
E-mail: chemosensors@mdpi.com
www.mdpi.com/journal/chemosensors

Disclaimer/Publisher's Note: The title and front matter of this reprint are at the discretion of the Guest Editors. The publisher is not responsible for their content or any associated concerns. The statements, opinions and data contained in all individual articles are solely those of the individual Editors and contributors and not of MDPI. MDPI disclaims responsibility for any injury to people or property resulting from any ideas, methods, instructions or products referred to in the content.

www.ingramcontent.com/pod-product-compliance
Lightning Source LLC
LaVergne TN
LVHW072358090526
838202LV00019B/2572